The Confinement of the Insane

The rise of the asylum constitutes one of the most profound, and controversial, events in the history of medicine. Recently, academics around the world have begun to direct their attention to the origins of the confinement of those deemed 'insane', exploring patient records in an attempt to understand the rise of the asylum within the wider context of social and economic change of nations undergoing modernization.

This edited volume brings together fourteen original research papers to answer key questions in the history of asylums. What forces led to the emergence of mental hospitals in different national contexts? To what extent did patient populations vary in terms of their psychiatric profile or their socio-economic background? What was the role of families, communities and the medical profession in the confinement process? This volume therefore represents a landmark study in the history of psychiatry by examining asylum confinement in a global context.

ROY PORTER was Professor of the Social History of Medicine at the Wellcome Trust Centre for the History of Medicine at University College London. Recent books included *Doctor of Society: Thomas Beddoes and the Sick Trade in Late Enlightenment England* (1991); *London: A Social History* (1994); *'The Greatest Benefit to Mankind': A Medical History of Humanity* (1997); *Enlightenment: Britain and the Creation of the Modern World* (2000) and *Bodies Politic: Disease, Death and the Doctors in Britain: 1650–1914* (2001). He was a co-author of *The History of Bethlem* (1997), and of *Gout: The Patrician Malady* (1998). He was widely published in a variety of fields, including eighteenth-century medicine, the history of psychiatry and the history of quackery. The author and editor of over a hundred books, Roy Porter died on 3 March 2002.

DAVID WRIGHT received his doctorate in economic and social history from the University of Oxford in 1993. In 1999, he returned to Canada, having been appointed the Jason A. Hannah Chair in the History of Medicine at McMaster University in Hamilton, Ontario. His books include *Mental Disability in Victorian England: The Earlswood Asylum, 1847–1901* (2001), and two edited volumes (with Anne Digby), *From Idiocy to Mental Deficiency: Historical Perspectives on People with Learning Disabilities* (1996) and (with Peter Bartlett) *Outside the Walls of the Asylum: The History of Care in the Community* (1999).

The Confinement of the Insane

International Perspectives, 1800–1965

Edited by

Roy Porter and David Wright

CAMBRIDGE UNIVERSITY PRESS
Cambridge, New York, Melbourne, Madrid, Cape Town, Singapore, São Paulo

Cambridge University Press
The Edinburgh Building, Cambridge CB2 2RU, UK

Published in the United States of America by Cambridge University Press, New York

www.cambridge.org
Information on this title: www.cambridge.org/9780521802062

First published 2003

A catalogue record for this publication is available from the British Library

ISBN-13 978-0-521-80206-2 hardback
ISBN-10 0-521-80206-7 hardback

Transferred to digital printing 2005

Dedicated to the memory of
Professor Roy Porter, 1946–2002

Contents

Figures

Tables

Notes on contributors

JONATHAN D. ABLARD is Assistant Professor of History at the State University of West Georgia, USA. He received his PhD from the University of New Mexico in 2000. He is the author of 'Law, Medicine and Commitment to Public Psychiatric Hospitals in Twentieth Century Argentina', in Mariano Ben Plotkin (ed.), *Argentina on the Couch: Psychiatry, the State and Society in Argentina, 1880–1970* (University of New Mexico Press, 2002).

THOMAS BEDDIES is currently an historian of medicine and staff member at the Institute for the History of Medicine at the University of Greifswald (Germany). His main interests include the history of psychiatry and the history of medicine in the era of National Socialism.

CATHARINE COLEBORNE lectures in History at the University of Waikato, New Zealand, and is the co-editor (with Diane Kirkby) of *Law, History, Colonialism: The Reach of Empire* (Manchester University Press, 2001) and the co-editor (with Dolly MacKinnon) of *Asylum: History, Heritage and 'Madness' in Australia* (University of Queensland Press, forthcoming). She has published widely on the history of madness in Australia. Her current work considers psychiatry in the museum and the construction of psychiatric histories in Australia and New Zealand.

HARRIET DEACON has been interested in understanding colonial institutions (prisons, leper and lunatic asylums) since working on her BA Honours thesis (on the Breakwater Prison in Cape Town, South Africa) at the University of Cape Town and her PhD thesis at the University of Cambridge (on the hospitals at Robben Island). Currently at Robben Island Museum in Cape Town, her publications encompass the history of racism in medicine, medical knowledge and exchanges between settler, scientist and *indigène*, cultural geographies of health and exclusion, and the notion of heritage at Robben Island.

ANDREA DÖRRIES received her medical degree from the University of Freiburg, Germany, with a specialization in paediatrics and human genetics (1983–93). She has conducted research projects in medical ethics and medical history at the universities of Freiburg and Berlin (1993–7). Since 1997, she has been Head of the Centre for Health Care Ethics (*Zentrum für Gesundheitsethik*) in Hanover. Her main interests include ethics in paediatrics and genetics, decision-making in medicine and health politics, and hospital ethics committees.

SEAN GOUGLAS is Assistant Professor of Humanities and Computing in the Department of History and Classics at the University of Alberta, Canada. He has received numerous prizes for his work on Geographical Information Systems in the history of agriculture and the history of medicine. He is currently researching the history of sudden and violent death in Western Canada and Australia.

SANJEEV JAIN is Additional Professor, Department of Psychiatry, National Institute of Mental Health and Neurosciences (NIMHANS), Bangalore, India. He completed his undergraduate medical training in Delhi, and residency in Psychiatry in Bangalore at the NIMHANS in 1985. He has been on the faculty since 1986. His clinical responsibilities at the hospital include the management of fifty in-patients, and an out-patient service. His research focuses on genetics of psychiatric disorders and the history of psychiatry in India.

CRISTINA RIVERA-GARZA is Associate Professor of Mexican History at San Diego State University, USA. She is the author of the award-winning historical novel, *Nadie me Verá Llorar* (Tusquets, 2000). She is currently completing the manuscript *Mad Narratives: Inmates and Psychiatrists Debate Gender, Class, and the Nation at the General Insane Asylum La Castañeda, Mexico 1910–1930.*

PETER MCCANDLESS is Professor of History at the College of Charleston, Charleston, South Carolina USA. He is author of *Moonlight, Magnolias, and Madness: Insanity in South Carolina from the Colonial Period to the Progressive Era* (1996). Currently, he is working on a history of disease and medicine in South Carolina from 1670 to 1820 and is an associate editor of the forthcoming *South Carolina Encyclopedia.*

ELIZABETH MALCOLM was educated at the universities of New South Wales, Sydney, Australia and Dublin, Ireland. She has worked in departments of Irish Studies at Queen's University, Belfast, and the University of Liverpool, and is currently Professor of Irish Studies at the University of Melbourne. She has published a history of St Patrick's Hospital, Dublin, Ireland's oldest psychiatric hospital, as well as a number of articles on the history of mental illness in Ireland and among the Irish abroad.

JAMES MORAN is currently a Canadian Institutes of Health Research Postdoctoral Research Fellow at McMaster University's History of Medicine Unit where he is working on an epidemiological history of madness in nineteenth-century Ontario. His interests in the history of madness have been recently published in his book, *Committed to the State Asylum: Insanity and Society in Nineteenth-Century Quebec and Ontario* (McGill: Queen's University Press, 2000).

ELAINE MURPHY is the Chairman of East London and the City Health Authority, London, UK; Visiting Professor in the Department of Psychiatry at Bart's Hospital and the London Hospital, Queen Mary Hospital, University of London and Honorary Senior Research Fellow at the Wellcome Trust Centre for History of Medicine at University College, London. She recently completed her doctorate in medical history, specializing on the Poor Law in Victorian England.

ROY PORTER was Professor of the Social History of Medicine at the Wellcome Trust Centre for the History of Medicine, University College, London. The author and editor of over seventy books, his publications on the history of psychiatry include *Mind-Forg'd Manacles: A History of Madness in England from the Restoration to the Regency* and *A Social History of Madness*. He was co-editor and co-founder of the journal *History of Psychiatry*.

PATRICIA PRESTWICH is a Professor of History in the Department of History and Classics at the University of Alberta, Canada. She has written extensively on the history of modern France, including *Drink and the Politics of Social Reform: Antialcoholism in France since 1870* (Palo Alto, California: Society for the Promotion of Science and Scholarship, 1988). She is currently completing a book manuscript on the Parisian asylum Saint-Anne, and is beginning a new project on the reform of psychiatric institutions in interwar France.

JONATHAN SADOWSKY is the Theodore J. Castele Associate Professor of the History of Medicine at Case Western Reserve University, Cleveland. His chapter is adapted from his 1999 monograph *Imperial Bedlam: Institutions of Madness and Colonialism in Southwest Nigeria*. He is currently working on the history of electro-convulsive therapy.

AKIHITO SUZUKI completed his doctorate at the Wellcome Institute for the History of Medicine, London, UK in 1992. He is now Associate Professor in History at the School of Economics at Keio University in Tokyo, Japan. He has published numerous articles on the history of psychiatry in England. He is now working on the history of psychiatry in Japan in the late nineteenth and early twentieth century.

DAVID WRIGHT holds the Jason A. Hannah Chair in the History of Medicine at McMaster University, and is Associate Professor in the Department of Psychiatry and Behavioural Neurosciences and the Department of History. He has published widely on the social history of developmental disability and psychiatry, including, most recently, a research monograph entitled *Mental Disability in Victorian England* (Oxford University Press, 2001). He is currently the principal investigator on a Canadian Institutes of Health Research project grant researching the historical epidemiology of mental illness in Canada, *c.*1850–1900.

Acknowledgements

This edited volume has evolved over several years and is the result of generous financial assistance of several foundations. The editors and authors are grateful for the assistance of Dr David Allen and the History of Medicine Grants panel of the Wellcome Trust, London, and from Associated Medical Services (Hannah Institute for the History of Medicine), Toronto. Anonymous referees of Cambridge University Press provided invaluable constructive criticism in the early stages of the book. We would also like to acknowledge the comments of Peter Bartlett, Jonathan Andrews and John Weaver. Editorial and administrative assistance was provided by Janna Bordonaro, Erika Dyck, Jessa Chupik, Angela Graham, Steve Bunn and James Moran of McMaster University, Hamilton. Marnie Houser and Jessa Chupik compiled the index for the book. Three chapters in this collection are partly based on previously published material. The contributors would like to acknowledge the Regents of the University of California, Alpha Academic, and the editors of the *Journal of Social History* for permission to republish the material included in the chapters by Jonathan Sadowsky, Andrea Dörries, and Patricia Prestwich respectively. The chapter by Gasser and Heller was translated from the French by David Wright and James Moran. Finally, the editors are grateful to William Davies and the staff at Cambridge University Press for their encouragement and support of this book. Special thanks must go to Maureen Leach for her excellent copy-editing of a challenging international collection of essays, and to Neil de Cort and the other members of the production and design team.

Introduction

Roy Porter

The closing decades of the twentieth century brought a rising and sustained critique of the welfare institutions of the modern state – one largely left-wing in origins but increasingly taken over and voiced by the radical right. Professions which professed to be 'enabling' were, claimed a rising chorus of critics, 'disabling'.[1] Social services which presented themselves as benign were, in reality, 'insidious', serving the interests of providers not consumers, promoting professional dominance, policing deviance and intensifying the social control required to ensure the smooth running of multinational capitalist corporations – or, in the right-wing version, such institutions were wasting tax-payers' money on scroungers and so encouraging malingering.[2]

Unsurprisingly, such political critiques of 'welfarism' (in its widest sense) spawned histories of their own. Replacing various kinds of Fabian, 'Whig' or celebratory historical interpretations which had treated the emergence of the 'caring professions' and social-security institutions as beneficial and progressive – as shifts from neglect to administrative attention, from cruelty to care, and from ignorance to expertise – a new brand of studies took altogether a more negative or jaundiced view of such social institutions and policies, and sought to blow their benevolent ideological cover.[3]

In no field were the new and critical histories more critical, indeed more indignantly impassioned, than the history of psychiatry. Traditional 'in-house' and Whig histories of the care of the insane had never been *particularly* triumphalist – after all, psychiatry had always been a house divided against itself, uneasy in its stance towards both the public and the medical profession at

[1] I. Illich, *Limits to Medicine: The Expropriation of Health* (Harmondsworth, 1977) and *Disabling Professions* (London, 1977).

[2] The literature here is so vast, it would be impossible to begin citing it. Of great importance, however, in clarifying the issues has been S. Cohen and A. Scull (eds.), *Social Control and the State* (New York, 1981).

[3] Once again, 'humanitarianism or control' is a topic on which the survey literature is too vast even to begin to cite, but see M. Micale and R. Porter (eds.), *Discovering the History of Psychiatry* (New York and Oxford, 1994), especially N. Dain, 'Psychiatry and Anti-Psychiatry in the United States', 415–44; G. Grob, 'The History of the Asylum Revisited: Personal Reflections', 260–81, and the substantial introduction.

large, and aware of its embarrassing want of 'magic bullets'.[4] But from the sixties, psychiatry and social policy towards the mad became subjected to intense historical analysis.

Perhaps most radically, and certainly most doggedly, the American (anti-psychiatrist) Thomas Szasz deemed mental illness a mythic and monstrous beast, and proclaimed that 'mental illness' was a fiction. Insanity, he has continued ever since to claim, is not a real disease, whose nature has been progressively scientifically unveiled; mental illness is rather a myth, forged by psychiatrists for their own greater glory. Over the centuries, medical men and their supporters have been involved, argues Szasz, in a self-serving 'manufacture of madness'. In this he indicts both the pretensions of organic psychiatry and the psychodynamic followers of Freud, whose notion of the 'unconscious' in effect breathed new life into the obsolete metaphysical Cartesian dualism. For Szasz, any expectation of finding the aetiology of mental illness in body or mind – above all in some mental underworld – must be a lost cause, a dead-end, a linguistic error, and even an exercise in bad faith. 'Mental illness' or the 'unconscious' are not realities but at best metaphors. In promoting such ideas psychiatrists have either been involved in improper cognitive imperialism or have rather naively pictorialized the psyche – reifying the fictive substance behind the substantive. Properly speaking, contends Szasz, insanity is not a disease with origins to be excavated, but a behaviour with meanings to be decoded. Social existence is a rule-governed game-playing ritual in which the mad person bends the rules and exploits the loopholes. Since the mad person is engaged in social performances that obey certain expectations so as to defy others, the pertinent questions are not about the origins, but about the conventions, of insanity. In this light, Szasz dismisses traditional approaches to the history of madness as *questions mal posés*, and aims to reformulate them.[5]

In some ways reinforcing and complementing Szasz's critique of the epistemological status of insanity, Michel Foucault's *Madness and Civilization*, first published in French in 1961, argued that mental illness must be understood not within the domain of positivist science but as inscribed within discursive formations. To be precise, 'madness' was a voice that, from Classical through Medieval times, spoke its truth and was listened to, within a Platonic philosophy

[4] J. G. Howells (ed.), *World History of Psychiatry* (New York, 1968). An important attempt at European comparative history is L. de Goei and J. Viselaar (eds.), *Proceedings: First European Congress on the History of Psychiatry and Mental Health Care* (Rotterdam, 1992). R. Porter, 'Madness and its Institutions', in A. Wear (ed.), *Medicine in Society* (Cambridge, 1992), 277–301, is a brief comparative study of institutions.

[5] T. S. Szasz, *The Myth of Mental Illness* (New York, 1961; London, 1972; revised edn, New York, 1974); and *The Manufacture of Madness* (New York, 1970; London, 1972). For discussion see R. E. Vatz and L. S. Weinberg, 'The Rhetorical Paradigm in Psychiatric History: Thomas Szasz and the Myth of Mental Illness', in Micale and Porter (eds.), *Discovering the History of Psychiatry*, 311–30.

of poetic *furor*, an Aristotelian assumption of the mad genius, or the Christian doctrine of divine or demonic possession inspiration. At a later stage as part of the developments dubbed by Foucault the 'great confinement', madness was 'shut up' (in both senses of the word), reduced to 'unreason' (a purely negative attribute), and rendered the object of supposed scientific investigation.[6] The critique of the 'great confinement' proved highly influential. Amongst the more conspicuous contributions, David Rothman applied the critical interpretation of the asylum (exposed as an engine of control) to the United States, and Andrew Scull saw madhouses serving a similar function in Britain, as well as being a vehicle of professional imperialism.[7]

The 'new historians' did not have it all their own way. Foucault's provocative formulations – which stood traditional history of psychiatry on its head, taking the heroes of the standard story and making villains of them – have been robustly rebutted by various professional psychiatrists. In *The Reality of Mental Illness*, Martin Roth and Jerome Kroll, for instance, counter-asserted that such have been the stability of psychiatric symptoms presented in recorded history that we may confidently affirm that madness is more than a label, a device for scapegoating deviants in the interests of social control: it is a real disease, probably with a biological basis.[8] For their part, traditionalist historians of social policy have continued to reiterate the progressivist view.[9] And historians of psychiatry of a socio-cultural bent have also taken issue with many of the empirical particulars of Foucault's reading of the transformations of madness and its treatment from Medieval times into the nineteenth century. *Discovering the History of Psychiatry*, edited by Mark S. Micale and Roy Porter, and *Rewriting the History of Madness: Studies in Foucault's 'Histoire de la Folie'*, edited by Arthur Still and Irving Velody, contain many essays offering detailed

[6] M. Foucault, *La Folie et la Déraison: Histoire de la Folie à l'Age Classique* (Paris, 1961); trans. and abridged as *Madness and Civilization: A History of Insanity in the Age of Reason*, by R. Howard (New York, 1965; London, 1967). C. Gordon, '*Histoire de la Folie*: An Unknown Book by Michel Foucault' and 'Rewriting the History of Misreading', in A. Still and I. Velody (eds.), *Rewriting the History of Madness: Studies in Foucault's 'Histoire de la Folie'* (London and New York, 1992), 19–43, 167–84.

[7] D. Rothman, *The Discovery of the Asylum: Social Order and Disorder in the New Republic* (Boston, Mass., 1971); A. Scull, *Museums of Madness: The Social Organization of Insanity in Nineteenth-Century England* (London and New York, 1979) – a much-revised version of this later appeared as *The Most Solitary of Afflictions: Madness and Society in Britain, 1700–1900* (New Haven, Conn., and London, 1993). The Castels' pioneering studies of France should also be mentioned: R. Castel, *L'Ordre Psychiatrique: L'Age d'Or d'Aliénisme* (Paris, 1973; and 1976); English trans. by W. D. Halls, *The Regulation of Madness: Origins of Incarceration in France* (Berkeley and Cambridge, 1988); F. and R. Castel and A. Lovell, *The Psychiatric Society* (New York, 1981).

[8] For instance M. Roth and J. Kroll, *The Reality of Mental Illness* (Cambridge, 1986).

[9] K. Jones: *Mental Health and Social Policy, 1845–1959* (London, 1960); *A History of the Mental Health Services* (London, 1972); and *Asylums and After: A Revised History of the Mental Health Services from the Early Eighteenth Century to the 1990s* (London, 1993).

critiques – rather than mere polemical bouquets or brickbats – of Foucault's views, producing many promising avenues of research.[10]

Overall it would seem that the Foucault who saw 'Reason' and society as involved in a joint mission (or even conspiracy) to control and silence madness did not offer a much more *sophisticated* historical view than traditional Whiggish and meliorist interpretations. But his emphasis upon the dialectic between 'Reason' and 'Madness' is surely valuable to historians. In some ways that is an insight which has been built upon by Sander Gilman and others who have examined madness as a particular mode of disease representation. Gilman had argued that the image of the insane forms part of wider construal of 'self' and 'other' whereby societies identify themselves by the projection of stigmatizing stereotypes. The mad form part of a world of the 'other' also populated by (for example) blacks, homosexuals, the criminal and other 'deviants'. Such an approach to the history of perceptions appears to offer a fruitful entry into the analysis of language, myth and metaphor respecting madness.[11]

The debates detonated by the works of Szasz, Foucault, Scull and others have been noisy, polemical and often angry.[12] Going beyond ideological, and sometimes personal, differences, new studies of the institutionalization of the mad, and the role of the psychiatric profession in it, have increasingly argued that the bait has been cast far too crudely – as if there were, for instance, a cut-and-dried choice between Whiggism and 'anti-psychiatry'. Closer scrutiny and more thoughtful analysis of the historical records, younger historians were claiming, revealed that the asylum was neither just a site for care and cure, nor just a convenient place for locking up inconvenient people ('custodialism').[13] It was many things all at once. And far from being a weapon securely under the control of the profession, or the state, it was a contested site, subject to continual negotiation amongst different parties, including families and the patients themselves. Monolithic and conspiratorial accounts are being replaced by ones

[10] Micale and Porter (eds.), *Discovering the History of Psychiatry*; Still and Velody (eds.), *Rewriting the History of Madness*.

[11] S. L. Gilman: *Difference and Pathology* (Ithaca and London, 1985); *Jewish Self-Hatred, Anti-Semitism and the Hidden Language of the Jews* (Baltimore, 1986); *Sexuality: An Illustrated History* (New York, 1989); *Inscribing the Other* (Lincoln, NE, 1991); *The Jew's Body* (New York and London, 1991); and *Health and Illness: Images of Difference* (London, 1995). See also related themes J. Hubert (ed.), *Madness, Disability and Social Exclusion: The Archaeology and Anthropology of 'Difference'* (London, 2000).

[12] For polemics see for instance J. L. Crammer, 'English Asylums and English Doctors: Where Scull is Wrong', *History of Psychiatry* 5 (1994), 103–15; K. Jones, 'Scull's Dilemma', *British Journal of Psychiatry* 141 (1982), 221–6. Scull has not been slow to hit back: 'Humanitarianism or Control? Some Observations on the Historiography of Anglo-American Psychiatry', *Rice University Studies* 67 (1981), 35–7; 'Psychiatry and its Historians', *History of Psychiatry* 2 (1991), 239–50; 'Psychiatrists and Historical 'Facts'. Part one: The Historiography of Somatic Treatments', *History of Psychiatry* 6 (1995), 225–42; and 'Psychiatrists and Historical 'Facts'. Part two: Re-Writing the History of Asylumdom', *History of Psychiatry* 6 (1995), 387–94.

[13] A. Scull, 'A Convenient Place to Get Rid of Inconvenient People: The Victorian Lunatic Asylum', in A. D. King (ed.), *Buildings and Society* (London, 1980), 37–60.

which emphasize the role of consumers ('purchasers') as well as suppliers, which highlight the market model, and give due weight to 'bottom up' as well as 'top down' history, histories of use as well as abuse, of resistance as well as domination – or which acknowledge (as in the later thinking of Foucault) the sheer complexity of the constitution of power.[14]

Different scholars have brought out different features of this more complex reading. Some, such as Peter Bartlett and James Moran, have emphasized how far the management of the mad remained outside psychiatric control.[15] Others, notably Len Smith, Elaine Murphy, and André Cellard stress that the handling of the insane should be seen not as monolithic and monopolistic but as a mixed economy of care provision, with inputs from the private sector, charity and the state.[16] Other scholars question the model of professional dominance and further argue that the active agency of the family in mediating forms of treatment and custody for a difficult relative was far more important than has hitherto been recognized.[17]

These debates provide the launching-point and the focus of inquiry for several of the studies in this book. Before teasing out some of their implications, it might be helpful at this point briefly to address each of the papers in this volume, to underscore key themes and potential points of comparison.

Cathy Coleborne, as part of a wider scholarly interest in gender and confinement, considers the role played by police in the institutionalization of 'insane' persons in lunatic asylums in Victoria, Australia. From the earliest days of the asylum in the colony, the police were involved in the detection, seizure and

[14] For one contribution amongst many see C. Jones and R. Porter (eds.), *Reassessing Foucault: Power, Medicine and the Body* (London, 1994).

[15] P. Bartlett, *The Poor Law of Lunacy: Administration of Pauper Lunatics in Nineteenth-Century England* (London, 1998); J. Moran, *Committed to the State Asylum: Madness and Society in Nineteenth-Century Ontario and Quebec* (Montreal, 2000). See also the studies in P. Bartlett and D. Wright (eds.), *Outside the Walls of the Asylum: The History of Care in the Community 1750–2000* (London and New Brunswick, NJ, 1999).

[16] L. D. Smith, *Cure, Comfort and Safe Custody: Public Lunatic Asylums in Early Nineteenth-Century England* (London, 1999); E. Murphy, 'The Administration of Insanity in East London 1800–1870', PhD thesis, University of London (2000). Pioneering was W. Llewellyn Parry-Jones, *The Trade in Lunacy: A Study of Private Madhouses in England in the Eighteenth and Nineteenth Centuries* (London, 1971).

[17] A. Suzuki, 'Lunacy in Seventeenth- and Eighteenth-Century England: Analysis of Quarter Sessions Records'. Part one, *History of Psychiatry* 2 (1991), 437–56. Part two, *History of Psychiatry* 3 (1992), 29–44; and see his 'Closing and Disclosing Lunatics within the Family Walls: Domestic Psychiatric Regime and the Public Sphere in Early Nineteenth-Century England', in Bartlett and Wright (eds.), *Outside the Walls of the Asylum*, 115–31; 'Framing Psychiatric Subjectivity: Doctor, Patient and Record-keeping at Bethlem in the Nineteenth Century', in B. Forsythe and J. Melling (eds.), *New Research in the Social History of Madness* (London, 1999); and his forthcoming book on family psychiatry in nineteenth-century Britain, provisionally entitled *Insanity at our Own Doors: Family, Patient and Psychiatry in Early Victorian London*. See also B. Forsythe and J. Melling (eds.), *Insanity, Institutions and Society: New Research in the Social History of Madness, 1800–1914* (London, 1999); Bartlett and Wright (eds.), *Outside the Walls of the Asylum*.

sequestration of women and men deemed to be 'insane'. As the nineteenth century progressed, the numbers of asylum inmates dramatically multiplied and the asylum system expanded rapidly. In order to cope with the spiralling patient numbers, a series of legal, medical and administrative measures was instituted in the latter part of the century, which shaped the meanings of both madness and the asylum.

Towards the end of the century, further legislation detailed the methods police were to follow in the committal of lunatics. They were asked to perform a role that was both medical and legal in nature, and were thus intimately involved in the forging of the asylum population. This has been taken as an indication that control of the insane was not primarily achieved through 'medicalization'. Yet, Coleborne argues, police *were* indeed central to the medicalization of madness, since they were asked to amass medical particulars of 'lunatics', and were also, in a variety of practical ways, the adjudicators of the boundary between sanity and insanity. Central to her chapter is the conviction that families, police, asylum authorities and the alleged insane all negotiated with each other, in the process producing definitions and experiences of both insanity and the asylum.[18]

In 'Ireland's crowded madhouses', Elizabeth Malcolm builds on the rather startling fact that Ireland was one of the first nations to construct a national asylum system.[19] The first purpose-built asylum was inaugurated in Dublin in 1814; a further nine were erected throughout the country in the 1820s and 1830s; and twelve more during the 1850s and 1860s. Subsequently, all these asylums were considerably enlarged or augmented with supplementary hospitals. Between 1851 and 1901, the asylum population rocketed by 337 per cent, to an astonishing 63.4 per 100,000. The United Kingdom and other European and colonial societies no doubt experienced enormous increases in their asylum populations during the latter part of the nineteenth century, but perhaps none on such a scale as Ireland.

Malcolm investigates this phenomenon through a meticulous study of the Irish asylum system. She shows that Irish asylums should not be seen as geriatric institutions, nor were their inmates the socially maladjusted or economically redundant 'misfits' supposed by certain historians to have been characteristic of late-Victorian asylums in England. A majority of the inmates at that time were 'ordinary' members of Irish society: persons under fifty – and many, particularly men, only in their twenties and thirties. The largest group among these men were rural labourers and farmers' sons. What, Malcolm asks, were the socio-economic origins of these patients? What conflicts within struggling rural Irish families led to institutional confinement?

[18] See also K. C. Kirkby, 'History of Psychiatry in Australia, pre-1960', *History of Psychiatry* 10 (1999), 191–204.

[19] For her earlier work see E. Malcolm, *Swift's Hospital: A History of St Patrick's, Dublin* (Dublin, 1988). See also P. M. Prior, 'Mad, not Bad: Crime, Mental Disorder and Gender in Nineteenth-Century Ireland', *History of Psychiatry* 8 (1997), 501–16.

Patricia Prestwich takes up the question of the dynamics of incarceration. Nineteenth-century psychiatrists, she notes, were fond of portraying their patients as 'fresh off the streets', without a medical identity until they came under the 'clinical gaze' and therapeutic control of the asylum physician – a professional view curiously echoed in Foucault's and Szasz's own formulations. Recent research in the history of institutional psychiatry, however, has been suggesting that the 'journey to the asylum' may be no less important than the clinical gaze for understanding the social composition and function of this contested institution – indeed scholars have recently been maintaining that admitting psychiatrists merely confirmed the diagnosis of insanity made by families, by neighbours, or by non-medical authorities. Such possibilities make it therefore essential to go beyond the concept of the asylum as an instrument of medical power and to examine the demands made on the asylum and its doctors by the community.

As examined in Patricia Prestwich's chapter, the records of the Parisian asylum of Sainte-Anne provide an opportunity to explore the complexities of the process of committal in France. Constructed in 1867 as the first of five new 'model' asylums in the Paris region, Sainte-Anne represented the hopes of Parisian psychiatrists for the scientific yet humane handling of the insane. It was specified as the teaching hospital for Paris, and its courses in psychiatric medicine were conducted by the most celebrated physicians of the period, including Valentin Magnan. The grounds of Sainte-Anne also housed the admissions office for all five asylums in the department of the Seine. There, from 1867 to 1912, Magnan examined and certified from 3,000 to 4,000 patients a year. Sainte-Anne was also the only public asylum situated within Paris itself, and was therefore the most convenient of these institutions for the Parisian population.

On the basis of quantitative and qualitative analysis of the admission records for over 7,000 patients treated at Sainte-Anne from 1873 to 1914, Prestwich examines three key questions of institutional confinement: first, how did inmates arrive at the asylum? Second, what kinds of people were committed? And third, what sorts of behaviour resulted in their confinement? Although the role of committal in maintaining public order is discussed, the accent is on what this committal process reveals about the motives and requirements of families. Prestwich then proceeds to examine the types of patients admitted to Sainte-Anne. Recent research on nineteenth-century asylums has established that they did not serve, as has often been suggested, as a 'dumping ground' for 'undesirables'. But it remains vital to analyse the diversity of the asylum population, in terms both of standard social characteristics (e.g. age, marital status and occupation) and of the types of behaviour leading to committal. Prestwich stresses the variety of medical and social problems faced by the community and, in consequence, the multiple demands for care and treatment placed on the asylum and its physicians. Gender, she suggests, was more important than the 'type of insanity' in distinguishing patients: women and men were frequently diagnosed as suffering from different afflictions and, as a result, had different experiences of the asylum.

In certain ways resembling one of the '*hopitaux généraux*' delineated by Foucault, Robben Island, off the southern coast of South Africa, accommodated lunatics, lepers and the chronic sick in a 'General Infirmary' for nearly a century after 1846. As Harriet Deacon shows in her chapter, the institution was established soon after the emancipation of slaves, at a time when the colonial government and a nascent Cape Town middle class were trying to impose a new order on the undisciplined urban underclass in preparation for self-rule. The Cape's most threatening insane were sent to the island asylum, which, until 1875, was the only such institution in the colony. Although it grew steadily after 1846, the total of insane isolated in the island asylum at any one time was relatively small, exceeding 200 only in the 1890s. The aggregate institutionalized population in the colony numbered only 645 in 1891; twice that sum of 'lunatics' and 'idiots' were kept in private houses. There was thus no 'great confinement', though the same pressures for institutionalization operated at the Cape as in Europe: the interruption of social networks of care, and a dominant-class horror of uncontrolled behaviour.

Deacon's analysis of the admissions registers for the asylum suggests how and why some of the Cape insane were confined at Robben Island. The selection and treatment of asylum inmates were related to social and economic patterns of change in society at large. At the time when the asylum was established, it took from country gaols and the overcrowded Cape Town hospital those regarded as disruptive to an institutional order on the mainland, which placed a new stress on the discharge of labour by gaol inmates and the rapid cure of patients in the hospital. Most of these patients were male, in a proportion which remained fairly steady throughout the century and demonstrated the primacy of gaol admissions and the overwhelming focus on detention rather than cure. During the first fifteen years, nearly half of the admissions were convicts.

In the 1860s and 1870s, however, the proportion of white paying patients rose fivefold, as the asylum underwent reforms along 'moral management' lines. New asylums were opened to take these middle-class patients (more of them being women) as pessimism over the curability of black lunatics coincided with a growing racism in colonial society. By the early twentieth century, four fifths of the island inmates were black (most of them deemed 'dangerous') and a third were convicts. The patient profile had come full circle, its function once again being to eject the most dangerous and threatening members of society from overcrowded prisons that made their black prisoners work at public works to prepare them for re-entry as disciplined labourers in a booming colonial economy resting on gold and diamonds.

Comparative history forms the analytic framework in two chapters – 'The Confinement of the insane in Switzerland, 1900–1970: Cery and Bel-Air

Asylums', by Jacques Gasser and Geneviève Heller, and 'The confinement of the insane in Victorian Canada' by David Wright, James Moran and Sean E. Gouglas. Both form part of larger interdisciplinary projects on the history of nineteenth-century Swiss and Canadian psychiatry. Cery and Bel-Air in Switzerland, and the Toronto and Hamilton asylums in Canada, were public asylums designed to receive both pauper and middle-class lunatics. Because these Swiss institutions were responsible to two *separate* cantons of the Swiss confederation (Geneva and Lausanne), they operated under *distinct* legislation governing the practice of confinement. The two Ontario asylums, by contrast, operated within the same provincial legislative framework. On the basis of a sample of patient records and other archival material, a socio-demographic analysis of inmates of both sets of asylums is offered, providing substantial new material on the social and 'medical' data (diagnoses, length of stay, number of stays) of patients.

Gasser and Heller provide a study of admission criteria to the Swiss asylums. They discuss the evolution of legislation which gave a relatively settled legal framework to each period, defining the type of admissions, those authorized to admit patients, and the reasons for confinement. In the canton of Geneva, until 1936, jurisdiction over the internment of the insane fell to the Department of Justice and Police; in the canton of Lausanne, by contrast, such jurisdiction lay under the control of the Department of the Interior (health and public welfare). Gasser and Heller also quantitatively analyse admission procedures during the whole period under consideration. They show the fluctuating proportions of patients who requested admission, were accepted on the request of others, or by civic authorities (generally by the intervention of a doctor outside the institution), or who were confined by judicial order. Finally, Gasser and Heller discuss the circumstances and processes of admission in specific situations, looking in particular at a handful of patient records, chosen from around 1930, which give some indication of the interweaving of medical and social criteria in the admission process.

Wright, Moran and Gouglas present detailed socio-demographic analyses that further question an older revisionist portrayal of the asylum as a 'dustbin' for the 'useless and unwanted' of industrial society. The Toronto and Hamilton asylums were not, according to them, populated by the fringe elements of industrial society, at least certainly not from a socio-demographic standpoint. Patients were admitted across the adult age spectrum. Men and women became patients in accordance with their representation in the general population.[20]

[20] The major studies suggesting women were disproportionately confined in asylums are: P. Chesler, *Women and Madness* (New York, 1973); E. Showalter, *The Female Malady: Women, Madness and English Culture, 1830–1980* (New York, 1985); Y. Ripa, *Women and Madness: The Incarceration of Women In Nineteenth-Century France* (Minnesota, 1990). For excellent summaries of feminist critiques of psychiatry and the history of psychiatry, see J. Busfield,

Indeed, the absence of sex as an important socio-demographic variable (when cross-referenced with age, occupation, length of stay, religion and geographical background) is striking. Wright, Moran and Gouglas use their statistical findings as a base upon which to reconsider the relationship between asylum admissions and wider patterns of employment, kinship networks, immigration and socio-economic growth in Victorian Ontario.

Andrea Dörries surveys the strengths and shortcomings for the historian of German psychiatry of surviving archival material. Founded in 1880, the Wittenauer Heilstätten psychiatric hospital in Berlin has preserved nearly all its patient records. In addition to an exhaustive run of patient files, documentation can be found on special treatments (e.g., malaria therapy for progressive paralysis due to tertiary syphilis) as well as material concerning hospital employees.

Aided by a computer-based analysis of patient records from the hospital, Dörries builds her paper with a view to describing patients' lives in the period from 1919 to 1960. Her database includes personal, medical, social, postmortem and admission data, as well as data on sterilizations performed during the 1930s and 40s. Using a random sample of 4,000 records (8 per cent of all surviving ones), the paper focuses on three topics: first, the social and demographic characteristics of admissions; second, the treatment and discharge of patients based on diagnosis, year and gender and the effect of the patients' social circumstances on their lengths-of-stay; and, third, the care and treatment of children who stayed at the hospital. Dörries takes particular account of changing political circumstances: the Weimar Republic, with its tremendous implications for most patients with physical disorders: and the post-war period in Germany, with the emerging new political systems in East and West Berlin. Her paper demonstrates continuities and discontinuities in the daily life of patients admitted to a municipal psychiatric hospital during the period of the first four decades in the twentieth century.

The records of the South Carolina Department of Mental Health form the foundation of Peter McCandless's chapter, an investigation of how an asylum in the southern United States changed as a result of radical changes in its patient population, and how those mutations were in turn related to major shifts in the society at large. Opened in 1828, the South Carolina Lunatic Asylum is the third-oldest state mental institution in the United States. American historians of mental illness have argued that it (and other early Southern asylums) began as custodial institutions caring for pauper lunatics. Its founders, however, hoped to create a curative establishment, grounded on moral treatment, for patients of

'Sexism and Psychiatry', *Sociology* 23 (1989), 343–64 and N. Tomes, 'Feminist Histories of Psychiatry', in Micale and Porter (eds.), *Discovering the History of Psychiatry*, 348–83. For a more recent discussion of the role gender played in the history of psychiatry, see the collected papers in J. Andrews and A. Digby (eds.), *Sex and Seclusion, Class and Custody: Perspectives on Gender and Class in the History of British and Irish Psychiatry* (Amsterdam, 2002).

all classes, and for several decades its officers struggled to achieve this goal. Before the 1860s, the number of patients remained small, never exceeding 200, and the total of paying patients (some from the wealthy planter elite) nearly equalled the number of pauper patients, whose care was funded largely by local governments. The patients were nearly all white – in a state which had a black majority population. Although the asylum could legally accept blacks after 1849, only five were resident when the Civil War ended in 1865.

After the Civil War, McCandless goes on to show, the nature of the asylum changed radically. Without explicitly abandoning its curative goals – indeed in 1895 it was renamed the South Carolina State Hospital for the Insane – it insensibly accepted a custodial function, handling large numbers of patients now deemed primarily chronic and incurable. The patient headcount grew rapidly, reaching 2,200 by 1920, and the social and racial nature of the patient population also changed markedly, as a result of significant changes in the situation of the state. One of the richest states in the 1820s, by the late nineteenth century South Carolina descended to become one of the poorest. Paying patients virtually disappeared, and the institution was inundated by impoverished 'beneficiary' patients whose costs were now paid by the state.

Emancipation fully opened the asylum to the black majority and the number of black patients skyrocketed to over 1,000 in 1920, ironically creating a biracial institution in a state increasingly devoted to rigid segregation. The changes in the institution's role and patient population accompanied a marked deterioration in its internal conditions, the result of grossly inadequate funding. To a large extent this development resulted from the state's steep economic decline and internecine political struggles. But inadequate funding also reflected changes in the institution's clientele. As a result, its officers were unable to provide even basic custodial care to patients increasingly marginalized by chronic disease, poverty and race.

The question of gender is central to Jonathan Ablard's examination of confinement in Argentina. As Ablard shows, by the early twentieth century Argentina had one of the most extensive and modern systems of public psychiatric care in Latin America. Despite the promise of these institutions and of plans to build new asylums in the interior of the country, however, by the 1930s all of Argentina's public facilities were in crisis, plagued by overcrowding, physical breakdown, legal irregularities, and impossible doctor-and-staff-to-patient ratios. Addressing one hospital in the city of Buenos Aires, the Hospital Nacional de Alienadas (National Hospital for the Female Insane), Ablard explores the decline of that hospital from two viewpoints. First he considers how the structural and ideological contradictions of public health policy condemned the National Hospital to overcrowding. From the 1870s to the 1930s, Argentina was a major destination for European immigrants. Argentine elites viewed those newcomers with ambivalence, believing that their presence disrupted the social and political order and that many were 'defective', and hence likely to require help from the state. Accordingly, public health, and particularly psychiatric hospitals, received paltry state subsidies.

As a further cost-saving device, hospitals for women such as the National Hospital were entrusted to charitable and religious organizations which performed the work for little or no recompense. The National Hospital was further burdened by the fact that it was Argentina's only psychiatric facility dedicated entirely to female patients. As a result, it received patients from all over the country. Many arrived with no identification, and often without the proper legal paperwork. Ablard also examines the hospital's relationship to everyday citizens and to public authorities. Despite conditions which were often harsh, families were the principal source of commitments until 1933. Thereafter the administration of the National Hospital, forced by overcrowding to refuse additional patients, established new admissions rules which led to a sharp increase in public authority commitments. This shift further undermined what many doctors had hoped was a trend towards a growing public confidence in the hospital.

The National Hospital, however, viewed its relationship to the general public with great ambivalence. On one hand, throughout the first half of the century, doctors continued to hope that commitments by family members would eventually lead to a growth in voluntary self-admissions. On the other hand, the hospital was constantly trying to restrict frivolous, inappropriate and medically unnecessary admissions. Yet its attempt to become a strictly medical institution was further impeded by socially grounded medical concepts that called for the protection and attentive control over women at large. As a result, doctors were ultra-vigilant for signs of mental illness in women who violated social norms, and also tended to be reluctant to release such women from the hospital's care.

The historiography of psychiatry in Argentina had relied hitherto almost exclusively on published medical literature. Ablard breaks new ground by examining previously untapped sources, including insanity proceedings, hospital annual reports, and the archive of the elite women's Society of Beneficence which ran the hospital from the 1850s until 1947. Much of the earlier literature assumed that Argentina followed European, and particularly French, models of hospital care. A careful study of primary documents reveals, however, that the Argentine psychiatric network reflected the peculiarities of national social and economic development, and particularly the relative weakness of the national state when it came to the implementation of health and social control policies.

Another Latin American nation here investigated in a parallel study is Mexico. Cristina Rivera-Garza traces the history of the General Insane Asylum, inaugurated in 1910 by General Porfirio Diaz – the flagship welfare institution devoted to the care of the mentally ill in early twentieth-century Mexico. One of the largest and most monumental projects of the modernizing agenda of the Porfirian regime, the asylum soon faced serious financial limitations as a result of the Mexican revolution, whose armed phase commenced just three months after the asylum's opening. Plagued by overcrowding, poor staffing and physical deterioration, the asylum authorities nevertheless kept detailed patient records. Using admission registers and medical files from 1910 through to 1930 – a year after the institution underwent medical and administrative reform – Rivera-Garza explores the continuities and discontinuities between the Porfirian strategies of confinement as set out in asylum regulations, and the actual procedures through which men and women became asylum inmates. In examining the social and demographic profiles of inmates, Rivera-Garza addresses the various ways in which both police and families used concepts of gender and class to detect mental illness in a rapidly changing social milieu. It is a kind of analysis especially important in an institution which, while serving a range of social classes, admitted great numbers of patients as free and indigent (100 per cent of women and 86 per cent of men in 1910) and on the strength of a government order (86 per cent of women and 68 per cent of men).

Drawing upon institutional reports and patient testimonies, she then examines the dynamic of life within the asylum grounds, paying special attention to the ways in which the layout and routine of the institution replicated and reinforced distinctive class and gender understandings of mental illness. In 'becoming mad', Rivera-Garza concludes, asylum inmate characterizations shed light on the negotiation through which state agents and family members distinguished mental illness, something of growing relevance in the context of the emergent revolutionary regimes concerned with the reconstruction of the nation. This social history of confinement in revolutionary Mexico City shows the contested origin of public health policies, variously interpreted as either a vertical imposition of an increasingly centralized state or the success of revolutionary welfare policies.[21]

In a welcome contribution to the history of colonial policy towards the mad in British Africa, Jonathan Sadowsky illustrates how the history of that colony's asylums re-enacted developments common in the comparative history of psychiatric institutions, while also illustrating themes peculiar to the politics and priorities of colonialism. Initially the institutions were, like many colonial imports, already obsolete by metropolitan standards, replicating virtually all the shortcomings British psychiatry had come to pride itself in overcoming. For most of the early twentieth century, administrators of the Nigerian colonial state ventilated a rhetoric of scandal and the need for reform, but when reform was at last achieved in the late 1950s and early 1960s, it was contemporary with Nigeria's gradual shift to independence, and the reform was largely accomplished through the initiatives of Nigerians.

[21] C. Rivera-Garza, *Mad Encounters: Psychiatrists and Inmates Debate Gender, Class, and the Nation in Mexico, 1910–1930* (Lincoln, NE, forthcoming).

Sadowsky appraises Nigeria's asylums under colonialism, by examining the material conditions in the asylum, the social processes of admission and discharge, the ideological conflicts raised by debates over asylum reform and, finally, the process of reform itself. Colonial asylum policy was driven by the theory of 'indirect rule', which held that interference in local practices should be kept to a minimum. It was a policy which, in the name of cultural preservation, justified parsimony, and which assailed reform of the institutions as an imposition on the 'native way of life'. Supporting this platform was the curious but common belief that one could, somehow, have colonialism without making any impositions. Westminster avoided a significant investment in treatment, and, at the same time, a strict 'hands-off' policy. By consequence Nigeria ended up with institutions which had none of the potential benefits of asylum care, and many of its most coercive traits. Colonial policies for the incarceration of the insane may thus be said to reflect in microcosm the contradictions of 'indirect rule'.

Examining provision for the insane in East London, Elaine Murphy emphasizes the gradual nature in the transition from private to public care. The capacity of an earlier private trade in lunacy to both adapt to market changes, and to impress the Lunacy Commissioners during their inspections, left the 'old mixed economy system' of care intact right to the end of the nineteenth century. Murphy notes that it was the force of state preference for public providers which ultimately placed them in a position of dominance. According to Murphy, the ultimate success of the public system thus lay more in the influence of families in the committal process and in the ability of the central government to control capital expenditure, than in the medical capture of madness by an incipient alienist profession or by the Lunacy Commission.

The relationship between (British) colonial psychiatry and indigenous practices of mental health services forms the backdrop to Sanjeev Jain's chapter on the history of psychiatry in India. Building on a burgeoning literature on the history of madness in the Subcontinent, Jain documents the influence of British psychiatric practice and the slow steps towards institutional treatment in colonial and later independence India. As Jain documents, rates of incarceration (as a percentage of the huge Indian population) were relatively speaking low, and thus Indian psychiatrists had to resort to a variety of techniques in providing mental-health services. Unique among these was the practice of the Amritsar Mental Hospital which invited patients' families to live on the grounds of the hospital in order to aid in the recovery of kin. Side-by-side with the continuance of traditional Ayurvedic and Unani medical practices, one is struck by the rapidity with which the few institutional based Indian psychiatrists employed the latest (European) treatments. Shock therapies, for instance, were reported to be used in Indian mental hospitals within months of being reported in European-based psychiatric journals. This 'internationalizing' aspect of psychiatry has only recently received serious historical attention.

In the other contribution focusing on Asia, Akihito Suzuki demonstrates how the development of psychiatric institutions in Japan from the Meiji Restoration (1868) to the Second World War presents a number of similarities to the experience of many European countries. These notably include the enactment of national legislation (from 1900) for regulating institutional provision for the insane, a series of *exposés* of glaring abuses in metropolitan asylums (1903), tensions between central and local government over psychiatric policy, and the involvement of police in the incarceration process.

His paper, however, spotlights one key difference. Before the Second World War, Japan did not witness the establishment of a publicly funded, nationwide network of large-scale asylums. As late as 1950, only 19,000 patients (from a national population of over 50 million) were institutionalized in about 150 psychiatric hospitals. Thus, instead of that dominance of vast long-stay mental hospitals which became prevalent in the west, a 'mixed economy of the care for the insane' persisted. Within this set-up, the style of psychiatric provision varied enormously.

Publicly funded psychiatric hospitals, modelled after English asylums, were outnumbered by private *Noh Byoin* ('brain hospitals'), many of which had origins in, and were situated on the sites of, Buddhist temples and Shintoist shrines. Psychiatric 'colonies' flourished in two cities (Kyoto and Kanazawa), both crystallized around ancient religious and philanthropic lore. A large number of certified patients (2,400 in 1950) remained within the household, and still more led their own troubled and dependent lives without being certified, let alone institutionalized. Thus, the history of Japanese psychiatry in the pre-1945 era provides a unique opportunity to examine some of the fundamental questions in the use of the asylum, and the Japanese experience of psychiatric provision must no longer be seen merely as a 'failure' to 'catch up' with the West.

Deploying a comparative perspective, Suzuki analyses the social history of confinement from around 1900 to 1945. In particular, using archival sources, he examines two different types of institutions: Matsuzawa Hospital in a Tokyo suburb, which was (and still is) a University-affiliated public psychiatric hospital with 1,000 beds; and Iwakura Hospital, a for-profit enterprise situated outside Kyoto, with its numerous satellite boarding-out houses for the insane. As well as adopting different therapeutic and managerial strategies, those two institutions catered for patients from different classes or social groups who came to the asylums through different paths and with different rationales: Matsuzawa pursued highly Westernized regimens on poorer patients from the metropolis who had usually been committed to the hospital by the police, while Iwakura provided self-consciously eclectic treatment for those from wealthier families from all over the country who sought refuge in the asylum. By examining these contrasts in the dynamics of institutionalization of the two facilities, and by putting them into the context of contemporary Japanese society, Suzuki illuminates the factors which shaped Japanese psychiatric provision into a system that was distinctively mixed and varied.

As will be evident, individually and collectively these studies address major questions in the history of psychiatry. Several seek to uncover the impulses behind the mighty movement accelerating in the nineteenth century and continuing in the twentieth to certify and incarcerate large populations of people (as Malcolm observes, a stunning one-in-a-hundred of the whole population in Ireland!). In some cases – Germany for instance – the movement seems to have been, as Scull classically argued, in part a response to capitalism, industrialism and urbanization.[22] But elsewhere the asylum grew in nations or regions (Ireland for instance, but also South Carolina) which were not only primarily rural but actually undergoing economic retrogression.

What were the forces behind the drive to confine the insane and to consolidate the procedures of incarceration? How important were doctors? Alternatively, were high-level political decisions and changing profiles of official public policy preponderant? This seems to have been the case in Mexico and Argentina. Elsewhere, the practice of 'street-sweeping', as described in Coleborne's study, indicates the prime role of the police in the committal of the insane in Victorian Australia. Or was the push to confine largely family-driven – a desire to get difficult relatives out of the domestic sphere? Malcolm intriguingly suggests that, although a nineteenth-century Foucauldian model of the 'great confinement' might seem attractive to some, in the early comprehensive system of state asylums which developed in Ireland, the urge to put people away actually came in large measure not from the agents of the state but from families and communities who were glad to have economic burdens shifted off domestic shoulders in circumstances of drastic economic crisis.[23] Prestwich for her part underlines the role of the bourgeois family in securing confinements in nineteenth-century Paris.

None of the authors in this book subscribes to the Foucauldian 'great confinement' model *tout court*. But that does not mean that they renounce the notion that institutionalization and the asylum served the ends of social control, of disciplining and punishing, however defined. Indeed several of the essays emphasize how the asylum particularly targeted certain groups. In Australia, Coleborne finds, women were disproportionately likely to be the victims of psychiatric labelling, stigmatization and asylumdom.[24] Moran and Wright's

[22] A. Scull, *The Most Solitary of Afflictions*. See also C.-R. Prull, 'City and Country in German Psychiatry in the Nineteenth and Twentieth Centuries', *History of Psychiatry* 10 (1999), 439–74.

[23] M. Finnane, *Insanity and the Insane in Post-Famine Ireland* (London, 1981).

[24] See also C. Siobhan Coleborne, 'Reading Madness: Bodily Difference and the Female Lunatic Patient in the History of the Asylum in Colonial Victoria 1848–1888', PhD thesis, La Trobe University, Melbourne (1997); for women as victims in Britain, Showalter, *The Female Malady*; and see also N. Tomes, 'Feminist Histories of Psychiatry', in Micale and Porter (eds.), *Discovering the History of Psychiatry*, 348–83.

study of lunatic asylums in Ontario by contrast finds that there it was the large floating population of young unmarried men who were often scooped up by the asylum. In antebellum South Carolina, lunatic slaves remained the problem of their masters, but after the Civil War, the asylums in that state increasingly were reduced to serving as custodial institutions for free blacks.[25] Conditions for black patients were worse in every way than for whites. Similar racial discrimination may be seen in Deacon's analysis of institutionalization in South Africa. Blacks formed the chief population of the Robben Island institution, while white South Africans were largely cared for in private facilities.[26] What all these studies point to is the need to specify, in their unique historical context, the distinctive range of factors which gave the psychiatry establishment its point; nor must we forget that these may have changed dramatically over time.

The 'new' history of psychiatry which emerged form the 1960s was, not surprisingly, centred on North America and western Europe.[27] Studies of other nations, regions and continents have followed, promoting the same or similar inquiries. We now know, for instance, much more about the development of psychiatry in Greece after its independence, in Central and South America, and in Russia.[28] This academic development inevitably led to the formulation of the question: what power role did psychiatry and the institutionalization of the insane play in colonial and quasi-imperial contexts? This is a matter particularly important in the light of the commonly held view that medicine – and by implication, psychiatry – are intrinsically *colonial* pursuits: they colonize the body, colonize the patient. If that is the nature of medicine, then must not psychiatry itself have served to promote the imperial mission? Indeed Megan Vaughan and other scholars have shown beyond doubt how the importation of western psychiatry into imperial contexts did not merely provide rationales for locking up troublesome indigenous individuals but supplied supposed psychological profiles of the 'native' at large, construed, for instance, as 'savage', or 'backward' or childlike, thereby rationalizing colonial rule.[29]

[25] See also P. McCandless, *Moonlight, Magnolias and Madness: Insanity in South Carolina from the Colonial Period to the Progressive Era* (Chapel Hill, NC, 1995).

[26] See also H. Jane Deacon, 'Madness, Race and Moral Treatment: Robben Island Lunatic Asylum, Cape Colony, 1846–1890', *History of Psychiatry* 7 (1996), 287–98; S. Swartz, 'Changing Diagnoses in Valkenberg Asylum, Cape Colony, 1891–1920: A Longitudinal View', *History of Psychiatry* 6 (1995), 431–52.

[27] Mention should also be made here of the comparative study: K. Doerner, *Madmen and the Bourgeoisie: A Social History of Insanity and Psychiatry* (Oxford, 1981).

[28] D. Ploumpidis, 'An Outline of the Development of Psychiatry in Greece', *History of Psychiatry* 4 (1992), 239–44; J. Brown, 'Heroes and Non-Heroes: Recurring Themes in the History of Russian-Soviet Psychiatry', in Micale and Porter (eds.), *Discovering the History of Psychiatry*, 297–307;

[29] See notably M. Vaughan, *Curing Their Ills: Colonial Power and African Illness* (Stanford, Calif., 1991).

These are lines of inquiry further explored in this book. What seems clear, however, is the sheer complexity of the colonial structures themselves. As is evident from the examples of Australia and South Africa, there was some bad feeling from settler groups in the colonies towards the mother country. Australians were never allowed to forget their convict origins and, partly for that reason, the handling of the 'insane' in that colonial milieu was never divorced from police business. In both places, the heavy arm of psychiatry fell disproportionately upon the disadvantaged native population.[30]

But once again the story proves more complex than it might seem. At first sight it might be expected that West Africa would provide a case in which psychiatrization would serve as an instrument of the rule of White over Black. But in fact Britain invested little faith or money in asylums in the biggest West African colony, Nigeria. Sadowsky shows why: Britain paradoxically, hypocritically but characteristically wanted to have its imperial cake and eat it – it wanted colonial rule, but chose to improve the colonized as little as possible.[31]

As is so often the case, the empire can offer a mirror for what was going on in the metropolitan domains. How far were the mad poor and other disadvantaged groups in London, Paris or New York being treated as colonized people? Certainly, as studies have shown, there was a very different 'psychiatry for the poor' than the 'psychiatry for the rich'.[32]

The case of Nigeria suggests that institutionalization and psychiatrization were less the bold and clear instruments of policy than the foci of uncertainties, muddles, conflicts and indecisions. As is revealed by the studies of European and American psychiatry contained in this book, much might be hoped and expected of the psychiatric enterprise, but in practice neither the bricks-and-mortar, nor the diagnostics, nor the therapeutics ever delivered the goods, and psychiatry never won the unambiguous respect of the politicians, the press, the pundits or the people.

Till recently, the history of psychiatry has had various obvious weaknesses. One relates to evidence. In many cases, important archives have been inaccessible to historians, or their astonishing richness of case material and documentation has overwhelmed the scholar. As the article below by Andrea Dörries in particular suggests, computer software is making it possible to overcome

[30] See also S. Garton, *Medicine and Madness: A Social History of Insanity in New South Wales, 1880–1940* (Kensington, Australia, 1988); M. James Lewis, *Managing Madness: Psychiatry and Society in Australia 1788–1980* (Canberra, 1988).

[31] See also J. Sadowsky: 'The Confinements of Isaac O.: A Case of 'Acute Mania' in Colonial Nigeria', *History of Psychiatry* 7 (1996), 91–112; 'Psychiatry and Colonial Ideology in Nigeria', *Bulletin of the History of Medicine* 71 (1997), 94–111; and *'Imperial Bedlam': Institutions of Madness in Colonial Southwest Nigeria* (Berkeley, 1999).

[32] R. Hunter and I. Macalpine, *Psychiatry for the Poor, 1851. Colney Hatch Asylum, Friern Hospital 1973: A Medical and Social History* (London, 1974); C. MacKenzie, *Psychiatry for the Rich: A History of Ticehurst Private Asylum, 1792–1917* (London and New York, 1993).

such obstacles.[33] Another is a consequence of parochialism. There have been too few comparative studies, and such 'worldwide' accounts that we have are mainly antiquated compilations.[34] Bringing together studies from all the continents, this book addresses differences and similarities in a variety of different societies, and hopefully points the way towards pioneering comparative studies.

[33] Valuable has been M. MacDonald, 'Madness, Suicide, and the Computer', in R. Porter and A. Wear (eds.), *Problems and Methods in the History of Medicine* (London, 1987), 207–29.
[34] See note 4, above.

1 Insanity, institutions and society: the case of the Robben Island Lunatic Asylum, 1846–1910

Harriet Deacon

Introduction

Robben Island, an island off the southern coast of South Africa barely six miles from Cape Town, the capital city of the Cape Colony in the nineteenth century, accommodated 'lunatics', 'lepers' and the 'chronic sick' for nearly a century after 1846. The 'General Infirmary' was established just eight years after the emancipation of slaves was finalized, at a time when the colonial government and a nascent middle class in Cape Town were trying to impose a new order on the undisciplined urban underclass in preparation for self-rule. The Cape's most dangerous insane were sent to the island asylum from 1846, that, until 1875, was the only asylum in the colony. By 1921, there were a number of other asylums established: Grahamstown (1875), Port Alfred (1889), and Fort Beaufort (1894) in the Eastern Cape, and Valkenberg (1891) near Cape Town.[1]

While Britain and some of her colonies provided extensive provision for the insane, the Cape did not. Most of the colonial insane were cared for at home or through private boarding arrangements: only the most desperate resorted to the asylum. In 1890, the proportion of registered white insane to the white population at the Cape was 1:1,180, about three times lower than that in Ireland, New Zealand, New South Wales, Victoria and Britain (from 1:294 to

The research on which this chapter is based was supported at Cambridge University by the Sir Henry Strakosch Memorial Scholarship, and the Patrick and Margaret Flanagan Scholarship. Completion of the chapter was supported by the Robben Island Museum.

[1] Current scholarship on Cape asylums includes H. J. Deacon: 'Racial Categories and Psychiatry in Africa: The Asylum on Robben Island in the Nineteenth Century', in W. Ernst and B. Harris (eds.), *Race, Science and Medicine* (London, 2000); and 'Madness, Race and Moral Treatment at Robben Island Lunatic Asylum, 1846–1910', *History of Psychiatry* 7 (1996), 287–97; S. Marks, '"Every Facility that Modern Science and Enlightened Humanity have Devised": Race and Progress in a Colonial Hospital, Valkenberg Mental Asylum, Cape Colony, 1894–1910', in J. Melling and W. Forsythe (eds.), *Insanity, Institutions and Society: A Social History of Madness in Comparative Perspective* (London, 1999); S. Swartz: 'The Black Insane at the Cape, 1891–1920', *Journal of Southern African Studies* 21 (1995), 399–415; 'Changing Diagnoses in Valkenberg Asylum, Cape Colony, 1891–1920: A Longitudinal View', *History of Psychiatry* 6 (1995), 431–52; and 'Colonialism and the Production of Psychiatric Knowledge in the Cape, 1891–1920', PhD thesis, University of Cape Town (1996); F. Swanson, 'Colonial Madness: The Construction of Gender in the Grahamstown Lunatic Asylum, 1875–1905', BA (Hons.) thesis, University of Cape Town (1994).

1:380).[2] There was also a much larger proportion of people classified as 'criminal' insane in the Cape than in Britain or New South Wales, although in New South Wales and elsewhere, police were still responsible for a large proportion of asylum committals before 1900.[3] Although it rose steadily after 1846, the number of insane confined in the Robben Island asylum at any one time was relatively small, only exceeding 200 in the 1890s. The total asylum population in the colony numbered only 645 in 1891; double the number of 'lunatics' and 'idiots' were kept in private houses. There was thus no 'Great Confinement' of the insane in the Cape Colony during the nineteenth century. Yet some of the same pressures for institutionalization operated at the Cape as in Europe: the disruption of social networks of care and a dominant-class fear of uncontrolled behaviour within an increasingly ordered urban society.

An analysis of admissions to the Robben Island asylum can illustrate the social dimensions of psychiatric practice at the Cape. Fox has suggested that patients committed to the San Francisco asylum in the early twentieth century were

> a strikingly heterogeneous [group, sharing] neither a common social background, a similar mental condition, nor even a customary 'route' to the asylum . . . What united them, instead, was a type of relationship to other people. The insane were disturbing, peculiar, or incomprehensible. They were in many cases out of touch with reality and in a small number of cases violent or destructive. But they became insane not when they crossed some well-defined boundary between health and sickness, between normality and abnormality. They became insane when other individuals decided they could no longer be tolerated.[4]

It is clear from the Robben Island records that the Cape asylum, unlike the San Francisco asylum,[5] was catering mainly for the 'dangerous' insane. This was partly a feature of the minimal institutional provision for the insane at the Cape and partly due to the legal strictures on admitting 'ordinary' lunatics before 1891. And yet within this framework the island admission records highlight interesting gender and racial variations in institutional use as well as changing patterns of admission and treatment that can be related to social and economic changes in the society at large.

Throughout the nineteenth century and into the twentieth, most of the patients in Cape asylums, including Robben Island, were male[6] and disproportionally many were white. When the Robben Island asylum was established, it took from country gaols and the overcrowded Cape Town hospital those who were

[2] 'Report of the Inspector of Asylums for 1890', Cape Parliamentary Papers (CPP), G37–1891, 14.

[3] S. Garton cited in C. Coleborne, 'Passage to the asylum', below.

[4] R. W. Fox, *So Far Disordered in Mind: Insanity in California, 1870–1930* (London and Berkeley, 1978), 79.

[5] *Ibid.*, 137–8.

[6] Swartz, 'Colonialism and the Production of Psychiatric Knowledge', 132; Valkenberg was an exception in having more women than men, 133.

considered most disruptive to an institutional order on the mainland that placed a new stress on the performance of work by gaol inmates and the speedy cure of patients in the hospital. It took time to develop a curative ethos on the island, however. During the first fifteen years three out of five asylum inmates were black, nearly half of whom had come through the criminal justice system. During the 1860s and 1870s, the proportion of white paying patients rose fivefold as the asylum underwent reforms along humanitarian 'moral management' lines. After 1875, new asylums were opened on the mainland to take these middle-class patients, more of whom were women than before. Greater pessimism over the curability of black 'lunatics' now coincided with increasing racism in colonial society.[7] Within the system of colonial asylums, Robben Island was marked for the most dangerous and threatening members of society. By the early twentieth century, four out of five of the island asylum inmates were black and a third were convicts. The asylum had come full circle, its function once again to remove troublesome black male prisoners from overcrowded prisons.

The process of admission

In order to analyse the process of admission to the asylum at Robben Island, a database was compiled from patient admission records. It is important to treat the statistical data with care, however. The records are incomplete before 1872, and systematically favour long-stay cases and those admitted through the Somerset Hospital, founded in 1818 as the first civilian hospital at the Cape. Early admission data have been gleaned from Old Somerset Hospital admission registers and correspondence files. Using the admissions database, the average admission rate for the period 1846–52 is 19.3 admissions per annum, while official statistics for the same period record 29.2 admissions per annum. It should be remembered too that categories such as 'nationality' changed over time, as did diagnostic terms and procedures.

The asylum population at Robben Island must be treated as a historically specific subset of those people who would today be defined as 'mentally ill' rather than as representative of the distribution of madness in the colony. Except possibly for middle-class British settlers after 1860, the major pressures for institutionalization at the Cape were poverty and fear of violence rather than the hope of a cure. There is little evidence that the establishment of an asylum in 1846 produced or reflected a change from home care to the use of the asylum as a therapeutic resource. Africans,[8] Dutch-Afrikaans settlers[9] and

[7] On racism in colonial psychiatry see Swartz, 'The Black Insane'.

[8] Although people who have been born in Africa, or lived most of their lives there, and certainly those who lived there before 1652, could all be termed Africans, in this paper I have used the term 'African' specifically to refer to those black indigenous inhabitants of the Cape who were probably not identified as Khoisan, 'Malay' or 'coloured'.

[9] The Dutch-speaking white settler community at the Cape came from a range of continental European countries, but predominantly from the Netherlands and Germany. Most immigrated to the Cape before 1806. Although there was considerable intermixing with the local slave and

Muslims[10] all continued to be reluctant to use the asylum. Admissions were dominated by those considered dangerous, by the friendless and the poor. I shall therefore start by examining the pressures for institutionalization and the alternatives to the asylum before exploring the processes through which the insane were identified and admitted to the asylum. Then I shall examine the social constitution of the asylum population.

The making of an asylum population

When the Robben Island asylum was opened in 1846 it provided an extra seventy hospital beds for lunatics in addition to the thirty or more in the Somerset Hospital in Cape Town. These places were soon filled by patients who had been in gaols, in the pauper asylum at Port Elizabeth and in home care. Few of these people had any alternative source of care. In 1861 the asylum keeper commented that if the lunatic men had 'any one to come for them . . . they would be sent away'.[11]

Ex-slaves, indigenous Khoisan[12] and Africans made up nearly 60 per cent (n = 87) of all first admissions given nationalities who were admitted to the Robben Island asylum in the period 1846–61.[13] Annual reports give a breakdown of lunatic numbers by race and nationality after 1859. In that year there were 156 lunatics, of whom 70 per cent (n = 110) were described as 'Hottentot',[14] 'African' (some of these were probably Dutch colonials) or 'Kafir',[15] and 23 per cent (n = 37) were from the United Kingdom.[16] Recent European immigrants, ex-slaves, African refugees from the frontier wars, and others without strong family networks were more likely to require state aid

Khoisan population, by the end of the nineteenth century those who saw themselves as 'white' had developed a strong racialized identity as Afrikaners.

10 In Cape Town at this time, Muslims were what nineteenth-century settlers called 'Malays', black descendants of slaves who had come from East Asia and parts of Africa, many of whom converted to Islam after their arrival at the Cape and intermarried with local settler and indigenous populations. A number of Cape Town Muslims were able to rise above the extreme poverty of the urban underclass.

11 Pierce, minutes of evidence, 'Report of the Commission of Inquiry into the General Infirmary and Lunatic Asylum on Robben Island', CPP, G31–1862, 109.

12 Some of the indigenous people who lived off the land around Cape Town and in the interior, mainly to the west and north, were hunter-gatherers and others were pastoralists. The Dutch called the former 'Bushmen' and the latter 'Hottentots'. Although later scholars have attempted to get away from the pejorative uses of these words by inventing new terms (San and Khoi or Khoekhoe respectively), which I have used in this paper, the distinction between the two is not always sustainable (hence the use of the term Khoisan).

13 The sample is small, and the large number of ex-slaves listed in the registers may be partly because they were admitted on government order. However, the general picture from official statistics is similar.

14 See explanation of the term 'Khoisan'.

15 A term used by settlers in the nineteenth century to refer to Xhosa-speaking Africans from the eastern Cape.

16 'Report on the General Infirmary, Robben Island for the year 1859', CPP, G11–1860, 4.

in times of distress. Emancipation and the transition to a market economy before mid-century, which reduced traditional family and employers' support for the mentally ill, and encouraged the use of the asylum in Britain,[17] may have increased the pressure on poor families to send cases to Cape Town for institutional care. Most of the Africans entering the Robben Island asylum were men – unemployed or migrant workers referred by employers or the criminal justice system rather than their own communities.[18] Similarly, Victorian asylums in Australia admitted few aboriginal patients.[19] In twentieth-century colonial Africa, governments still considered the care of the African insane to be the concern of their communities rather than the state.[20]

There were options outside the asylum that even the poor could utilize in the absence of family care. The Dutch Reformed Church, to which most Dutch-Afrikaans settlers belonged, provided boarding-out care for non-violent lunatics. There were Cape Dutch home remedies for hysteria and epilepsy[21] and some patent medicines, very popular among the Dutch-Afrikaans community, like 'Dr Forsyth's Chemic Health Restorer', were said to cure 'nervousness'.[22] For those without church or financial resources there were other options. The insane orphan Maggie K was kept at the Salvation Army Rescue Home for some time before being taken to the police by a friend of the family.[23] In 1883 a Woodstock resident called a 'Malay doctor' to attend to her servant girl who had 'gone mad or was in a fit'. 'Brutus', the doctor, chanted and sang in the 'vernacular' to ward off the presence of the Devil who, he said, had been in the room the previous night.[24] 'Vertical' charity thus assisted some of the mad just as it assisted the poor in general.

Institutional provision for the insane in the Cape Colony was limited and in demand. The Robben Island Surgeon-Superintendent, Dr Edmunds, attested in 1871 to the 'numerous applications' for admission.[25] Even in the 1880s, the asylums at Robben Island and Grahamstown were usually so full that cases had to be kept in gaols for many months where they were 'aggravated by becoming the butt and amusement of the prisoners'.[26] These cases included convicts who

[17] A. Scull, *The Most Solitary of Afflictions: Madness and Society in Britain 1700–1900* (London, 1993), 26–32.
[18] On the paths of black people into asylums see Swartz, 'The Black Insane', 408.
[19] Coleborne, 'Passage to the asylum', below.
[20] M. Vaughan, *Curing their Ills: Colonial Power and African Illness* (Cambridge, 1991), 120.
[21] D. G. Steyn *et. al.* (eds.), *Volksgeneeskuns in Suid-Afrika: 'n Kultuurhistoriese Oorsig, Benewens 'n Uitgebreide Versameling Boererate* (Pretoria, 1966), 258; L. Pappe, *Florae Capensis Medicae Prodromus: An Enumeration of South African Plants used as Remedies by the Colonists of the Cape of Good Hope* (Cape Town, 1868), 16, 46.
[22] *The Lantern*, 18 May 1878.
[23] Sub-Inspector Barnes to Resident Magistrate of Cape Town, 10 July 1898, Cape Town Municipality Papers: Lunacy 1878–1910, 1/CT 12/53, Cape Archives, Cape Town (CA).
[24] *The Lantern*, 25 August 1883.
[25] 'Report on the General Infirmary, Robben Island for the year 1871', CPP, G18–1872, 8.
[26] 'Reports of the Civil Commissioners, Resident Magistrates and District Surgeons for 1882', East London, CPP, G91–1883.

had become insane as well as non-criminal cases.[27] Dangerous or violent cases usually got precedence for admission to the island because of the shortage of asylum provision. The Old and the New Somerset Hospitals in Cape Town were also used to house lunatics. One Jan du P, from the country town of Paarl, having threatened his family with a knife, thinking that he was being poisoned, was kept in the Old Somerset Hospital for twelve years with 'chronic mania' before his transfer to Valkenberg in 1891.[28] The more modern New Somerset Hospital was used for patients like Miss S, who in 1864 was transferred from the Old Somerset Hospital where it was deemed 'quite impossible... to pay proper attention to lunatic females' of her class.[29]

The wealthy insane had a wider choice of options than the poor: home care, boarding-out, private asylums and state institutions. Initially, however, the island asylum had a poor public image that discouraged all but the most desperate applicants. There was also a general aversion to hospitals among the middle classes at the Cape. Growth in middle-class use of asylums abroad and their increasing association with cure made the institutional option more popular during the latter half of the nineteenth century. Walter E, a colonial-born Englishman who worked as a clerk in the attorney general's office, was sent to Robben Island in 1855 and again in 1880. He was described as 'an imbecile... harmless and quite childish', but had delusions of persecution by 'Malays'. After Walter E's discharge from the island, Dr Beck suggested a 'complete change' to cure his 'loss of memory and general nerve depression'. But when he returned from his holiday abusive, threatening suicide and sexual assault, he was sent to Robben Island again before being transferred to Valkenberg in 1891.[30]

Private care of the wealthier insane continued to play an important role even after middle-class facilities were made available in Cape asylums. Most of the propertied insane were admitted to the Valkenberg and Grahamstown asylums, who cultivated a more elitist image than the Robben Island asylum.[31] During the 1880s some private practitioners dissuaded relatives from sending patients to Robben Island or the Somerset Hospital.[32] Cape Town doctors continued to treat some 'better class' lunatics privately in 1879, patients for whom the 'existing arrangements' on Robben Island were said to be 'quite unfit'.[33] As

[27] R. Southey to Edmunds, 19 Feb. 1868, letters despatched by Colonial Office, CO 6861, CA.
[28] Valkenberg Asylum casebook 1, 1891–4, University of Cape Town (UCT) Manuscripts Collection, Cape Town.
[29] J. Laing to Colonial Secretary, 23 Dec. 1864, letters received by Colonial Office, CO 827, CA.
[30] Valkenberg Asylum casebook 1, 1891–4, UCT Manuscripts Collection, Cape Town.
[31] Report on Robben Island in 'Reports on the Government-aided Hospitals and Asylums and Report of the Inspector of Asylums for 1892', CPP, G17–1893, 135.
[32] D. Moyle, 'Laying down the Line: The Emergence of a Racial Psychiatric Practice in the Cape Colony During the Nineteenth Century', unpublished paper, Psychology Department, UCT, 1988, 10.
[33] Dr J. F. Manikus, Minutes of Evidence, 'Report of the Commission appointed to inquire into and report upon the best means of moving the asylum at Robben Island to the mainland', CPP, G64–1880, 21.

late as 1898, the *Cape Argus* reported that poorer patients were sent to the asylum sooner, as rich families 'will do anything rather than send [their insane relatives] to a hospital'.[34] In 1890, only thirteen of the thirty-nine propertied insane placed under curatorship by the Supreme Court were accommodated in asylums – the rest were kept in private homes.[35]

Although most of those recognized as insane were not sent to asylums, private asylums never loomed as large at the Cape as they did in England.[36] In 1845, Harriet O complained that there were no private houses for the treatment of the insane in Cape Town. Her father was forced to go either to the Somerset Hospital or to Robben Island.[37] In 1905, only a Miss Durr's in Mowbray was licensed under the 1897 Act as a private lunatic asylum. It housed three uncertified European women patients as voluntary boarders.[38] Informal boarding houses were more common. Thomas McS, an English hotel keeper in Caledon, was boarded with a family after the death of his mother in 1890, three years after he began to get violent. He was admitted to Valkenberg in 1891. Ebenezer K, declared 'of unsound mind' in the Supreme Court in 1843, was boarded out for ten months before going to England where he was in fact certified sane.[39] Sending the insane 'home' to England was not general practice. In 1889, Robben Island surgeon-superintendent Ross had to make a special case of C, a 'dipsomanic with strong leading delusions', whom he wanted to send to relatives in England at government expense.[40]

The 1891 census provides the first accurate estimates of the relative balance between different forms of provision for lunatics. It shows that 1,281 lunatics were being maintained in private dwellings, as opposed to 120 in jails and approximately 645 in Cape asylums.[41] Males were significantly over-represented among those certified as insane, but whites were only slightly over-represented. While the male–female ratio approached unity in the colony,[42] men represented nearly two-thirds of the insane. While whites represented about a third of the

[34] Quoted in the *South African Medical Journal* (1898), 48.

[35] Report of the Inspector of Asylums in 'Reports of the Medical Committee . . . for 1890', CPP, G37–1891, 10.

[36] See W. L. Parry-Jones, *The Trade in Lunacy: A Study of Private Madhouses in England in the Eighteenth and Nineteenth Centuries* (London, 1971).

[37] Memorial from Harriet O, 6 Dec. 1845, memorials received by Colonial Office, CO 4026, doc.468, CA.

[38] Report of the Inspector of Asylums in 'Reports on the Government-aided Hospitals . . . for 1905', CPP, G32–1906, 54.

[39] Valkenberg Asylum casebook 1, 1891–4, UCT Manuscripts Collection, Cape Town; and 'Supreme Court', *Cape Town Mail*, 28 February 1846.

[40] W. Ross to Under Colonial Secretary, 11 March 1889, Colonial Office, letters received, CO 1438, CA.

[41] Report of the Inspector of Asylums in 'Reports on the Government-aided Hospitals and Asylums and Report of the Inspector of Asylums for 1892', CPP, G17–1893, 126, 134.

[42] Census of 1891, CPP, G6–1892, 22.

Table 1.1 *The number of lunatics and idiots in the colony in 1891, as recorded in the 1891 census*

	Asylums	Outdoor lunatics	Outdoor idiots	Gaols
Male	396	244	450	
Female	249	253	334	
White	350	131	266	
Black	295	366	518	
Total	645	497	784	120

colonial population in 1891,[43] they comprised two-fifths of the insane. In a situation of scarcity of asylum accommodation, admission was granted more often to white men (whose insanity threatened white supremacy and raised the spectre of degeneration and hereditary insanity) and black men (whose insanity threatened white society by disrupting employment relations or the taboo on sexual contact with white women).[44]

Cape asylums were also admitting only a small proportion of those who were recognized by their communities and the authorities as insane, and admitting these patients very selectively (see Table 1.1). Although the number of asylum patients had nearly doubled in the previous decade,[45] it still represented only about half of the number outside asylums and gaols. Nearly two thirds of certified mental patients in private dwellings ('outdoor' lunatics and idiots) were classified as 'idiots' – cases who were probably considered less violent or dangerous than 'lunatics'. This trend was reversed within the asylums, where most inmates were certified as 'lunatics'. Yet more of the black and slightly more of the female insane were 'outdoor lunatics' and more of the white and male insane were institutionalized.[46]

Admission procedures

Defining someone as insane was a necessary condition for admission into the asylum. The doctor was only called to ratify the definition if the person had already been labelled as insane in social terms and had also become socially or

[43] Census of 1891, CPP, G6–1892, 15 (in that section of the colony defined by the 1875 census, and thus excluding recently annexed territories).

[44] H. J. Deacon, 'A History of the Medical Institutions on Robben Island, 1846–1910', PhD thesis, University of Cambridge (1994), chapter five; Swartz, 'Colonialism and the Production of Psychiatric Knowledge', 113–16.

[45] Report of the Inspector of Asylums in 'Reports on the Government-aided Hospitals and Asylums and Report of the Inspector of Asylums for 1892', CPP, G17–1893, 126.

[46] Moyle, 'Laying Down the Line', 11.

economically problematic.[47] The relative scarcity of complaints about wrongful detention at the Cape testifies to the use of the asylum mainly for those cases who were seriously mentally ill, for whom there were few viable alternatives or whose families approved of their detention. The 'Lunacy Panic' about wrongful detention of alleged lunatics in England during the 1850s,[48] which caused a flurry of legislation in India,[49] hardly touched the Cape.

At the nineteenth-century Cape, where in legal terms until 1891 all lunatic admissions had to be either criminal or potentially so, the boundaries between the lunatic and the criminal lunatic were vague. Because gaols were used to accommodate paupers and the insane as well as the criminal, and to police 'vagrancy' too, there was little pressure to sharpen the boundaries between the various groups. A prisoner called Rachel N in the House of Correction in Cape Town in the mid-1870s became too violent to control and was sent to the Somerset Hospital lunatic wards on the order of the Under-Colonial Secretary. 'Convalescent', she was returned to the House of Correction six months later to complete her sentence of imprisonment.[50]

Not all lunatic cases stood trial: brought to the gaols for some offence or 'nuisance', some were certified insane by the local doctor or district surgeon and sent to the asylum when there was a vacancy. Michael P, reportedly a source of 'annoyance' and 'violence' to his friends in Cape Town, was several times imprisoned in the Cape Town gaol, where he 'forced conversation upon other prisoners of the most beastly and unnatural description'. A medical board asked to determine whether he was insane, decided that as a temporary measure he should be sent to the Robben Island pauper wards.[51] Henry I, a similar case, had run after the young women in his master's house, lit a fire in the stable and scribbled nonsense on the fence. He was sent to Grahamstown Asylum in 1880 as a criminal lunatic without trial.[52] The separation of criminal and ordinary lunatics only became an issue in the 1890s, when numbers of the latter increased, and a large proportion were sent to Robben Island from where, it was argued, they could not easily escape.

[47] See Swartz, 'Colonialism and the Production of Psychiatric Knowledge', 76–97 for a discussion of the certificates required by the committal process at the Cape during this period. See Coleborne, 'Passage to the asylum', below, for a discussion of New South Wales' certificates.

[48] T. Turner, 'Not Worth Powder and Shot: The Public Profile of the Medico-Psychological Association, *c.*1851–1914', in G. Berrios and H. Freeman (eds.), *One Hundred and Fifty Years of British Psychiatry 1841–1991* (London, 1991), 3.

[49] W. Ernst, *Mad Tales from the Raj: The European Insane in British India, 1800–1858* (London, 1991), 45–6.

[50] Under-Colonial Secretary to Surgeon of Old Somerset Hospital, 16 Dec. 1876, Old Somerset Hospital Papers, letters received: 1876–1888, HOS 1, CA.

[51] Report by the Superintendent of Police, memorial of P, March/April 1859, memorials received by Colonial Office, CO 4110, doc. P36, CA.

[52] Case of Henry I, n.d., Health Branch: Criminal lunatics 1893–9, CO 8050, CA.

Until the first quarter of the nineteenth century in England, notions of culpability were centred around obvious signs of behavioural disturbance (e.g. violence), and required proof that the insane did not know wrong from right, for if they did they were not insane and could control their actions. After 1825, the defence of partial insanity, or monomania (delusion), began to be accepted in English courts and was accompanied by a far greater amount of medical testimony because the signs of insanity were only discernible by expert eyes.[53] In the colony, these ideas took root as well. Elliot, a lunatic on Robben Island in 1848, had been accused of stealing clothing in Cape Town and had subsequently destroyed the clothing given to him in the asylum. Although he had been rejected by the legal system as insane, his lack of violence and his apparent consciousness of his misdeeds were commented on by Dr Hall, who said, 'we cannot avoid thinking that some degree of knavery is mixed up with his lunacy, which a little gentle discipline would in all probability correct'.[54] The idea of partial insanity was also clumsily suggested in evidence before the Robben Island Commission in 1861. The assistant lunatic keeper, appropriately named Mr Nutt, complained that the lunatics who refused to work, fought each other and stole from the boat, knew that they were doing wrong: 'They are not quite right [he said], but some are only a little wrong.'[55]

Psychiatric assessment of dangerousness and the use of the diminished responsibility defence are now crucial in the sentencing of those who are deemed mentally disordered in South Africa.[56] In dealing with the forensic patient, the relevance of the crime to sentencing and duration of asylum care remains a serious issue today.[57] 'Dangerousness' played an important role in justifying asylum admission during the nineteenth century. Besides family applications, the courts and police networks were the major screening mechanisms for asylum admissions during the nineteenth century, and often invoked the notion of dangerousness. Whether criminal or not, a large proportion of the patients sent to Robben Island were perceived as dangerous. In the period 1846–1910, 406 out of 1,141 first admissions (36 per cent) entered in the database are listed as dangerous. Suicidal cases made up about 9 per cent of first admissions in this period.

[53] J. P. Eigen, 'Delusion in the Courtroom: The Role of Partial Insanity in Early Forensic Testimony', *Medical History* 35 (1991), 27, 29.

[54] Report by J. Hall on Robben Island, 19 April 1848, minutes of evidence, 'Report of the Commission of Inquiry', CPP, G31–1862, 133–4.

[55] Nutt, minutes of evidence, 'Report of the Commission of Inquiry', CPP, G31–1862, 169–70.

[56] Henning cited in A. Cohen, 'The Psychiatric Assessment of Dangerousness at Valkenberg Hospital', MA thesis, UCT (1991), 27, suggests that in South Africa today, although most Attorneys General feel that the duration of a patient's detention in an asylum should be directly related to the seriousness of their crime, the therapeutic policy of the Department of Health relates length of detention to cure.

[57] Cohen, 'The Psychiatric Assessment of Dangerousness', 28.

Because of the bias towards long-stay patients in the pre-1872 records, and the scarcity of accommodation for the mentally ill in this period, one might expect that dangerousness would be a major criterion for admission to Robben Island before 1872. Indeed, fourteen (nearly a third) of the forty-nine patients sent from Somerset Hospital to Robben Island in 1846, the year the latter hospital opened, were described as 'violent' or 'treacherous'.[58] The cases of Joseph O and Cornelia S, both held in the lunatic wards of the Old Somerset Hospital in 1845, and earmarked by the authorities for transfer to Robben Island, illustrate the influence of assessments of dangerousness in sending patients to the Island. The relatives of both cases did not want them transferred, as Robben Island was too far away, and was already stigmatized. Joseph O was an epileptic who had been cared for at home by his daughter for six years until he became violent, when he was put into the Old Somerset Hospital. Cornelia S was a 'peaceful' lunatic who had been kept in the Old Somerset Hospital for fifteen years, visited by her sister whose husband could not afford to keep Cornelia at their home. Cornelia was allowed to remain in the Old Somerset Hospital while Joseph, who was considered too disruptive for the pauper wards, was transferred to the island.[59]

For the whole period before 1872, however, only a tenth of first admissions to Robben Island were described as dangerous in the Somerset Hospital registers (see Table 1.2). On the Island in 1848, the surgeon-superintendent reported that 'with two or three exceptions the lunatics [were] tranquil'.[60] In 1861 the chaplain, Revd J. A. Kuster complained that he visited the lunatics only once a month, as '[s]peaking with them affects my nerves very much, there being much disturbance from the noisy ones'.[61] Noisy or disruptive behaviour, in the wards, at work, or in church, was reported as the major disciplinary problem in the asylum, although every year there were a few cases of violent assault.[62] This suggests that, although always important as a justification for admission or transfer, the notion of 'dangerousness' was used far less before 1872 than thereafter in admission registers for the island asylum. This may have been because in 1879,[63] the first mental health legislation concerned with institutionalization

58 H. Bickersteth to Acting Secretary to Government, 23 June 1852, in 'Report of the Select Committee on and documents connected with, the Robben Island Establishment', CPP, A37–1855, 41.

59 Memorial of H.O., 6 Dec. 1845, memorials received by Colonial Office, CO 4026, doc.468, CA; memorial of J.S., 3 Dec. 1845, memorials received by Colonial Office, CO 4024, doc.127, CA.

60 Report on Robben Island, 19 April 1848, minutes of evidence, 'Report of the Commission of Inquiry', CPP, G31–1862, 133.

61 Revd J. A. Kuster, Minutes of evidence, 'Report of the Commission of Inquiry', CPP, G31–1862, 49.

62 For example, J. Verreaux, minutes of evidence, 'Report of the Commission of Inquiry', CPP, G31–1862, 191.

63 See A. Kruger, *Mental Health Law in South Africa* (Durban, 1980), pp. 12–13 for a discussion of the Act.

Table 1.2 *Proportion of first admissions to the Robben Island Lunatic Asylum recorded as dangerous, 1846–1910*

Date	'Dangerous' lunatics
1846–1871	10% (n = 45)
1872–1890	61% (n = 234)
1895–1910	41% (n = 127)

at the Cape was passed to provide a legal basis for the detention of 'criminal' and 'dangerous' lunatics.

By the time the detention of 'ordinary' lunatics was provided for in the 1891 Act,[64] Robben Island was earmarked for dangerous and criminal lunatics anyway. As more mainland asylums opened, a greater proportion of 'dangerous' and 'criminal' patients were sent to the island asylum. By 1881, Grahamstown asylum accommodated a significantly smaller proportion of 'maniacal and dangerous' cases than did Robben Island.[65] In the period 1872–90, two-thirds of first admissions to the Robben Island asylum were characterized as dangerous. This proportion dropped to two-fifths in the period 1895–1910 (see Table 1.2). By this time more of the Robben Island patients were criminal lunatics (34 per cent rather than 28 per cent) whose detention was already justified by the court and legislative changes that had increased the range of patients who could be admitted, including 'ordinary' and 'voluntary' patients, which reduced the burden on dangerousness as a justification for admission.

Men were consistently more likely than women to be designated 'dangerous' in the Robben Island registers, in line with contemporary gender stereotypes. Although there was a long association in colonial discourse between blackness and dangerousness too, and black mental patients are perceived as especially dangerous in Britain and America today,[66] 'dangerous' admissions at the Robben Island asylum were not disproportionately black. This may have been because far more of the black admissions came through the criminal justice system, and their detention was already justified on those grounds (see above).

The definitions of insanity

The process of defining madness was not uniform. Disagreement over who was insane illustrates the fluid nature of boundaries of deviance. There were cases

[64] *Ibid.*, p. 14. [65] *Cape of Good Hope Blue Book for 1881*, 228.
[66] P. Moodley and G. Thornycroft, 'Ethnic Group and Compulsory Detention', *Medicine, Science and Law* 28 (1988), 324–28; M. Sabshin, H. Diesenhaus and R. Wilkison, 'Dimensions of Institutional Racism in Psychiatry', *American Journal of Psychiatry* 127 (1970), 787–93.

at Robben Island thought not to be insane, either by their families or by the staff. In 1860, for example, the attendant Pierce felt that the violent actions of a Mr S, employed in the colonial auditing department in Grahamstown, towards a fellow lodger who had stolen something from him, were used against him in 'some foul play' resulting in his admission to Robben Island.[67] S had no trial, and was apparently sent to the island without being told what had happened.[68] He was not considered insane, but nevertheless spent some time on the island.

There were more fundamental debates about the meaning of insanity, however. In 1873, the island chaplain, Baker, commented on a case of 'religious mania' as follows:

> [De V is now] quite sane, may be a little 'eccentric', and doubtless too practically religious to be regarded as 'quite right' by ordinary people. I counselled him not to bring religion *openly* to bear on trifling matters of daily routine or domestic life. I wish many were like him.[69]

He admitted nevertheless that the Bible could be misused, commenting in 1869 that

> From conversations . . . with one or two of the patients, I have been convinced that it would be better not to give the Bible generally to them, but a book of suitable readings from the Scriptures, with forms of prayer. They morbidly turn to unprofitable expressions, and find food for their diseased minds [in the full Bible text].[70]

Baker was tolerant of 'eccentricities', as long as they coincided with his moral viewpoint. His emphasis on religious instruction and morality as the only good way of living gave his psychological counselling a particular emphasis. He recognized the need to speak less 'plainly' to 'a sensitive Lunatic',[71] but did not agree with the surgeon-superintendent, Dr Biccard, who quoted, in 1876, 'a medical man of 12 years' experience' who made it a rule 'never to discuss or allow to be spoken of, matters of Religion and Politics in his Asylum'. Only convalescents should be allowed to attend church, said Biccard.[72] In other ways, Baker's view of the insane was more inclusive than the medical definition, possibly due to the fact that he was more concerned with the content of utterances than with the pathological form.

In general, Baker and the other island chaplains seem to have relied on the usual visual and audible indications of the 'ordinary features of insanity', such

67 Pierce, minutes of evidence, 'Report of the Commission of Inquiry', CPP, G31–1862, 108.
68 G. M. S., minutes of evidence, 'Report of the Commission of Inquiry', CPP, G31–1862, 224.
69 Baker, 11 Nov. 1873, Chaplains' Diaries, AB 1162/G2, University of the Witwatersrand Manuscripts Collection (UWMC), Johannesburg.
70 Baker, 10 Sept. 1869, Chaplains' Diaries, UWMC, AB 1162/G2.
71 Baker, 30 April 1877, Chaplains' Diaries, UWMC, AB 1162/G3.
72 F. L. C. Biccard to Under-Colonial Secretary, 6 July 1876, letters received by Colonial Office, CO 1027, CA.

as a lack of 'rational' conversation, over-excitement, and strange appearance or lack of composure.[73] In 1870 he urged a Miss P to be calm, in spite of her excitement at being discharged, in order that this not be interpreted as insanity.[74] In 1874, using a popular etiology, he linked the violence of some of the lunatics on a Sunday to the presence of a new moon.[75] In 1876, he was concerned about the over-hasty discharge of a certain Mrs S, who he felt ought to be removed from the 'society of insane people' but that 'from her expressions about her husband and her feeling towards myself, etc., I fear she should not retake the full status of a wife'.[76] Baker supported the asylum doctors' decisions regarding institutionalization, however. For example, Baker wrote to the brother of a patient, Dr P. J. van B, whose family was agitating for his release on the grounds of temporary insanity, noting that the patient had 'conducted himself as a gentleman'. But Baker then wrote in his journal that he had 'made no remark as to mental disease – nothing to be used as proof of sanity' in their attempts to free the patient against medical advice.[77]

Medical diagnoses

Nineteenth-century doctors' definitions of who was insane tallied closely with social definitions. Almost all they added to the process was a medical diagnosis. During the early nineteenth century in Europe, doctors' classifications of the insane centred around gross behavioural signs, and simple putative causes: major categories were mania, melancholia, phrenzy, dementia and lethargy.[78] The Robben Island doctors used a similar classification, centred around mania, dementia, melancholia and idiocy or imbecility. More detailed diagnoses were given as the century wore on. S. Swartz has suggested that nineteenth-century medical certificates for the insane in the Cape were legal documents justifying institutionalization, rather than medical diagnoses with implications for treatment. She indicates that in the latter part of the century, these justifications hinged on evidence that the patient was becoming childish (dementing); that a patient was passive (lazy, lethargic, mute, withdrawn) or violent and hyperactive; and/or that patients were immoral (including all sexual behaviour such as masturbation).[79]

Many asylum patients in Britain before mid-century suffered serious bouts of psychosis, were suicidal or suffered from serious mental disability. By the

[73] Baker, 28 Dec. 1869, Chaplains' Diaries, UWMC, AB 1162/G2.
[74] Baker, 29 July 1870, Chaplains' Diaries, UWMC, AB 1162/G2.
[75] Baker, 20 Oct. 1874, Chaplains' Diaries, UWMC, AB 1162/G3.
[76] Baker, 11 Aug. 1876, Chaplains' Diaries, UWMC, AB 1162/G3.
[77] Baker, 5 July 1872, Chaplains' Diaries, UWMC, AB 1162/G2.
[78] G. E. Berrios, 'Historical Background to Abnormal Psychology', in F. Miller and P. J. Cooper (eds.), *Adult Abnormal Psychology* (Cambridge, 1988), 30.
[79] Swartz, 'Colonialism and the Production of Psychiatric Knowledge', 78–9.

Table 1.3 *First diagnoses of first admissions to the Robben Island Lunatic Asylum who were given diagnoses, 1846–1910 (percentage)*

Date	Mania	Dementia	Idiocy Imb.[a]	Melancholia	General[b]	Other	Total
1846–61	52.7 (n = 137)	19.6 (n = 51)	1.6 (n = 4)	0.4 (n = 1)	24.6 (n = 64)	1.2 (n = 3)	100 (n = 260)
1862–71	28.0 (n = 52)	30.6 (n = 57)	6.5 (n = 12)	3.2 (n = 6)	30.6 (n = 57)	1.1 (n = 2)	100 (n = 186)
1872–90	53.5 (n = 204)	26.8 (n = 102)	9.7 (n = 37)	6.6 (n = 25)	2.4 (n = 9)	1.0 (n = 4)	100 (n = 381)
1895–1910	52.8 (n = 161)	13.4 (n = 41)	7.2 (n = 22)	13.8 (n = 42)	9.5 (n = 29)	3.3 (n = 10)	100 (n = 305)

[a] Idiocy and imbecility were not formally distinguished although the general trend was towards defining more severe cases as idiots.
[b] I have invented this category to describe non-specific diagnoses such as 'insanity' or 'lunacy '.

mid- to late nineteenth century, British alienists admitted more patients with less serious disorders.[80] It is difficult to ascertain reliably the extent of severe dysfunctional behaviour from the diagnoses in the Robben Island admission registers, however, as there are no surviving case books detailing behaviour.[81] The frequency of the appellation 'dangerous' (see above) does perhaps indicate that aggression and behavioural dysfunction were very common. Throughout the period, the most common diagnoses for the Robben Island admissions were 'mania' of various types, and 'dementia' (see Table 1.3). The less disruptive forms of insanity ('idiocy', 'imbecility' and 'melancholia') were diagnosed slightly more frequently as the century wore on, perhaps indicating a greater willingness among doctors to venture out of general descriptions such as 'insanity' or a greater preponderance of mental deficiency and depression among patients. As diagnoses became more sophisticated and 'scientific', general descriptions like 'insanity' or 'lunacy' were used less often. Diagnoses of epilepsy remained constant, representing about 10 per cent of first admissions throughout the period 1846–1910, often coupled with other diagnoses.

Delusions were a clear identifier of the insane, by doctors and lay people alike. It has been argued that delusions cannot tell us much about the social fabric of life for the population at large, but delusional content may nevertheless reflect

[80] See L. Smith, *Cure, Comfort and Safe Custody: Public Lunatic Asylums in Early Nineteenth-Century England* (London, 1999).

[81] Swartz, 'Changing Diagnoses in Valkenberg Asylum', 451 has suggested that because diagnoses and descriptions of symptoms changed so often in asylum records, information about first diagnoses in admission registers does not represent the complexity of the system of psychiatric diagnosis. This means that while we can compare diagnostic patterns to social prejudices, we cannot simply translate nineteenth-century diagnoses into modern ones.

general social tensions.[82] The 1880s was a time of increasing concern about the 'Malay' (Muslim) threat in Cape Town, as the white middle classes believed that the smallpox epidemic of 1882 was exacerbated by the burial practices of the Muslim community. Muslims were associated with magic, poisoning and dirtiness.[83] One man who was sent to Robben Island in 1886 complained that he could not fish because the 'Malay men' hid under the water and took the fish off his line.[84] Another patient refused to apologize for beating his wife and children because he said 'a Malay had bewitched him'.[85] The idea of being 'Malay tricked' was also a feature of delusional content among Valkenberg patients.[86]

Causes of insanity

Lay and medical explanations of insanity commonly emphasized the inability of the insane person to cope with the trials and temptations of life, or the adverse effects of excess. This 'social' explanation of insanity was older than, and existed alongside, physical explanations (referring to brain lesions) and later physiological ones (referring to brain function) advanced by alienists in Europe and America.[87] But the assignation of etiology was not a priority for the asylum doctor. Only just over a quarter of admissions to the Robben Island asylum between 1872 and 1890 are given etiologies, while the register is complete in most other respects. Of the cases given etiologies, 37 per cent (n = 39) were deemed hereditary, 31 per cent (n = 33) due to physical causes and 26 per cent (n = 28) to moral causes. The latter included 'adverse circumstances', 'disappointed affections', 'religious enthusiasm' and 'temper'.[88] 'Physical' etiologies, that also had moral or social dimensions, included climate, 'deviant' sexual behaviour such as masturbation or promiscuity, and alcohol abuse.[89] By the late nineteenth century, doctors saw heredity as the primary etiology. Dodds suggested in 1891 that female insanity was due mostly to hereditary factors or 'other bodily diseases'.[90] Dr Greenlees of the Grahamstown Asylum

[82] J. C. Burnham, 'Psychotic Delusions as a Key to Historical Cultures: Tasmania, 1830–1940', *Journal of Social History* 13 (1980), 373.
[83] J. V. Bickford-Smith, *Ethnic Pride and Racial Prejudice in Victorian Cape Town: Group Identity and Social Practice, 1875–1902* (Cambridge, 1995), 71–4.
[84] Case of Jan, n.d., Health Branch, Criminal lunatics 1893–1899, CO 8050, CA.
[85] Case of Bekker, 25 Nov. 1895, Attorney General's Papers, Lunatics 1894–5, AG 1932, CA.
[86] Swartz, 'The Black Insane', 404–5.
[87] Berrios, 'Historical Background', 29.
[88] These etiological terms were not invented by colonial doctors (see Sankey cited in R. Russell, 'Mental Physicians and their Patients: Psychological Medicine in the English Pauper Lunatic Asylums of the later Nineteenth Century', PhD thesis, University of Sheffield (1983), 41).
[89] For an example of self-diagnosis see A. Simons to W. J. Dodds, 22 December 1894, Valkenberg Asylum casebook 1, 1891–4, UCT Manuscripts Collection, Cape Town.
[90] Report of Inspector of Asylums in 'Reports of the Medical Committee . . . for 1891', CPP, G36–1892, 10.

argued that heredity was a more important cause of insanity among whites in the Colony than in England.[91] There was a marked drop in diagnostic interest at Robben Island after 1890, due probably to the large number of supposedly 'incurable' black and criminal cases. Etiologies were given to only 11 per cent (n = 34) of first admissions in the period 1895–1910, compared to 28 per cent (n = 106) in the period 1872–1890.

Throughout the period 1846 to 1910, alcoholism was advanced as a cause of insanity in just over a tenth (n = 18) of all cases given etiologies. This is the second largest category after 'heredity'. In evidence from the 1850s, it is clear that although addiction to drink was seen as a cause of insanity, being drunk was not conflated with being insane, and the insane alcoholic could be cured by abstinence. K was said by his brother to have 'destroyed' his 'mental faculties' through drink in 1849;[92] Birtwhistle suggested in 1850 that Hugh G be sent on a sea voyage to avoid temptation from drink.[93] In 1855 Birtwhistle said that Mr V, admitted to Robben Island with 'mania', had merely been 'suffering from the effects of drink' when examined by Dr Frankel on the mainland, and was not therefore showing further signs of insanity.[94] Epileptic cases, said the chaplain in 1876, would be improved by 'the withholding of intoxicating drinks'.[95] Alcohol was nevertheless provided for patients as part of their asylum diet because it was a central part of nineteenth-century medical treatments.[96] In the 1860s, a patient with *delirium tremens* was turned away from the Somerset Hospital with the advice to go home and drink some whisky.[97]

The social profile of the Robben Island patient

This section examines the social profile of first admissions to the Robben Island asylum between 1846 and 1910, focusing on the period 1872–1890, for which there is most information. This period differs from earlier and later periods at the asylum because non-criminal cases make up a large proportion of the intake, and white lunatics predominate. The figures in Table 1.4 show that the Robben Island admissions were disproportionately likely to be middle-aged

[91] Quoted in R. C. Warwick, 'Mental Health Care at Valkenberg Asylum 1891–1909', BA (Hons.) thesis, UCT (1989), 57.
[92] Memorial of K, 29 October 1849, memorials received by Colonial Office, CO 4047, doc., 3, CA.
[93] J. Birtwhistle to Colonial Secretary, 23 Oct. 1850, Robben Island Letterbook, RI 1, CA. See also Birtwhistle to Colonial Secretary, 2 Nov. 1854, Robben Island Letterbook, RI 1, CA.
[94] Birtwhistle to Colonial Secretary, 18 March 1855, Robben Island Letterbook, RI 1, CA.
[95] Chaplain's Report in 'Report on the General Infirmary, Robben Island for the year 1876', CPP, G20–1877, 6.
[96] W. Edmunds in P. E. Wodehouse to the Duke of Newcastle, 7 Aug. 1863 in 'Correspondence between the Colonial Government and the Authorities at Home Relative to the Robben Island Institution 1863–1865', CPP, A9–1865, 18.
[97] P. Landsberg, minutes of evidence, 'Report of the Select Committee appointed to take into consideration the papers laid on the table referring to Somerset Hospital', CPP, A27–1865, 20.

Table 1.4 *Social profile of first admissions to the Robben Island Lunatic Asylum, 1846–1910*

Date	Total in sample	White %	Male %	Paying %	Mean age	Median age
1846–61	261	37.8	61.8	0.8	33.5	30.0
1862–71	186	56.5	68.5	4.8	33.8	32.0
1872–90	384	60.8	65.0	24.0	35.2	34.5
1895–1910	310	19.4	63.9	7.1	34.9	32.0

males between 1862 and 1890. Paying patients represented about a quarter of first admissions between 1872 and 1890.

Different admission or diagnostic patterns for racial and gender-defined groups at the Robben Island asylum could be caused by the race or gender-bias of colonial officials or doctors, or by systematic differences between these groups in terms of family circumstance, culture and incentives or opportunities for seeking care. The relative role of these factors in diagnosis could be established by looking at individual case records. These are however absent from the Robben Island archive, the only detailed case records coming from records of those transferred to Valkenberg or Grahamstown or the Old Somerset Hospital. An analysis of the patient profile can nevertheless inform our understanding of the way in which the asylum was used by psychiatrists, their clients and the community.

Fox shows that admissions to the San Francisco Asylum (1906–29) were mostly lower-class, single adult males.[98] Black admissions to Robben Island were largely single adult males but among white admissions there was an increasing tendency to use the asylum for middle-class, married white females in the period from 1860 to 1890. Both black and white women continued to be underrepresented at Robben Island compared to the colonial population, however, possibly for different reasons. Black women, especially Africans, were not fully urbanized and therefore avoided contact with white employers or agents of the state. More white families could afford private care to avoid the stigma of institutionalization, and they were more likely to keep mentally ill women at home.

Recent historians of gender and psychiatry have argued that women have suffered the brunt of psychiatric intervention as they are represented in greater numbers both in Victorian asylums and in the more diffuse psychiatric patient population today.[99] This feminization of psychiatry is not evident in South

[98] Fox, *So Far Disordered in Mind*, 105.
[99] E. Showalter, 'Victorian Women and Insanity', *Victorian Studies* 23 (1980), 161.

Africa: neither at Robben Island during the nineteenth century, nor today.[100] During the nineteenth century the gender ratio in Cape asylums remained stubbornly favourable to men. In the early twentieth century, the preponderance of male patients at Robben Island can be partly ascribed to the increasing proportion of criminal insane patients (largely men), and possibly also to the increasing proportion of black patients, for which group there may have been some gender specific recruitment because of the initial predominance of males among African migrant labourers in the urban areas. By the 1890s only Valkenberg Asylum attracted a significant proportion of long-stay female patients whose middle-class families found the asylum acceptable.[101]

External factors and the allocation of institutional beds in segregated asylums can also influence gender ratios, however. The dominant use of the Victorian asylum for pauper cases (women were more likely to be recipients of poor relief), and the provision of more ward space for women in asylums built after the 1830s were important factors in creating the consistently high ratios of women to men in Victorian asylums.[102] And although admission ratios are valuable in detecting inequalities, they do not tell the whole story. As Fox has pointed out, inequalities in admission ratios, or the lack of such inequalities, does not automatically imply the absence of gendered inequalities associated with psychiatric care.[103] In fact, he shows that in San Francisco between 1906 and 1930, although gender ratios on admission approached unity, women admitted to the state asylums suffered from longer attacks, were more likely to have had previous commitments and attacks, and were overrepresented in the age group sixty-five and over, compared to men.[104] Both variation in length of stay (women stayed longer) and allocation of bed spaces (men had more bed spaces) played a role in the gendering of psychiatric provision at Robben Island.

Compared to the general population, proportionally fewer black people than white were admitted to Robben Island asylum. As in the past, black South Africans today have different admission figures for certain psychiatric conditions,[105] there are racial differences in the type and form of some mental diseases,[106] and some have argued that there are different intra-racial profiles depending on experiences of urbanization and what has been termed 'transculturation'.[107] In modern South Africa where racial differences are bound closely to class and cultural divides, different patterns of aid-seeking,[108]

[100] Swartz, 'Colonialism and the Production of Psychiatric Knowledge', 21.
[101] *Ibid.*, 35, 45, 46. [102] Showalter, 'Victorian Women and Insanity', 162, 164.
[103] Fox, *So Far Disordered in Mind*, 105, 124. [104] *Ibid.*, 127–29, 131.
[105] Freed and Bishop cited in L. Swartz, 'Aspects of Culture in South African Psychiatry', PhD thesis, UCT (1989), 39.
[106] Bartocci cited in Swartz, 'Aspects of Culture', 41.
[107] Cheetham *et al.* cited in Swartz, 'Aspects of Culture', 47.
[108] Swartz, 'Aspects of Culture', 40.

different cosmologies,[109] complex culturally based communication failures between psychiatrist and patient and other variables can systematically influence psychiatric profiles of the various racial categories. Leslie Swartz argues that different incentives and opportunities for admission may by themselves produce different psychiatric profiles for various racial groups in South Africa today.[110] He has also criticized modern cross-cultural studies in South African psychiatry for treating black and white patients as culturally separate.[111] In assessing degree and amount of mental illness among Xhosa speakers in South Africa using the Present State Examination (PSE) in translation as a standard diagnostic tool may be problematic.[112] Many studies continue, explicitly or implicitly, to use racial stereotypes (e.g. the African personality), or a passive, static representation of African culture, in order to explain the extent and type of mental illnesses among black people.[113] Similar patterns of racism within psychiatry have been documented in Britain, where schizophrenia, for example, is more commonly diagnosed among blacks and the Irish.[114]

Cultural factors would certainly have affected the admission process at Robben Island. Fox argues that in early twentieth-century San Francisco, insane foreigners were relatively unlikely to have family in the area who could refer them to the asylum and were therefore more likely to be picked up by police. But by comparing police referrals for foreigners and for locally born cases without family in the area, Fox can conclude that cultural factors played a large role in determining which people were taken in by police. Foreigners wandering about the city in inappropriate areas would be picked up and, if unable to give a good account of themselves in English, appearing confused or hostile, would often be sent to the courts on a charge of insanity.[115] In the nineteenth-century Cape, recent arrivals to urban areas, especially those who did not speak English, sometimes ended up in the asylum.[116] Recent arrivals from rural African communities in the eastern Cape may have experienced similar problems, as they did in Argentina and Australia.[117] Cultural 'boundaries' were not always contiguous with nineteenth-century racial categories, however. Cultural similarities between rural Cape 'coloured' and

[109] H. Ngubane cited in Swartz, 'Aspects of Culture', 22.
[110] Swartz, 'Aspects of Culture', 40. [111] *Ibid.*, 52.
[112] Gillis *et al.* cited in Swartz, 'Aspects of Culture', 36.
[113] Swartz, 'Aspects of Culture', 17, 23.
[114] M. Lipsedge and R. Littlewood, *Aliens and Alienists: Ethnic Minorities in Psychiatry* (London, 1989).
[115] Fox, *So Far Disordered in Mind*, 87–89.
[116] For example, see the report of a Russian detained at Robben Island who could not speak any English, Health Branch Correspondence, 1901, CO 7258, CA.
[117] J. Ablard, 'The limits of psychiatric reform in Argentina, 1890–1946', below, notes a similar pattern for long-distance peasant admissions to asylums in Buenos Aires; see also Coleborne, 'Passage to the asylum', below, on Chinese immigrants to Australia.

Dutch-Afrikaans 'white' populations (especially in the eyes of British-born policemen) may sometimes have outweighed 'racial' difference between the two groups.

Although admission processes favoured the admission of new arrivals to the city, they also reflected general patterns in an already racialized social order. In the colony as a whole, blacks were more likely to be poor and non-Christian than whites. Common-law marriages were also more common among blacks. These patterns were echoed in the Robben Island intake. Between 1872 and 1890 there were only four black paying lunatic patients at Robben Island (3 per cent of 150 black first admissions), all 'mixed-race', while paying patients made up 37 per cent (n = 79) of white admissions. Blacks were more likely than white admissions to be non-Christian or Dutch Reformed than from Anglican, Catholic or other Christian denominations. A greater proportion of white admissions were married and more of the black admissions were single. Black admissions to Robben Island were also more likely to have come from the eastern Cape and western Cape hinterland, although not from outside the colony, than whites. This is not surprising – far more Africans lived in the eastern Cape and far fewer Africans were urbanized, even in 1904.[118]

One would expect that racialized theories about insanity would produce racialized etiologies and diagnoses, but this was less marked in the analysis of admission registers than expected. Black patients admitted to Robben Island were given substantially fewer etiologies than whites, an indication of greater medical concern about white insanity and the perception that black insanity was less complex.[119] There was also a disproportionate number of blacks among idiotic and imbecile admissions and fewer blacks among those given general diagnoses like 'insanity'. This may reflect a different patient profile (due to different pressures for admission) rather than racism in diagnosis, however. Diagnoses of melancholia (then based partly on considerations of mood) certainly had a racial connotation. This diagnosis was restricted to twenty white patients before 1888, of whom fifteen were paying, and thirteen were either married or widowed. A disproportionate number of white, married and paying patients were thus given diagnoses of melancholia. The diagnostic category may have been used to decrease the stigma attached to institutionalization for better-class white patients with family attachments. Racial differentiation is also evident in the diagnosis of melancholia at other asylums. Black patients were believed to be unable to experience depression because they had insufficiently

[118] The 1904 census shows that 8 per cent of Africans, 48 per cent of 'coloureds' and 52 per cent of whites were urbanized; C. Simkins and E. van Heyningen, 'Fertility, Mortality and Migration in the Cape Colony, 1891–1904', *International Journal of African Historical Studies* 22 (1989), 94.

[119] For the same pattern in other Cape asylums see Swartz, 'Colonialism and the Production of Psychiatric Knowledge', 102.

evolved nervous systems. Even today, it is often assumed that black patients suffer less from melancholia than whites.[120]

Admission patterns are associated with a complex set of influences, which are difficult to disaggregate. This can be demonstrated with regard to age of first admissions. For the entire period 1846–1910, the mean age at first admission was 34.5 years, with the median at 32.0 years (n = 1114). In the first seventeen years of the asylum, the median age of first admissions was 30.0 (n = 260), but between 1872 and 1890, the median age was 34.5 (n = 362). The data before 1872 is skewed towards long-stay cases who would have been younger on admission. The increase in median age was probably due to an increase in recording of admissions in the forty to fifty-nine age group. By the 1870s, psychiatric services were expanding to take in more demented and mentally retarded patients (diagnoses of dementia, idiocy and imbecility had increased) but these were concentrated in the younger age groups rather than in the age-bracket providing senile cases today.[121] When Robben Island accommodated more criminal cases, the median age decreased again (to 32.0 years, n = 307), as the penal system mainly accommodated and referred younger men.

Throughout the period, the Robben Island first admissions cluster around the twenty to thirty-nine age group, that represents over 60 per cent (n = 686) of all first admissions, although this age group represented only about 20 per cent in the general population (and 40 per cent of the population aged over twenty) in 1875.[122] Most nineteenth-century asylums in Britain also accommodated mainly young adults and those in early middle age.[123] The middle-age range of the Robben Island admissions illustrates its use as a place of detention for those cases who could be expected to work but did not, or for those who presented a particularly dangerous aspect if violent, but it may also be a product of the asylum's use by immigrants. There are very few first Robben Island admissions under twenty years (6 per cent, n = 70) and over sixty years (7 per cent, n = 64). Comparing these proportions to those in the colonial population in 1875 (about 50 per cent and 5 per cent respectively),[124] one can see that very young people were grossly underrepresented but old people were not. Local communities may have been more reluctant to use the asylum for the very young, who might have been more economically useful and easier to control than the very old. A slightly greater proportion of older people was admitted later in the century, but this is probably a product of sample bias. Without much institutional provision

[120] Swartz, 'Changing Diagnoses in Valkenberg Asylum', 441–3.

[121] Before Kraepelin's work in the late nineteenth century, no distinction was made between dementia due to 'dementia praecox' (later called schizophrenia) and senile dementia. Many of the dementia cases were therefore under sixty.

[122] Census of 1891, CPP, G6–1892, 48.

[123] A. Digby, *Madness, Morality and Medicine: A Study of the York Retreat 1796–1914* (Cambridge, 1985), 176–7.

[124] Census of 1891, CPP, G6–1892, 48.

for the elderly at the Cape, it is not surprising that the old were not under-represented.

In a contemporary American study, sociologists have found that married people have consistently lower incidence of mental illness.[125] For those who suffer mental illness before marriage, marriage may not be an option. Also, families are able to provide some care for the insane. In asylums in nineteenth-century Britain, more patients were single than married.[126] One would therefore expect to have fewer married admissions to Robben Island than married people in the general population. In the years 1872–90, of those Robben Island lunatic admissions whose marital status was listed, 40 per cent (n = 129) were listed as 'married' and 54 per cent as 'single' (n = 175). By contrast, in 1875 about 56 per cent of the colonial urban population aged twenty and over were married.[127] The assumption thus holds true. There was a gender dimension to this pattern, however. Proportionally more of the female and white admissions to Robben Island were married. Between 1872 and 1890 half (n = 60) of female first admissions at the Robben Island asylum given marital status were married while only a third (n = 69) of male first admissions were married. Just over half (n = 65) of female first admissions were single or widowed compared to two-thirds (n = 141) of male first admissions. Fox suggests that women in early twentieth-century San Francisco were more likely than men to be perceived as insane within the family, and more likely to be sent to an asylum through family intervention. If subject to a dementing process, women had fewer socially acceptable options than men for care outside the home.[128] The Robben Island case may indicate a similar pattern in the Cape.

The increase in 'better class' patients

When the first 'better-class' patient arrived at Robben Island in a whale boat from the Old Somerset Hospital in 1850, surgeon-superintendent Dr Birtwhistle failed to treat him better than the others and was berated by the Colonial Secretary for this.[129] Most of the lunatics during the 1840s and 1850s were paupers, leaving little behind them when they died, with a few exceptions.[130] The favoured patient V earned £7 2*s*. 6*d*. from selling property in King William's Town in 1851.[131] D, admitted as a paying patient in April 1854, had exhausted

125 D. R. Williams, D. T. Takeuchi and R. K. Adair, 'Marital Status and Psychiatric Disorders among Blacks and Whites', *Journal of Health and Social Behaviour* 33 (1992), 140–1.

126 Digby, *Madness, Morality and Medicine*, 175.

127 Census of 1875, CPP, G42–1876, 15.

128 Fox, *So Far Disordered in Mind*, 97, 131.

129 Birtwhistle to Secretary to Government, 12 Sept. 1850, CA, Robben Island Letterbook, RI 1.

130 Birtwhistle to Colonial Secretary, 21 April 1851 and 14 March 1853, CA, Robben Island Letterbook, RI 1.

131 Birtwhistle to Secretary to Government, 21 April 1851, CA, Robben Island Letterbook, RI 1.

his deposit by August and was demoted to non-paying status.[132] By 1855, just over a tenth of the lunatics were 'better-class', but not all of these were paying patients.[133] In 1861 surgeon-superintendent Minto differentiated between 'better-class' lunatics (one third of the patients) and 'educated' lunatics (very few).[134] The standard rate for paying patients (the same as at Old Somerset Hospital) was two shillings per day[135] – four times the bare minimum needed to maintain oneself in Cape Town during the 1830s, and beyond the reach of all but the skilled worker or subsidized employee.[136]

The reform of the asylum in the 1860s attracted more middle-class patients to the Robben Island asylum in search of a socially acceptable therapeutic and custodial solution. Between 1872 and 1890, a quarter (n = 92) of first admissions were paying, in contrast to less than 5 per cent (n = 11) before 1872. Between 1846 and 1890, the proportion of white patients (who would probably have been wealthier) increased from about two-fifths to nearly three-fifths. British immigrants and other immigrant Europeans made up a much higher proportion of first admissions than colonial-born whites in the period before 1872. But by the 1870s, a growing number of colonial-born whites were admitted, including some from 'respectable' colonial families. By 1880 half of the approximately 200 Robben Island lunatics were white, compared to a third of the 100 patients at Somerset Hospital.[137] Edmunds's success in improving the public image of the Robben Island asylum attracted more paying patients, some with well-respected colonial names, although even in the 1870s better-class families still sent relatives 'reluctantly' to Robben Island.[138] During the 1890s, when more mainland asylums opened, however, the proportion of white, non-earning and paying patients at the Robben Island asylum declined dramatically.

Although more wealthy patients entered the asylum in the 1860s the percentage of first admissions from the professional, self-employed or skilled classes remained fairly stable (at 11–13 per cent) before 1890 (see Table 1.5) because many of the new middle-class admissions were non-earning women. The proportion of non-earners (mainly housewives) admitted to Robben Island had tripled by the 1860s and 1870s, but declined thereafter. As convict and unskilled

[132] *Ibid.*, 15 Aug. 1854, CA, Robben Island Letterbook, RI 1.

[133] 'Return of Chronic Sick, Lunatics and Lepers in the General Infirmary', in 'Report of the Select Committee Appointed to take into Consideration the Papers Laid on the Table Referring to Somerset Hospital', CPP, A9–1855, 17.

[134] J. C. Minto, Minutes of evidence, 'Report of the Commission of Inquiry', CPP, G31–1862, 15.

[135] 'Reports on the Somerset Hospital and the General Infirmary, Robben Island, for the Year 1855', CPP, G12–1856, 2.

[136] S. Judges, 'Poverty, Living Conditions and Social Relations: Aspects of Life in Cape Town in the 1830s', MA thesis, UCT (1977), 3.

[137] Captain Mills, minutes of evidence, 'Report of the Commission', CPP, G64–1880, 1.

[138] Robben Island Lunatic register 1846–1904, RI 181, CA; Ebden, minutes of evidence, 'Report of the Select Committee . . . on Removing the Lunatics and Lepers from Robben Island', CPP, A3–1871, 3.

Table 1.5 *Occupations of first admissions to the Robben Island Lunatic Asylum whose occupations are given, 1846–1910 (percentage)*

Date	Middle class	Artisans/ skilled	Semi-skilled	Unskilled	Not earning	N/A (Convict)	Total
1846–61	11.4	5.7	17.1	14.3	5.7	45.7	100
	(n = 4)	(n = 2)	(n = 6)	(n = 5)	(n = 2)	(n = 16)	(n = 35)
1862–71	14.7	12.0	17.3	14.7	17.3	24.0	100
	(n = 11)	(n = 9)	(n = 13)	(n = 11)	(n = 13)	(n = 18)	(n = 75)
1872–90	12.9	9.9	9.6	19.8	20.1	27.8	100
	(n = 43)	(n = 33)	(n = 32)	(n = 66)	(n = 67)	(n = 93)	(n = 334)
1895–1910	1.0	4.1	8.1	36.9	15.9	33.9	100
	(n = 3)	(n = 12)	(n = 24)	(n = 109)	(n = 47)	(n = 100)	(n = 295)

admissions (mainly black men) increased again in the 1890s, middle-class admissions declined in general.

The increase in paying patients admitted to Robben Island during the 1870s was most dramatic among white females (see Figure 1.1). The percentage of white admissions among the female insane is a particularly sensitive indicator of the use of the asylum for those who were not just poor or considered dangerous. White families were generally wealthier and therefore able to choose whether to admit their relatives to the asylum or keep them at home. Black families, by contrast, seem to have played only a minor role in selecting the asylum as a place of custody for insane relatives. The colony's penal system, that tended to target blacks in any case, was the major referral agent for black admissions to the lunatic asylum. In the period 1872–90 nearly 40 per cent (n = 17) of African first admissions were convict lunatics as opposed to 13 per cent (n = 12) of British-born first admissions and 24 per cent (n = 20) of white colonials. But although middle-class whites of European extraction were perhaps more willing to resort to an asylum for insane relatives, this willingness was tempered by considerations of gender. They were less likely to see female lunatics as dangerous than men. The stigma of institutionalization also rested more heavily on women, who were traditionally seen as occupying the private sphere of the home. The admission of greater proportions of white females to an asylum therefore indicates a reduction in the stigma attached to institutionalization and a recognition of the asylum's therapeutic role. (It was only after this shift in perception that married women would be more likely than married men to enter the asylum, as Fox has observed above.)

In 1859 there were as many white as black men in the Robben Island lunatic asylum, but more black women than white.[139] The gap between the proportions

[139] 'Report on the General Infirmary, Robben Island for the Year 1859', CPP, G11–1860, 4.

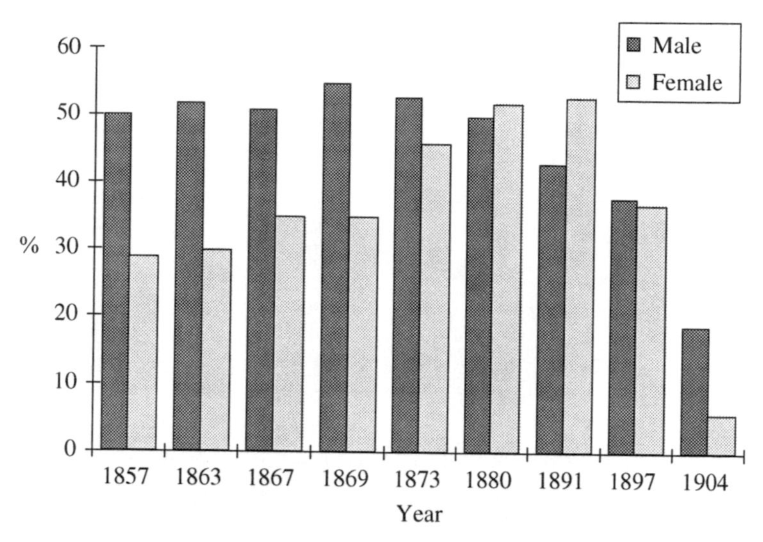

Figure 1.1 White lunatics resident in the Robben Island Lunatic Asylum as a percentage of total lunatics resident, by gender, 1857–1904

of white and black female lunatics narrowed gradually during the 1860s (see Figure 1.1 above). By 1873 there were sixty-eight male and thirty-four female white lunatics and sixty male and forty female black lunatics in the asylum.[140] Segregated accommodation at Robben Island was used to reserve places for better-class patients and encourage applications. But white female admissions were encouraged even more actively than white male admissions. A new building was constructed for the female asylum in 1867, a separate wing was set aside for better-class female patients on the island by 1873[141] and paying female patients had their own ward by 1880.[142] In 1871, surgeon-superintendent Edmunds claimed that although arrangements for the males were 'incomplete', better-class female patients were well enough catered for.[143] Female inmates formed a disproportionate number of the paying patients, most of whom were admitted between 1870 and 1890. They were overrepresented in the non-earning category, mainly as 'housewives', and underrepresented in the employed petty bourgeois and semi-skilled categories. Grahamstown asylum admitted some 'better-class' women when it opened in 1875, but the long distance from Cape Town may have encouraged some families to use Robben Island instead. Disproportionally more women than men were admitted to Robben Island from

[140] 'Report on the General Infirmary, Robben Island for the Year 1873', CPP, G23–1874, 3.
[141] *Ibid.*, 6.
[142] Lunacy Inspectors' Report in 'Report on the General Infirmary, Robben Island for the Year 1880', CPP, G22–1881, 9.
[143] Edmunds, minutes of evidence, 'Report of the Select Committee', CPP, A3–1871, 3.

Table 1.6 *Outcomes of all first admissions to the Robben Island Lunatic Asylum, 1846–1910 (percentage)*

Date	Discharged	Died	Transferred	Escaped	Not given	Total
1846–61	3.4 (n = 9)	20.7 (n = 54)	9.6 (n = 25)	–	66.3 (n = 173)	100 (n = 261)
1862–71	12.4 (n = 23)	41.4 (n = 77)	16.1 (n = 30)	–	30.1 (n = 56)	100 (n = 186)
1872–90	25.0 (n = 96)	39.3 (n = 151)	32.6 (n = 125)	0.3 (n = 1)	2.9 (n = 11)	100 (n = 384)
1895–1910	31.6 (n = 98)	32.3 (n = 100)	20.3 (n = 63)	–	15.8 (n = 49)	100 (n = 310)

the Boland,[144] and slightly fewer from the more distant western Cape hinterland. Families were possibly less willing to send their female relatives too far from home.

Leaving the asylum

In order to assess the functioning of the institution, some indication has to be given of length of stay and therefore the composition of the resident population as opposed to admissions. An analysis of length of stay and outcome can reveal much about the social profile of patients and the reasons for their institutionalization. For this purpose the data from the years 1872–90 is the most reliable. Of those given discharge dates before 1872, only one case stayed less than eleven years in the asylum. This is because most early cases in the database are gleaned from the Old Somerset Hospital register that does not list outcome after the patient's stay at Robben Island. Tables 1.6 and 1.7 show how the admissions database is skewed towards patients who were not given outcomes in the years before 1861 and those who remained in the asylum in 1872 and died there. A significant number of cases admitted in the period after 1895 were not given outcomes.

There has been some debate as to whether British asylums filled up with chronic cases during the last third of the nineteenth century.[145] This did not happen at Robben Island in the long term, although there were a number of long-stay patients, especially before 1891. Of first admissions to the asylum in the period 1872–90,[146] only 16 per cent stayed a year or less. About two-

[144] The Boland, or uplands, refers to an area encompassing the wine- and wheat-farming towns in the immediate vicinity of Cape Town such as Stellenbosch and Paarl.

[145] G. N. Grob, 'Marxian Analysis and Mental Illness', *History of Psychiatry* 1 (1990), 229–30.

[146] That is, all those who were given outcomes, or dates and reasons why they left the asylum. The earlier cases in the database either have no outcome listed (those from the Somerset Hospital

Table 1.7 *Length of stay of first admissions to the Robben Island Lunatic Asylum whose outcome is given, 1846–1910 (percentage)*

Date	<1 year	1–2 years	2–4 years	4–9 years	9–18 years	>18 years	Total
1846–61	1.1 (n = 1)	–	–	–	13.6 (n = 12)	85.2 (n = 75)	100 (n = 88)
1862–71	4.6 (n = 6)	6.9 (n = 9)	3.8 (n = 5)	22.3 (n = 29)	23.1 (n = 30)	39.2 (n = 51)	100 (n = 130)
1872–90	15.8 (n = 59)	18.8 (n = 70)	22.7 (n = 59)	21.2 (n = 55)	18.0 (n = 67)	7.8 (n = 29)	100 (n = 339)
1895–1910	25.0 (n = 65)	27.3 (n = 71)	22.7 (n = 59)	21.2 (n = 55)	3.8 (n = 10)	–	100 (n = 260)

thirds of the cases stayed over two years, and a quarter stayed longer than nine years (see Table 1.7). This indicates the relative severity of the Robben Island cases or the slow discharge process. In two state asylums in Harrisburg in America between 1880 and 1910, about half of the patients were discharged after one year, and this proportion rose to 60 per cent in private asylums.[147] With the establishment of other mainland asylums in the 1890s, the number of patients who were transferred or discharged from the Robben Island asylum increased,[148] and the number of long-stay patients at Robben Island decreased accordingly.

Outcomes varied according to length of stay. Of all first admissions in the period 1872–90 who were given outcomes, a quarter (n = 96) were finally discharged, two-fifths (n = 151) died in the asylum and a third (n = 125) were transferred to other asylums. But these outcomes were unevenly distributed between long- and short-stay patients. Over half of the discharges (56 per cent, n = 54) occurred within the first two years of admission. Discharge became very unlikely after four years in the asylum (82 per cent (n = 79) of all discharged cases had left by then). This was similar to the pattern at the Retreat asylum in York (1796–1910) although discharge happened even more rapidly at York.[149] Death rates at Robben Island were higher than discharge rates for every length-of-stay group, except for those staying between one and two years in the asylum, when the weak and debilitated cases had already perished. Naturally, the longer a patient stayed in the asylum, the more likely s/he was to die there. But the high death rate in the initial intake can be contrasted with the pattern at York, where

registers) or are the long-stay cases that were in the asylum in 1872 when the register was started. The best data on length of stay thus exists for admissions after 1872.

[147] C. McGovern, 'The Myths of Social Control and Custodial Oppression: Patterns of Psychiatric Medicine in Late Nineteenth-Century Institutions', *Journal of Social History* 20 (1986), 6, 19.

[148] This is evident from the official reports rather than the database figures as outcomes were not reported as often after 1890.

[149] Digby, *Madness, Morality and Medicine*, 227.

death rates only exceeded discharge rates for those staying over two years in the asylum.[150] This indicates the more extensive use of the island asylum for patients who were seriously ill or debilitated from poverty. As more asylums were opened during the course of the century, the reorganization of asylum provision on the basis of race, violence and chronicity necessitated some movement of patients between asylums. At Robben Island, therefore, long-stay patients were very likely to be transferred. Indeed, after four years in the asylum, patients were more likely to be transferred than to die. The most common reasons for transfer or non-transfer were difficulty of control or suitability for work. After 1891, difficult cases were increasingly transferred out of Grahamstown and Valkenberg asylums to Robben Island or Fort Beaufort, the asylum for chronic cases. In 1896, for example, a patient admitted to Grahamstown and described there as not dangerous but a 'voluble, spiteful fussy little Irishwoman who, from the violent use she makes of her tongue, is always getting into trouble', was transferred 'not improved' to Robben Island.[151] The use of black patients as manual labourers within asylums affected transfer patterns.[152] When the Grahamstown Asylum became solely for whites in 1908 'the withdrawal of the native labour supply was rather severely felt' and black male patients were again housed nearby two years later.[153] In 1916 Valkenberg admitted its first black patients (housed in a separate building) for similar purposes.[154] Manual labour provided by black patients was essential in reducing the expense of running asylums at the Cape.[155]

A closer analysis of length-of-stay and outcome can provide some ideas as to social and economic factors prompting a resort to the asylum. The data can be usefully disaggregated by diagnosis, area of origin, socio-economic group, gender and race (see Table 1.8). The severity of a patient's illness affected the length of time s/he stayed in the asylum. The few melancholic cases (this was the socially acceptable face of insanity for the middle classes),[156] showed a

[150] *Ibid.*, 228. The statistics Digby gives in Table 9.6 have been used to get a figure of 43.2 per cent for deaths in the group staying two to ten years and 35.4 per cent for those discharged recovered or improved. The discharge rate would in fact exceed the death rate if one included the cases discharged unimproved, but this would perhaps be more comparable to the transfer figures for Robben Island than the discharge figures as York did not transfer cases to other asylums.

[151] Admissions for 1896, Grahamstown Asylum Papers, casebook 1875–93, HGM 16, CA.

[152] Work was a common theme of institutional conditions for working-class patients. See Ablard, 'The Limits of Psychiatric Reform in Argentina, 1890–1946', below, for a discussion of manual labour in Argentinean institutions.

[153] Report of the Commissioner of mentally deranged and defective persons in 'Report of the Department of the Interior for 1916–1918', Union of South Africa Parliamentary Papers (UPP), UG31–1920, 21.

[154] *Ibid.*, 19. [155] Swartz, 'The Black Insane', 412.

[156] Melancholia was presumably acceptable to the middle class because it involved introversion, a return to privacy, rather than florid extroversion. The same romantic idea that associated middle-class depression with civilization and creativity played out in the arena of tuberculosis, the wasting disease.

Table 1.8 *Social profile of first admissions to the Robben Island Lunatic Asylum by length of stay, 1872–1890*[a] *(percentage)*

Length of stay	Cases in Sample	Male[b] %	White %	Married[c] %	Paying %	Convicts %
< 1 year	59	67.8 (n = 40)	72.9 (n = 43)	55.6 (n = 30)	28.8 (n = 17)	8.5 (n = 5)
1–2 years	70	62.9 (n = 44)	75.4 (n = 52)	47.5 (n = 28)	37.1 (n = 26)	7.1 (n = 5)
2–4 years	75	56.0 (n = 42)	58.7 (n = 44)	34.9 (n = 22)	28.0 (n = 21)	9.3 (n = 7)
4–9 years	73	79.5 (n = 58)	42.5 (n = 31)	30.5 (n = 18)	15.1 (n = 11)	20.5 (n = 15)
9–18 years	67	59.7 (n = 40)	52.2 (n = 35)	41.4 (n = 24)	13.4 (n = 9)	52.2 (n = 35)
Over 18 years	29	58.6 (n = 17)	62.1 (n = 18)	23.1 (n = 6)	27.6 (n = 8)	75.9 (n = 22)
All lengths	373	64.6 (n = 241)	59.9 (n = 223)	40.1 (n = 128)	24.7 (n = 92)	23.9 (n = 89)

[a] Because there are no data on sex, race, marital status, etc. for some cases within the length-of-stay group, the percentages listed here are valid percentages, excluding missing data.
[b] One missing case: total valid cases 372 of 373.
[c] Fifty-four missing cases: total valid cases 319 of 373.

high turnover, as 52 per cent (n = 13) of first admissions with this diagnosis left the asylum within two years. Over a third of first admissions between 1872 and 1890 with diagnoses of general insanity, idiocy or imbecility tended to leave the asylum within two years of admission, in contrast to a fifth of those with diagnoses of mania and dementia. The latter cases formed the bulk of long-stay patients. Long-stay cases had mostly been legally detained as criminal insane. Three-quarters (n = 22) of those cases staying over eighteen years were committed under Act 20 of 1879 (which was until 1891 the only legal means of detention, but was not used for the majority of admissions). While about 10 per cent of cases in most occupational groups stayed for over ten years, 64 per cent (n = 57) of convicts stayed for over ten years. 'Dangerousness' (as recorded in the asylum admission registers) does not seem to have implied long-term detention, however. Its primary role was to justify initial detention.

But the accessibility and viability of alternatives to asylum care also affected the length of stay. Those coming from the Boland or Greater Cape Town were, not surprisingly, more likely to leave the asylum within less than two years than cases from the eastern Cape or the western Cape hinterland. Cases with some chance of recovery and with family or friends in contact, were often transferred from Robben Island to mainland asylums.[157] The declining public image of the Island asylum after 1880 also encouraged middle-class patient mobility. The patient T asked to be transferred from Robben Island to Grahamstown in 1906, as 'I am mixed up with all kinds of criminal patients here whereas in Grahamstown I shall be among my own class.'[158] Married, white or paying patients were more likely than single, black or non-paying patients to leave the asylum in the first two years, perhaps partly because of the greater number of melancholia diagnoses among the former group.

Gender differences in length of stay can be partly explained by gendered notions of the necessity for institutional care. In the first two years of their stay in the asylum, men and women were equally likely to leave it by death, transfer or discharge. But later rates of leaving varied somewhat by gender. In the period 1872–90, a third (n = 39) of the female first admissions with outcome given stayed over nine years at the Robben Island asylum, compared to only a quarter of the men (n = 57). This was not due to differential outcomes. Men and women died, were transferred and discharged in similar proportions, although more men than women were admitted overall. Disproportionally more men than women were admitted to Robben Island with imbecility and dementia, which diagnoses tended to produce longer length of stays. There was also a group of long-stay women patients in the Valkenberg asylum after 1891. Swartz has argued that women admitted to Valkenberg were often represented, both before and after admission, as violent and troublesome as a sign of their insanity (it is interesting to note they were not more likely to be marked as 'dangerous' however). She attributes this partly to the cultural expectation that women should be more passive and sedentary than men.[159] Certain kinds of disruptive or violent behaviour by women in the asylum may thus have been more of a barrier to discharge than among men.

Patterns of asylum use also differed along racial lines. Black patients in general were more likely than white admissions to stay longer than three years in the asylum. They were slightly more likely to die in the asylum rather than be discharged or transferred. It is therefore unlikely that black families were using the asylum for the temporary detention of insane relatives. The black admissions to Robben Island were usually there under coercion rather than by relatives' request. The situation at the Cape parallels that in India where, with growing pressure on accommodation after legislation enforcing the detention of criminal lunatics in asylums in 1851, the basic criterion 'for the admission of Indians was to become their danger to the community at large rather than

157 Robben Island Commissioner to Under Colonial Secretary, 12 March 1901, Robben Island Letterbook, RI 61, CA.

158 W. H. T. to Colonial Secretary, 21 April 1906, Health Branch, letters received, CO 7810, CA.

159 Swartz, 'Colonialism and the Production of Psychiatric Knowledge', 46, 197–9.

their state of mind'.[160] In colonial Malawi too, the main motivation behind the establishment of an asylum in 1910 was the need to take criminal lunatics out of prisons, where they were disruptive of the institutional order.[161]

Conclusions

How and why 'lunatics' entered mental institutions is related closely to the social function of the institutions and the class, race or gender-specific meanings attached to insanity. The asylum was established in 1846 primarily as a dumping ground for those who were clogging up the colonial gaols and hospitals as chronic, disruptive or disabled cases. It was no coincidence that the establishment of the Robben Island asylum coincided with a programme to utilize convict labour on public works. During the 1860s and 1870s, middle-class white families began to use the asylum more, especially for female relatives. The reformed asylum was an icon of civilization at a time when the colonial middle class was attempting to establish itself on both the colonial and the international stage.[162] More 'dangerous' and criminal cases (mainly black men) were sent to the island asylum after two new mainland asylums for white middle-class patients were opened in 1875 and 1891. By the beginning of the twentieth century, when the colony's newly exploited mineral riches relied heavily on the control of black migrant labour, Robben Island inmates were again mainly poor, black, male and criminal.

The social profile of the Robben Island patient was determined by a variety of complex factors. In the Cape Colony, where there was no great confinement of the insane, facilities for institutionalization were minimal. These scarce resources were allocated and used in specific ways by the state, the medical profession, communities and families to create an asylum population dominated by single, urbanized men of working age. White patients were overrepresented in Cape asylum populations as a whole, but at Robben Island they dominated numerically only between 1862 and 1891. Black women were strongly underrepresented. This differential allocation of bed-space was related to patterns of mental illness, availability of alternative sources of medical care for the insane, community definitions of mental illness, community attitudes towards western medical care, colonial state priorities, medical practices and medical prejudices.

Although the asylum inmates at Robben Island, like those in the San Francisco asylum, had all been defined by society as inconvenient or intolerable, they

[160] Ernst, *Mad Tales from the Raj*, 50.

[161] M. Vaughan, 'Idioms of Madness: Zomba Lunatic Asylum, Nyasaland, in the Colonial Period', *Journal of Southern African Studies* 9 (1983), 220.

[162] See H. J. Deacon, 'Remembering Tragedy, Constructing Modernity: Robben Island as a National Monument' in C. Coetzee and S. Nuttall (eds.), *Negotiating the Past: the Making of Memory in South Africa* (Oxford, 1998) and Ablard, 'The Limits of Psychiatric Reform in Argentina, 1890–1946', below.

formed a specific subset of the colonial population who had travelled several well-worn paths to the asylum door. Patients admitted to Robben Island from 1846 to 1910 were disproportionally single, as are mental patients today. Almost all were in the middle-age range, over twenty and under sixty years, indicating the importance of inability to work or conduct family life in encouraging institutionalization. Most of the patients in Cape asylums, including Robben Island, were male. This was partly because of the legal emphasis on dangerousness as a criterion (more often applied to men) for admission to scarce asylum beds, and the way in which asylums were used by the state to empty gaols of disruptive criminals (who were predominantly men), especially at Robben Island. Married people were less likely to enter the asylum in general, as they had support networks at home, but married and well-off white women were more likely to enter the asylum than similarly placed men, probably because of their lower status and power within the household. While women were in general less likely to be institutionalized than men, they were more likely to become long-stay patients.

Robben Island asylum was at the custodial end of the psychiatric spectrum, except for a brief interlude in the 1860s and 1870s, and housed numerically more black than white patients for most of the period 1846–1921. But even at Robben Island, white patients were overrepresented in asylum admissions compared to the general population. This was explained at the time by racist theories that represented the white brain as more evolved and civilized, and therefore more susceptible to, and requiring more protection from, insanity than the black brain.[163] Better explanations can be found, however. White insanity was feared within the colonial order because it connoted degeneration and threatened hereditary insanity. Black insanity was feared mainly in its contact with white communities – potentially disrupting employment relations or breaking the taboo on sexual contact with white women. The black insane sequestered in relatively independent or isolated communities were thus not a major concern of the colonial state. Africans and Khoisan communities continued to use indigenous healing methods and to resist western medical treatment systems for mental illness well into the twentieth century. African women were particularly underrepresented among admissions because of their limited exposure to white employment and the criminal justice system, the major admission route of their male counterparts.[164] Many black male admissions were either seriously physically ill or debilitated on admission, suggesting both poorer general health, possibly due to distance from community networks of assistance, and a long route to the asylum through the criminal justice system.

163 Swartz, 'Colonialism and the Production of Psychiatric Knowledge', 113–16.
164 Swartz, 'The Black Insane', 409.

Those without family networks who could not afford home care were most likely to depend on the asylum, but cultural factors and alternative sources of aid kept many out of the asylum. This included recent immigrants (mainly British), ex-slaves and poor black inhabitants of Cape Town and the surrounding area. British settlers, mainly recent immigrants without other options, at least before the asylum became respectable in the 1870s, formed the majority of the voluntary patients in Cape asylums. Christian churches and associated organizations seem to have provided an asylum entry route for some poor mixed-race people and ex-slaves, but not, it seems for those from poor white Dutch-Afrikaans communities. Community attitudes towards institutionalization and western medicine played an important role in determining admission patterns, even among relatively poor communities. Like Africans, Dutch-Afrikaans settlers and Muslims in Cape Town were reluctant to use the asylum. Well-organized religious charity and community assistance programmes in urban areas gave them more options outside the asylum. Besides the western treatment approach, inflexibility regarding diet, the lack of proper provision for Muslim burials and the association of the island asylum with Christianity probably discouraged black Cape Town Muslims from using the asylum extensively.

2 The confinement of the insane in Switzerland, 1900–1970: Cery (Vaud) and Bel-Air (Geneva) asylums

Jacques Gasser and Geneviève Heller

Introduction

The two asylums upon which this comparative study of patient records is based share many similarities. Situated only sixty kilometres apart, both are public teaching hospitals of two neighbouring cantons – Vaud and Geneva – in the French region of Switzerland, the Swiss Romande. In Switzerland, which is a confederation of states (cantons), there is little centralization of power. Thus, the responsibility for the mentally ill lies under cantonal jurisdiction. This explains the fact that there were different laws for different cantons, and that there were no massive 'national' mental hospitals. Over the course of the nineteenth and beginning of the twentieth century, most of the cantons established one or two public asylums for a variable, though not numerous population. In 1930,[1] the largest of the twenty-five public institutions of Switzerland, Zurich's *Rheinau*, had 1,200 beds. The principal private institutions numbered twenty-one and catered mostly to members of the domestic and foreign middle class.

The cantonal asylum of Vaud, named Cery, was established in 1873. It was an imposing building, corresponding to the type, popular in that era, of monumental u-shaped structures. It succeeded the first public asylum which began welcoming pauper lunatics in 1811. The asylum of Bel-Air, in the canton of Geneva, was established in 1900, replacing the first cantonal asylum, which had been constructed in 1838. Its composition of several pavilions represented a break from the u-system of buildings. Both Cery and Bel-Air were situated outside of town, surrounded by a park and a farm or market garden. For most of the twentieth century, these institutions were under the direction of resident medical officers who were also, simultaneously, professors of psychiatry in their respective medical faculties of the University of Lausanne and the University of Geneva.

Translated from French by David Wright and James Moran.

[1] There is, however, a comparative study of the 1930s, that of H. Bersot, *Que fait-on en Suisse pour les malades nerveux et mentaux?*(Berne, 1936, collection *Contributions à l'étude des problèmes hospitaliers*, chapter nine).

In 1900, the canton of Vaud (whose capital is Lausanne) was a farming and, to a lesser degree, an industrial region, numbering 280,000 habitants in an area of 2,000 square kilometres. For this population, Cery had 500 beds. The canton of Geneva, by contrast, constituted an urban and tertiary sector economy boasting 130,000 inhabitants for an area of 100 square kilometres. Bel-Air was correspondingly smaller than its sister institution, with 300 beds. This comparative study examines the period from 1900 to 1970. During this epoch, the buildings were modified, enlarged and modernized. By 1970, there were 570 beds in Cery (the highest density in Cery exceeded 800 beds in 1960) and 725 beds in Bel-Air (which reached a maximum density, then diminished to 657 beds in 1975). These remain, to this day, the principal public and teaching mental hospitals of the two Swiss cantons.

It is necessary to consider the period before the establishment of each of the institutions to understand their development. This question will be considered later in a discussion of legislative matters. As public institutions, the asylums of Cery and Bel-Air prioritized the admission of *les malades assistés* and *les malades internés* by administrative and judicial request, without excluding completely the admission of a less unfortunate and less restricted population.

Social and medical characteristics of patients

This project examined the records of nearly 50,000 individuals in order to study the criteria for confinement of psychiatric patients in the Swiss Romande, from the beginning of the twentieth century to 1970. This figure corresponds to the number of case files which were preserved in the archives of the two mental hospitals. From this group of records, we selected a 10 per cent random sample for each decade in order to conduct a statistical analysis.[2] We have thus retained 1,865 case files from Bel-Air and 2,844 case files from Cery (see Table 2.1). One-third of the total number of files analysed concern the 1960s, and half (47.8 per cent from Bel-Air and 51.5 per cent from Cery) relate to the last two decades (1950–69), reflecting a net increase in psychiatric admissions since the 1950s.

Moreover, the population that we studied was distributed nearly equally by sex. At Bel-Air, one sees slightly more women than men (itself a reflection of the general population). At Cery, by contrast, one observes a surplus of men

[2] The collection of data and a preliminary analysis was conducted by George Klein in 'Evolution des modalités d'admission non volontaire en hôpital psychiatrique; L'exemple de Cery: 1873–1949', thesis of the Faculty of Medicine, University of Lausanne (1996). Jacques Caspary and Philippe Rossignon also contributed to this project within the framework of research financed by the *Projets de développement de l'Association Vaud-Genève*, directed by Jacques Gasser and entitled 'Le passé dans le présent de la psychiatrie'.

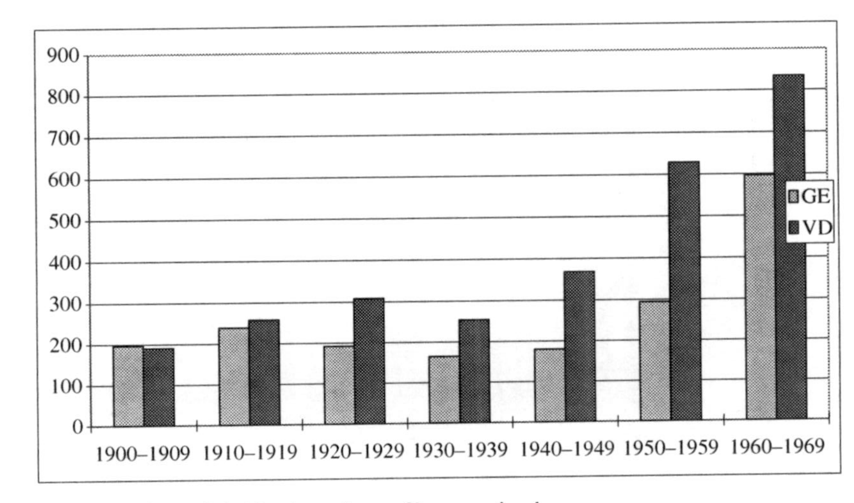

Figure 2.1 Number of case files examined

Table 2.1 *Number of case files examined*

Years	GE[a]	VD
1900–9	197	191
1910–19	239	259
1920–9	193	308
1930–9	165	254
1940–9	180	368
1950–9	294	629
1960–9	597	835
Total	1,865	2,844

[a] In all the tables and figures in this chapter, the cantonal asylum of Geneva, Bel-Air, is indicated by the initials 'GE', and that of Vaud, Cery, by the initials 'VD'.

Table 2.2 *The distribution of case files by sex (percentage)*

	GE		VD	
Years	Female	Male	Female	Male
1900–9	46.7	53.3	41.1	58.6
1910–19	53.6	46.4	43.2	56.8
1920–9	48.7	51.3	46.1	53.9
1930–9	55.8	44.2	37.4	62.6
1940–9	55.6	44.4	40.8	59.2
1950–9	53.4	46.6	43.9	56.1
1960–9	52.8	47.2	48.0	52.0

(about 60 per cent until the 1960s) during the entire period under study (see Table 2.2). This disparity can be explained by a differing attitude amongst psychiatrists and local authorities as to the question of inebriety. In effect, in the canton of Vaud, individuals suffering from alcoholism were mostly incarcerated in a psychiatric institution; in Geneva, they were often accepted into a general hospital. In addition, we were able to calculate that 90 per cent of those individuals hospitalized in the psychiatric institution for alcohol-abuse problems were men. Thus, in our group of patients at Cery, if one subtracts alcoholics, we find a distribution of the sexes reflective of their composition in the general population.

The distribution of patients studied by age reflects that of the general population, with an increase in aged patients (those over sixty) over several decades

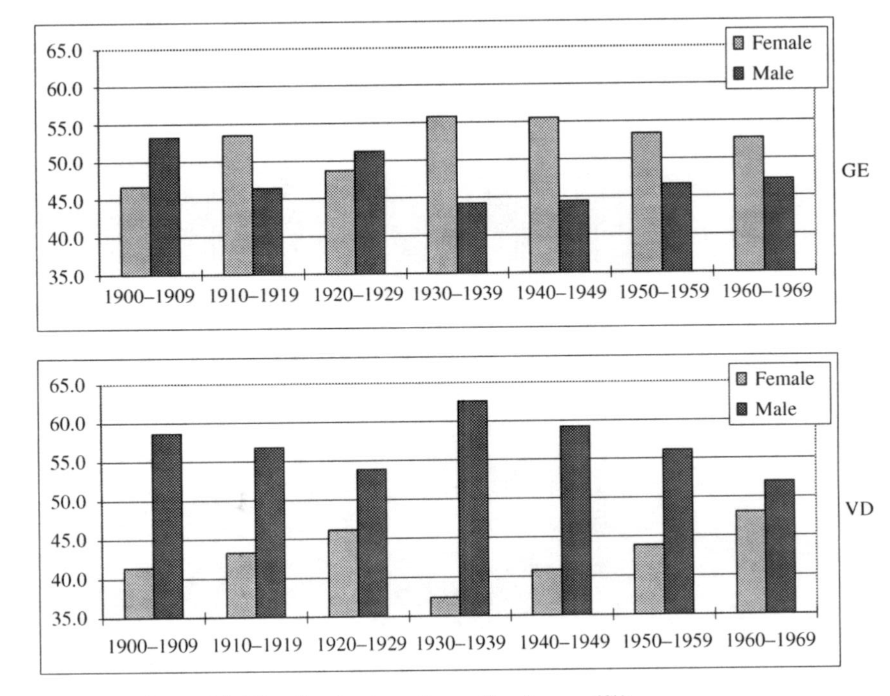

Figure 2.2 The distribution of case files by sex (%)

Table 2.3 *Number of case files by age and by decade (percentage)*

	1900–9		1910–19		1920–9		1930–9		1940–9		1950–9		1960–9	
	GE	VD	GE	VD	GE	VD	GE	VD	GE	VD	GE	VD	GE	VD
<20	6.6	6.8	6.3	9.0	6.7	8.8	3.7	8.7	7.2	5.4	4.4	4.3	4.5	5.1
20–29	13.7	17.4	18.4	17.6	16.6	15.9	11.0	14.2	10.6	16.3	12.6	16.4	14.9	18.0
30–39	18.8	23.7	21.3	26.6	22.3	17.2	20.1	22.4	17.8	19.0	12.6	18.1	14.3	15.1
40–49	23.9	16.3	20.9	17.2	18.1	19.5	21.3	20.1	17.2	17.4	11.6	20.2	11.7	14.9
50–59	13.2	20.0	12.1	11.7	11.4	17.5	20.1	15.0	12.2	16.6	13.9	15.1	10.4	13.3
>59	23.9	15.8	20.9	18.0	24.9	21.1	23.8	19.7	35.0	25.3	44.9	25.9	44.1	33.7

(this increase is most pronounced in Bel-Air from 1940 onwards: between 35 and 45 per cent of patients were over sixty years); the modest number of those under the age of twenty (between 4 per cent and 9 per cent) illustrates a presence of this age-group clearly underrepresented in the asylum compared to the general population. For the remainder of the patients, we see a relatively homogeneous and stable distribution across the age groups (see Table 2.3).

The evolution of the marital status of patients demonstrates that unmarried people were in the majority until the 1930s, in both Cery and in Bel-Air, even if, compared to the general population, this majority was underrepresented (approximately 10 per cent fewer unmarried persons in the asylum). This underrepresentation is probably explained by the fact that those under the age of twenty were seemingly absent from the asylum and were for the most part unmarried.[3] Next, there are the married patients who are more important numerically after 1930 (see Table 2.4). Those divorced and widowed are relatively stable during the different decades and form together about one-quarter of the entire asylum population. For those who were married, the rates are comparable to the general population; as for those divorced and widowed, they are strongly overrepresented in the asylum (as many as five times for those divorced and two times for those widowed).

There is further evidence that the vast majority of those hospitalized during this period under study were institutionalized only once (see Table 2.5). In effect, in an average of all of the decades in our study, more than three-quarters (76.8 per cent) of patients were in this situation, and the level was relatively stable, with the exception of Cery in the 1950s where the proportion of admissions who were first admissions dropped to 59.4 per cent before rebounding to 70 per cent in the 1970s. These numbers are explained partly by the appearance of new neuroleptic therapies which began to be used widely, in Cery at least, from the middle of the 1950s, and which permitted psychotic patients to leave hospital (but which did not avoid more numerous relapses in the community).

[3] The authors would like to thank David Wright for alerting us to this point.

Table 2.4 *Number of case files by marital status and by decade (percentage)*

	1900–9		1910–19		1920–9		1930–9		1940–9		1950–9		1960–9	
	GE	VD	GE	VD	GE	VD	GE	VD	GE	VD	GE	VD	GE	VD
Unmarried	47.9	42.9	42.2	48.6	45.1	44.5	30.9	39.0	37.2	38.6	32.3	40.9	31.2	36.1
Married	29.9	42.4	33.8	34.4	32.1	34.4	44.8	40.2	37.8	41.0	37.1	42.6	37.4	40.5
Divorced	4.6	2.6	3.8	3.9	8.3	5.5	12.1	9.8	7.2	9.8	12.2	7.8	9.7	9.4
Widowed	17.5	12.0	20.3	13.1	14.5	14.6	12.1	11.0	17.8	10.6	18.4	8.7	21.6	14.0

Table 2.5 *Number of admissions/readmissions by decade (percentage)*

	1900–9		1910–19		1920–9		1930–9		1940–9		1950–9		1960–9	
	GE	VD	GE	VD	GE	VD	GE	VD	GE	VD	GE	VD	GE	VD
1	81.7	75.4	79.1	85.7	78.8	74.0	75.8	81.5	75.6	76.1	79.3	59.4	83.2	69.9
2	13.2	16.8	14.6	10.4	10.9	15.9	18.8	13.8	14.4	16.3	11.6	18.2	10.4	17.2
3	2.0	5.2	4.2	2.3	5.7	4.9	4.2	3.5	5.6	4.1	4.1	9.6	3.9	6.5
4	0.5	1.0	0.4	1.2	2.1	1.0	0.6	0.4	2.8	1.1	1.7	3.2	1.7	2.4
5	0.5	0.0	0.8	0.0	1.0	1.3	0.6	0.8	0.6	1.1	1.0	2.6	0.3	1.7
>5	2.0	1.6	0.8	0.4	1.6	2.9	0.0	0.0	1.1	1.4	2.4	7.0	0.5	2.3

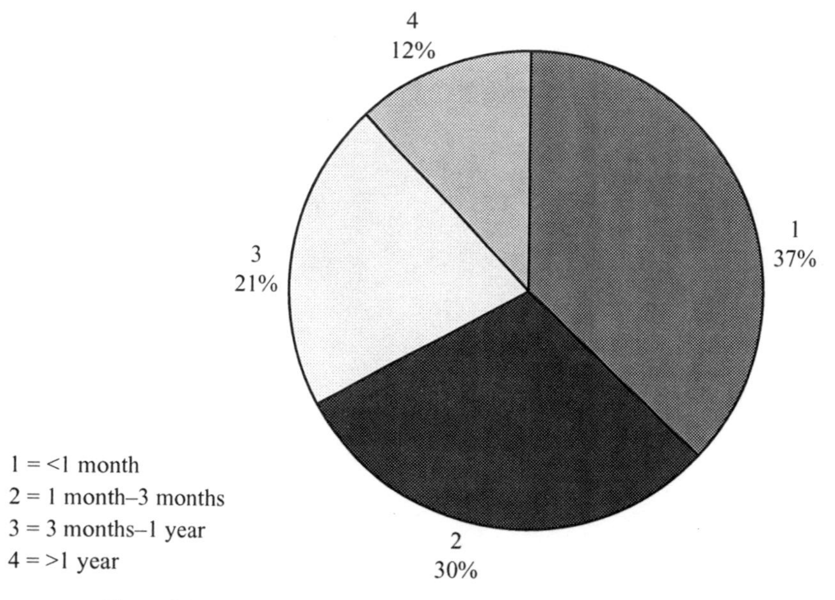

Figure 2.3 Length of stay of hospitalization

We also conclude that, on average, more than 90 per cent of patients of our group were hospitalized one or two times and only about 4 per cent more than four times (we also note the exception of Cery during the 1950s when 7 per cent of patients – forty-four people in the group studied – were hospitalized more than five times). What should not be forgotten in the various quantitative results is the constant rate over seventy years of psychiatric hospitalization; in effect, if we except the Lausanne situation from the 1950s, the number of hospitalizations per patient only varied by a few per cent.

The last variable that we analysed is that of the duration of hospitalization. On average, one third of patients (37 per cent) were hospitalized for less than a month, with Bel-Air having a relatively constant rate across the seven decades

Table 2.6 *Duration of hospitalization by decades (percentage)*

	1900–9		1910–19		1920–9		1930–9		1940–9		1950–9		1960–9	
	GE	VD	GE	VD	GE	VD	GE	VD	GE	VD	GE	VD	GE	VD
<one week	8.2	4.2	9.7	4.2	13.1	7.8	5.5	6.3	5.1	6.8	8.8	7.0	10.4	11.0
1 to 2 weeks	12.8	6.8	14.7	7.3	12.0	9.7	7.3	8.7	7.3	9.5	9.5	12.6	10.2	9.1
2 to 3 weeks	5.1	5.8	11.3	6.2	6.8	4.9	5.5	8.7	7.9	9.5	7.5	11.1	8.9	10.4
3 weeks to 1 month	11.2	13.6	9.7	12.4	11.5	11.0	9.1	12.6	9.0	12.8	11.6	13.5	10.7	16.6
1 to 2 months	12.8	10.5	15.1	24.3	17.3	23.1	20.6	22.4	17.4	23.4	22.4	22.9	20.8	26.5
2 to 3 months	9.7	8.9	8.4	7.3	7.3	8.1	8.5	8.3	10.7	12.8	8.8	11.3	11.4	10.2
3 months to 1 year	25.5	27.7	19.7	25.1	16.8	17.2	24.2	18.1	23.6	18.5	17.3	15.3	18.6	11.4
1 to 5 years	7.1	15.7	7.1	10.0	8.9	12.3	10.3	10.2	12.4	4.1	12.6	4.9	7.4	3.8
>5 years	7.7	6.8	4.2	3.1	6.3	5.8	9.1	4.7	6.7	2.7	1.4	1.3	1.5	1.0

(except during the two decades of the 1930s and the 1940s, when one sees a decline to 27 per cent and 29 per cent) and Cery showing a progressive increase (from 30 per cent in 1900–19 to 47 per cent in 1960–9).

Two-thirds (67 per cent) of the patients in our group stayed in the asylum for less than three months and the remaining third can be divided into stays of longer duration (about 20 per cent additional patients stayed less than one year and another 10 per cent patients more than one year). Finally, on average, 5 per cent of the patients stayed hospitalized more than five years, with a progressive decline during the decades, reaching a rate of about 1.5 per cent from the 1950s onwards at Cery and at Bel-Air. In absolute numbers, a small number of people (ten to fifteen in our group per decade) since the beginning of the century spent a very long time in the asylum or psychiatric hospital. We may ask ourselves whether the relatively stable but small number of very long stays reflects an institutional reality in psychiatry.

Conditions of admission

The social process of admission to the psychiatric hospital in Switzerland will be examined in three ways. First, we will analyse the legal requirements concerning the motives and the methods of admission,[4] as well as their evolution in the two cantons of Geneva and of Vaud over the course of the last century. The legal conditions of admission specified, on the one hand, the type of person vulnerable to being institutionalized and the motives behind their confinement, and, on the other hand, they distinguished the types of confinement. Further they indicated which people were liable to request the confinement and under what circumstances people were empowered to authorize or to impose it. Second, we will analyse statistically certain general data relating to the admission of patients. This takes into account most notably, whether or not the individual was considered dangerous, and the type of authority (medical, administrative, judicial) under which admission was requested. Finally, we will examine several case files with close attention to detail. Such a reading will lead to a greater and more nuanced understanding of concrete situations. We will examine, above all, why individuals pleaded for the confinement of a particular person.

Legislation

The first lunatic asylums of the cantons of Vaud and of Geneva, created in the first half of the nineteenth century, responded both to the phenomenon of asylum construction during this era, and to political, administrative and medical imperatives. The reorganization of the administration of a number of Swiss

[4] In order to protect lunatics' property, the laws also governed other aspects concerning the management of institutions and modes of surveillance found therein.

cantons including Vaud (to this date a dependent territory of the canton of Berne) and Geneva (which had been temporarily annexed to France) reflected a concern for better management of services and the need for public order. These new institutions also represented a wish on behalf of authorities to distinguish those whom one could simply care for (incurables), those persons guilty of crimes and/or of begging (thereby necessitating punishment or rehabilitation), and those persons who were victims of physical or mental malady (necessitating care, protection, but also, for certain lunatics, confinement). Thus, whereas the mentally ill were previously found intermixed with other categories of the assisted (beggars, the disabled, and prisoners) in institutions created in the seventeenth and eighteenth centuries (such as the Grand Hôpital or at the Prison de l'Evéché in Lausanne, at the Petite Maison des aliénés in Berne, or at the Maison de Discipline de l'Hôpital général in Geneva), the creation of institutions reserved specifically for lunatics became oriented little by little to a distinct and specific responsibility separate from those of prisons and the physically ill.

The authorities of the Canton of Vaud (a new Swiss canton from 1803) advocated by decree in 1910, the creation, in addition to the existing cantonal hospice for physical illness, the building of an establishment for incurables and of a House of Lunatics (Maison des aliénés). All three institutions were supported by public funds. This decree specified the criteria for confinement: 'In the House of Lunatics, lunatics of both sexes are received, whose existence in their families and in society have become painful and dangerous, or who have a probable hope of recovery.'[5] These criteria were formulated later on with more precision, but the principles upon which they were based remained: distress (pénibilité – an interesting notion which was later abandoned), dangerousness, and curability.

The House of Lunatics was instituted in 1811 in an existing building for an expected forty patients. Sixty years later, there were 170 inmates. The number of patients continued to increase because of the confluence of several phenomena: the increase in the general population, the organisation of public assistance, public order control, and the more effective identification of mental pathologies. The construction of a new building was thus thrust upon the authorities. The style of architecture of the lunatic asylums, which was largely disseminated and discussed at the international level, drove the conceptions and the compromises that the authorities and doctors made in asylum construction. The asylum of Cery, and thirty years later, the asylum of Bel-Air, did not escape this pattern.[6]

The canton of Geneva adopted its first lunacy law (*Loi sur le placement et la surveillance des aliénés)* on 5 February 1838,[7] a piece of legislation which was

[5] Canton de Vaud, decree of 18 May 1810, art. 2.

[6] C. Fussinger, D. Tevaearai, *Lieux de folie. Monuments de raison. Architecture et psychiatrie en Suisse romande 1830–1930* (Lausanne, 1998).

[7] Completed by the law of 7 April 1838, this legislation revisited elements of a provisionary rule on the goverance of lunatics of 11 November 1829 and adopted, word for word, certain articles of the French law.

inspired by the French law of 1837 (in effect from 30 June 1838). The Geneva law was contemporaneous with the creation of the cantonal House of Lunatics, Les Vernets, and concerned all those lunatics who were sequestered in public or private institutions. Considerably more comprehensive than the legislation regarding the protection of lunatics for the canton of Vaud, it nevertheless did not specify the reasons for committal. Whereas confinement depended on the welfare and health authorities in the canton of Vaud, it was the perogative of the authorities of justice and police in the canton of Geneva. This notable difference in authority over confinement has been criticized. The obligation of Geneva to request a confinement from the police appears to have been a humiliating procedure, treating lunatics as if they were accused criminals. This practice, however, did not disappear until 1936. The admission of the mentally ill to asylums was at that time authorized by the Department of Hygiene and Public Assistance, a procedure consistent with the other hospitals. One cannot conclude, however, that this had an effect on the hospitalized population. The admissions in the asylum were submitted for authorization by the executive authorities of each canton (the Welfare Department or the Police Department in these cases) because it was a type of confinement that could infringe upon the penal code.

Partial modifications occurred at different times during the nineteenth and twentieth centuries. To simplify, we are going to discuss here the principal laws. For the canton of Vaud, there was a law of 14 February 1901 (concerning the administration of persons suffering from mental illness) and a law of 23 May 1939 (concerning mental illness and other psychopathies) which would be replaced by a new law in 1985. For the canton of Geneva, there was the law concerning the administration of lunatics from 25 May 1895, and the law concerning the administration of persons suffering from mental disorder from 14 March 1936 which would be replaced by a new law in 1979. There was not any fundamental difference between the new laws of the two cantons. We note rather a parallel evolution concerning the three areas considered here, namely, the categories of lunatics, the types of admission and the causes of admission.

Categories of lunatics These laws concerned lunatics (*aliénés* is the only term used in the text of the first generation of laws), a global category in the nineteenth century to which was added, little by little, other patient groups, such as epileptics, alcoholics, the mentally disabled, and addicts. The asylums, however, already welcomed these categories of patients before the terminology became diversified. In effect, the diagnostic statistics in the annual reports or the examination of the case files indicate that these types of patients were already hospitalized in the asylum during the nineteenth century. This modification in the laws signifies a clearer differentiation in patients' mental pathologies. The epileptic or alcoholic was not conflated with the lunatic, although it was convenient, in certain circumstances, to intern them in lunatic asylums. We note, however, that at the same time (at the end of the nineteenth and the beginning of the twentieth century), specific institutions were created for epileptics, alcoholics, and persons suffering from mental disability.

In Geneva, the law of 1895 concerned 'the resident lunatics in the canton of Geneva' and 'the lunatics of Genevan nationality' (art. 1) and 'the epileptics, alcoholics, and in general, all the other patients whose mental state compromises public security' (art. 2). In the canton of Vaud, the law of 1901 was extended in 1921 'to persons suffering from addictions (morphine addiction, cocaine addiction, alcoholism, etc.)', and further, in 1928, to persons suffering from mental infirmity (*infirmité mentale*). We note that this modification is tied to the legalization of sterilization in the canton of Vaud (a unique situation in Switzerland at this time). Persons suffering from mental infirmity were vulnerable to being sterilized rather than incarcerated.[8] Certain admissions were made for an expert evaluation of such cases. In the law of 1939 (Vaud), all of these categories were subsumed under the categories, 'the mentally ill and other psychopaths'.

One is able also to remark that the term lunatic '*aliéné*' was abandoned; 'the limits of madness were far from being always clearly demarcated... we wish to try to lessen the stigma that results from a stay in a *maison de santé*: we propose to eliminate the term *aliénés* from the title of the law'.[9] A less pejorative formulation, more nuanced, and suggesting a gradation and diversity of mental pathologies, was adopted. Thus the legislation on lunatics *aliéné* became a law on the government of persons suffering from mental illness (Vaud, 1901) and the law on the government of persons suffering from mental afflictions (Geneva, 1936). It is interesting to note that the transition from the notion of the lunatic to that of the mentally ill occurred in the first third of the twentieth century in the Romande legislation, whereas one is able to situate this medicalization of madness in the 1850s in psychiatric texts.[10]

Admission types The principal distinctions between different types of admission over the two centuries are, respectively, requests for asylum required by different people and, secondly, confinements ordered by an administrative or judicial official. The terminology was changeable and was thus ripe for confusion. One must note in particular that the *admission volontaire* of the nineteenth century was not the same as 'voluntary admission' in the twentieth century. Thus, according to the Geneva law of 1895, voluntary admissions

[8] G. Jeanmonod, G. Heller, J. Gasser, 'Déficience mentale et sexualité. La stérilisation légale dans le canton de Vaud entre 1928 et 1985', *Médecine et Hygiène* 57 (1999), 2050–4. This article gives the first results of a research project, financed by the National Swiss Foundation for Scientific Research, on Eugenics in the Swiss Romande: 'A study of the sterilization of the sick and developmentally disabled at the end of the 19th century to the present'.

[9] E. Grandjean, *Considérations sur la Loi genèvoise du 25 mai 1895 sur le régime des aliénés* (Yverdon, 1902), 10.

[10] See G. Lantéri-Laura, *Psychiatrie et connaissance* (Paris, 1991).

included all admissions requested by anybody (with the exception of ordered admissions), although now they are only concerned with admissions requested by the patient him/herself.[11] The notion of an emergency admission (*admission d'urgence*),[12] and that of a provisional admission (already in the law of 1862 in the canton of Vaud), were introduced at the end of the nineteenth century. In certain cases this facilitated the preliminary procedure of confinement.

The requests for admission could be made by different people, a list of whom would be too copious to detail. Requests were made not only by the parents or the authorities, but 'by such persons who, in the circumstances, could authorise a similar request',[13] 'by the sick person himself, by his legal representative, [and] by all the members of his family who agree to such a procedure';[14] one could also have 'a request ordered in the interest of public safety and the safety of the person himself' by the authorities of the commune, the district, or the canton. The admission requested by the patient did not yet have a specific status. In the law of 1936[15] as with that of 1939,[16] the admissions requested by the patient, called free admissions (*admissions 'libres'*), are distinct from the admissions requested by other people to the *Departement*. The procedure for free admissions was simplified, and discharge was made easy. Free admission had an autonomous and entirely medical status. Thus, the expressed wish by the psychiatrists since the beginning of the century to free the care of the mentally ill from the double humiliation of administrative interference and dependence on the authorities, was realized in the case of the consenting patient.

The medical certificate, an optional requirement in the first laws of the nineteenth century, became obligatory for all admissions.[17] This was not the case for the admissions by judicial authority in the law of 1901,[18] nor for the free admissions.[19] It seemed to be required in all the cases in Geneva. The modalities of establishing a medical certificate were often specified (doctors, skilfulness in completing them, delays in their availability, direct observation of the patient, content of the certificate) and were of various degrees of urgency.

Motives behind admission The motives for admission were formulated in a very simple manner in the laws of the nineteenth century. It was the patients whose 'existence in their families and in society became painful and dangerous, or who had a probable hope of recovery'[20] and the institutions were destinations for the 'treatment and the supervision of lunatics'.[21] The mental illness itself, care, danger and protection were the grounds for confinement. One finds these elements (which did not necessarily have to be all present at one time) in the Vaud law of 1901: 'The present law applies to all those persons suffering

[11] Geneva law of 1979. [12] Geneva, 1895, art. 22. [13] Geneva, 1895, art. 25.
[14] Vaud, 1901, arts. 10 and 11. [15] Geneva, art. 21. [16] Vaud, art. 18.
[17] Vaud, from the rules of 1862, Geneva, with the law of 1895. [18] Vaud, art. 35.
[19] Vaud, 1939, art. 18. [20] Vaud, 1810. [21] Geneva, 1838.

from mental illness found in the Canton of Vaud, for all whose state necessitates care or threatens danger to others.'[22] Note that there was no specific mention of danger to the lunatic him/herself. However, the medical certificate had to attest to the 'existence of a mental illness and the necessity for treatment in an institution'.[23] With the law of 1939, one loses the priority accorded until then to care and treatment; the emphasis is shifted to dangerousness. This shift is evidenced in legal texts relating to the authority of police (public order and public safety): 'The present law applies to the mentally ill and other psychopaths (the mentally infirm, drug addicts, alcoholics, etc.) whose state presents a danger to themselves, to others, to the public order, or public morals'.[24] The dominant tone of the law given by this first section was that of better social control rather than that of assistance and care. One thus returns to the priorities of the Geneva legislation on lunatics during the nineteenth century owing to its connection to the Department of Justice and Police. However, the medical certificate did have to mention the necessity for treatment or supervision.[25]

From the law of 1895 in Geneva, the category of dangerousness as grounds for committal was diversified. There was a distinction between dangerousness (of a varied nature) with respect to others, and danger to the patient him/herself. In effect, the law concerned all those patients 'whose mental state was of a nature that compromises security, decency, or public tranquillity or their proper security'.[26] The need for treatment was not made explicit, neither in the law of 1895 nor in the law of 1936, but the medical certificate had to 'establish the sickness or the necessity for confinement'[27] and, 'reveal the symptoms of the sickness and the reasons necessitating admission'.[28] By contrast, in the law of 1979, the notion of dangerousness disappeared in the explicit motives, giving way to a need for treatment: 'The present law applies to persons... suffering from mental illness and whose mental state requires care in a psychiatric institution.'[29] The notion of dangerousness was resumed for involuntary admissions; it even became an indispensable condition. In effect the request for admission formulated by a doctor had to attest to 'mental troubles', to 'a grave danger to themselves or others', and to the necessity for 'treatment and care in a psychiatric hospital'.[30] Dangerousness was no longer mentioned in the first section of article 1, but lower in article 24. It is necessary to note still that the motive for admitting the accused was generally that of putting them under the observation of experts.[31]

The examination of the evolution of the different laws allows us to observe changes wrought by experience, using more precise, or more exact formulations, and occasionally more liberal and realistic formulations. Except for the often

[22] Vaud, 1901, art. 1. [23] Vaud, 1901, art. 12b. [24] Vaud, 1939, art. 1.
[25] Vaud, 1939, art. 16. [26] Geneva, 1895, art. 2, Geneva, 1936.
[27] Geneva, 1895, art. 25. [28] Geneva, 1936, art. 22. [29] Geneva, 1979, art. 1.
[30] Geneva, 1979, art. 24. [31] Vaud, 1901, art. 35 and Geneva, 1936, art. 21.

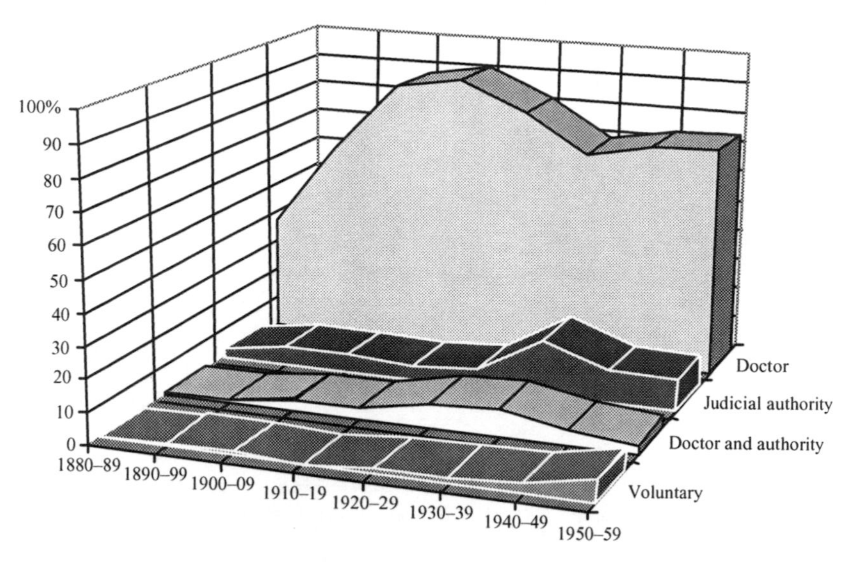

Figure 2.4 The individual(s) demanding admission (Cery, 1880–1960)

very detailed and complex dispositions (relating to the protection of the sick and the security of society), we are able to remark that a new formulation is often precipitated by a change in the spirit of this law echoing a change in the evolution of practice: the accentuation of the role of doctors relevant to the administration, the desire to reassure public opinion, and to 'dedramatize' admission.

Quantitative results on the admission of patients

At Cery, we are able to distinguish who requested admission quite easily from the case files, be it a doctor (with a medical certificate), a representative of political, administrative, or judicial authority (with a document written on the letterhead of one of the authorities), or, conjointly between the two (a medical certificate *and* an official document). It is possible, moreover, to discover if the patient requested his own admission.

All through our period, one can observe that in the majority of case files and, conforming to the law of 1901 (Vaud), a doctor wrote a medical certificate of admission. However, in the 1920s, there was a rise in admissions supported by an official and a doctor, and in the 1930s, a rise in admissions demanded by a judicial authority alone. Both decline in the 1940s. That corresponds, in the causes of admission, to an increase and then a decrease in 'expert' requests. This fluctuation suggests that the 1920s and 1930s saw a hardening of the social

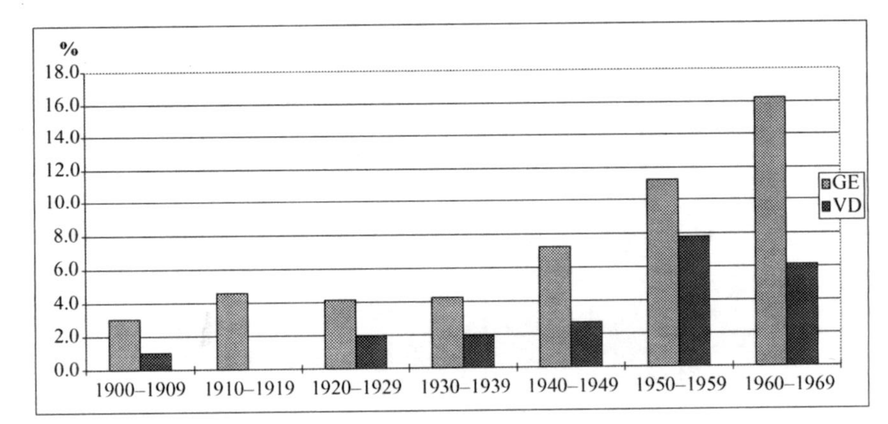

Figure 2.5 Number of voluntary patients (in percentage)

Table 2.7 *Number of voluntary patients (percentage)*

	GE	VD
1900–9	3.0	1.0
1910–19	4.6	0
1920–9	4.1	1.9
1930–9	4.2	2.0
1940–9	7.2	2.7
1950–9	11.2	7.8
1960–9	16.2	6.1

order, which declined in Switzerland during the war. In confronting the results concerning the (juridical) requests and the sex of patients, one is able to remark that 70 per cent to 100 per cent (according to the year concerned) of admissions requested by juridical entreaties concerned men.

The admissions requested by the sick persons themselves are, so to speak, non-existent until the end of the 1950s (see Table 2.7/Figure 2.5), when we see a rise, however modest (about 10 per cent) in this category. Today, they represent fully half of total admissions. It is necessary to point out however, that the results of a quantitative analysis are not very trustworthy when the numbers are quite small. We note however that in our research which concerns one case file in every ten, we analysed forty-nine admissions cases requested by a sick person himself amongst 629 case files examined between 1950 and 1959, and that there were only twenty such cases admitted to the institution in the entire first half of the twentieth century. Some printed forms, signed by the patients themselves were recovered after 1935. It is important here to put into perspective the notion of patient-driven admissions. A qualitative analysis makes it appear that certain patients presented themselves (in a state of delirium) at the asylum for refuge, and that others came under the pressure of others, still others self-admitted because they asked (with complete lucidity) for help, notably in the case of depression or drug dependence.

One must recall here that the Vaud law of 1901 recognized the admission by the patients themselves and the law of 1939 dispensed with the need to present a medical certificate and facilitated their discharge. The law of 1901 already authorized admission without medical certificate, whilst it was still required by a judicial authority.

The categories utilized in the case files were not identical at Cery and Bel-Air. Whereas the specificity of the laws which did not require medical certificates for admissions by juridical authority (Vaud, 1901) or by the patient himself (Vaud, 1939) facilitated, in the quantitative analysis, the discovery of such instances at Cery; at Bel-Air such a distinction was not easily identifiable. All of the admissions had to be accompanied by a medical certificate and, until the law of 1936 (Geneva), were picked up by the Department of Justice and Police which introduced a certain confusion between the admissions authorized by it or requested by it.

Qualitative analysis concerning social and medical motivations behind admission

Certain indications concerning the motives behind admission were easily retrievable in the case files; others more entangled or scattered. Due to the gaps in evidence, one can not usefully or comprehensively determine motivation because one can not fully evaluate or speculate on missing information. Indeed, the motivations behind admission are often not explicitly articulated. One must specify that the indications furnished in the case files offer only a very imperfect conclusion in individual cases. However, one is able to find some mention of official motivation for admission (mental affection, danger, supervision, placement under observation, request for expert opinion), and often mention of the recent circumstances which precipitated the demand for admission. Besides, older events and circumstances were reported by one or other of the testifiers, parents or witnesses, or by the sick person himself. One also finds mention of the actual person making the request, and occasionally one can identify the name of the person who made the request. What can be elucidated here, thanks to different parts of the case files, is at what point the threshold of tolerance was passed and what recent events, added to other circumstances, led eventually for a request for admission.

This research focuses on several case files chosen in an arbitrary manner concerning the admissions made in the 1930s. The individual situations were inscribed in the personal history of the patient but also in the social history of the era.

Betty: 'schizophrenic' At the age of twenty-nine, Betty was brought by taxi to the asylum at Cery. She had escaped from her parents' house during the night in a fit of anger, and, the next day, had caused a scandal in the street as her father tried to take Betty home; but she escaped him. She then escaped a policeman who had come to help out, but he ultimately detained her. Examined by a doctor, at the request of the local police, she was admitted as an emergency case. Betty spent close to two months in Cery before being transferred to another asylum. Over the course of the previous years, there had been several troubling incidents because of Betty's strange behaviour. The last scandal, a most violent act against her father in the street, seemed to push the family to seek this solution of confinement first suggested by the general practitioner.

However, she and her parents had endured the effects of her sickness for a long time. She was anxious, crying, hardly eating anything, becoming thin, tired, and frightening others; she was not able to get by, did not work and could not stand people speaking in her presence. According to her parents, they witnessed the troubled life of their daughter; her suffering added to the daily difficulties of living. They had her cared for by a private doctor and accompanied her to her consultations. They did not spare time or their money. It was the public manifestation of the sickness, rather than its domestic face, which led to her confinement in an asylum. The first scandals, which did not have the intensity of the last ones, had been surmounted because her parents managed to supervise and control her at home.

Bertrand: 'chronic alcoholism, psycopathy' A forty-seven-year-old doctor, Bertrand spent five months at Cery for observation and detoxification. Following a complaint against Bertrand over a serious professional mistake, and after several warnings, the cantonal health service forced its appointed physician to 'take him to the Cery asylum' because 'his state of health must be considered a public danger'. The health service physician asked for an emergency confinement 'for chronic alcoholism, and attempted suicide' and Bertrand was taken away by a policeman. Two days after admission, the council of the state took away Bertrand's licence to practise medicine. One week later the Justice of the Peace gave the divorce demanded by his wife; at the end of three months, the Justice of the Peace asked for an expert in psychiatry because Bertrand made a lawsuit for false imprisonment. He left Cery to present himself for judgement relative to his lawsuit. Kept in prison during the night preceding the judgement against the recommendations of a psychiatrist, Bertrand committed suicide.

Thus, professional fault aggravated by a lapse in judgement, precipitated his admission. Alcoholism and the attempt of suicide were the explicit reasons (following an order to pay, one month before his confinement, he had attempted suicide). The situation worsened progressively, one source suggesting that the sickness had lasted ten years, others indicating only two or three years. From all sides grief followed him: from lawsuits over his debts, from the bank, from his wife for ill-considered spending, and from drunkenness. Successive criminal offences, inadequate behaviour in the practice of medicine, and failure in the financial domain led finally to his admission and the request for psychiatric expertise. Confinement was a measure of protection both against new suicide attempts, and against the public danger that his mental state represented in his professional capacity. The confinement also allowed for administrative, civil, and penal measures (professional interdiction, divorce, court proceedings) to be taken against Bertrand. The only available therapy in such cases was a regimen of abstinence.

Louis: 'constitutional psychopathy, opiomania' A man of thirty-one years, without employment and a resident of Berlin, Louis was admitted to Cery for observation. He was there a month. He was taken by a policeman to the prison by order of a judge who requested a medical examination to see if the young man was an insane person, or if he demonstrated an 'incomplete mental development' that would diminish his responsibility for what he had been accused of doing. He was arrested following a complaint and risked a penalty of ten years for several swindles. The complaint was lifted thanks to the influence of his father, a high level German bureaucrat working in Berne, who promised to ask the Justice of the Peace in Switzerland for the control of his son. As soon as this was established, the father drove his son to the border.

Charles: 'not a lunatic but an antisocial character' Charles, a fifty-seven-year-old clockmaker, without a regular job, spent eight months in Cery. Transported by the Prefect in person to be placed under observation, Charles came from the prison where he had finished his term. He was imprisoned for not paying a fine for poaching. The Prefect did not want to release Charles at the end of his prison term because the police feared him. Charles was an avid hunter who possessed weapons and who promised revenge on those who imprisoned him. He was a man who lived a solitary life in a cabin, raised dogs, doing small work tasks to maintain his existence. One week after the admission of Charles to Cery, the prefecture, on the advice of the sanitary service, requested from Cery a report, which was not made until seven months later. On the basis of this report, the sanitary service, under the advice of the council of health which held authority over psychiatric institutions, ordered the immediate release of Charles.

Alain: 'paranoid schizophrenic' At the age of thirty-three Alain, in a very distressed state, was brought to Cery by his father and his maternal uncle. His condition worsened irremediably and he died after a very painful three-year stay at Cery. Entries relating to Alain's critical mental state were frequent (in contrast to the file of Charles), and included references to his medical treatment as well as descriptions of the disruptions to asylum life caused by his behaviour.

His admission to Cery came at a critical moment when Alain's life had become too painful with the aggravation of his condition. Difficulties during childhood, during adolescence, from the death of his mother, the remarriage of his father, were cumulative steps in his worsening mental state. Hallucinations, growing anguish, and increasing attempts at suicide for several days made matters urgent. His family seemed to have used up much of their patience and attempted other avenues before confinement, including successive consultations with different physicians. As Alain's suffering grew, so did that of his family, and the danger level increased. 'The treatment of a similar case like this could only be done at a specialized clinic.' It was in these terms that the medical certificate for admission justified the need for confinement.

Lydie: 'erotic pubescent reaction' A young woman of fifteen years, Lydie was brought to Cery by a parent. Her mother did not know what to do and worried about the consequences resulting from the liberties that Lydie was taking with the boys. Lydie had quit school one year before graduation for indiscipline, stayed at home without occupation and did not learn a trade. Lydie spent forty-one days at Cery after which she was released on her own. It appears that 'the lesson served her well'. Her file does not indicate whether the medical certificate leading to her committal was made as the result of a specific incident relating to her condition. 'Mental troubles characterized by an exaltation of mood, a tendency to linguistic excess, excitation of the senses, the risk of "fugue" and indecency. She reacts against all authority, whether scholastic or familial . . . I request her placement under observation at the asylum of Cery.' It was probably the mother of Lydia who appealed to the doctor.

Daisy: 'non-alienated intellectual retard' Daisy, aged sixteen years, was admitted to Cery upon the order of the Sanitary Service after requests for abortion and sterilization were made by her guardian and by a general physician. These measures were considered necessary 'because of heredity from her mother', and 'because the child resulting from her pregnancy would be a burden to the community'. After twenty-five days, she was released on her own without the requests having been carried out.

Assessing the case files

The files that we have examined enable us to observe in concrete fashion the different modalities of admission. Individual files also make evident the extent to which institutionalized psychiatry was inextricably linked with, and even indistinguishable from, medical, social and penal contexts. It is difficult to appreciate the role of different actors in the unfolding of the process which led to an institutional admission. For example, in the case of Daisy, if it was her guardian who seemed to make the request for sterilization, the patient file nevertheless does not allow us to appreciate the role of each intermediary, between the family physician, the physician of the infirmary where Daisy's pregnancy was discovered, the guardian, the Justice of the Peace and the Sanitary Service.

In most of the files, we encountered the polymorphic idea of danger (present/future, physical/moral, private/public). We see this overriding theme in attempts at suicide (Alain), great loneliness (Betty), crumbling family fortune, compromised reputation of family members, dupery (Louis), attempted suicide, professional mistakes endangering the health of others, debts, falsehoods (Bertrand), threats to a policeman by an armed individual (Charles), risk of falling pregnant (Lydie), risk of bringing one or more children into the world at the expense of public assistance (Daisy).

To explicit danger can be added motives, ephemeral or lasting, like amorality, asocial behaviour, and disruption to public order. These are evidenced by the young man who did not work and 'who knew nothing but enjoyment as the sole goal of life' (Louis), the doctor for whom alcoholism brought about moral decay (Bertrand), the hunter with no regular employment and no proper home, who lived like a savage (Charles), the young woman who fought and roused the curiosity of onlookers (Betty), another young woman who refused to go to school, or to work (Lydie), a third woman who was behind at school (Daisy). In each case we can uncover issues relating to a lack of social integration, to a threat, real or felt, like that related to the autonomy and the capacity to provide for their needs, producing dependence or a disturbance in the relations with others. Finally, it is worth noting in a minority of the admissions examined, a great suffering, which was not however the pivotal event leading to committal. This could be seen in the distressing nature of the situation for the individual in question, and for his/her social contacts who witnessed the suffering, who were impotent to alleviate the situation, and who were exhausted by it (Betty, Alain, Bertrand).

No single argument seems to suffice in explaining requests for asylum committal, and in deciding on prophylactic (sterilization), protective (confinement) and therapeutic (hospitalization and treatment) measures. The mental illness itself, it seems, does not suffice if it does not encroach on the public sphere, exhaust relatives, manifest itself in attempted suicide, etc. Betty and Alain were ill long before they were admitted to Cery. The possibility of consultations in private offices or in 'policlinique' (public out-patient clinics) with specialists permitted the delay in their confinement. These two examples suggest that the public psychiatric institution was one aspect of a multi-faceted network. The fluctuations and the development of this network offered alternative solutions to that of admission to a public asylum. Thus, for example, we can imagine that twenty years later Lydie's situation could be resolved through a specialized consultation for adolescents. Many examples make manifest the role of buffer played by temporary confinement which offered a respite for the presumed patient or for his/her social circle, permitting a crisis situation to dissipate. Certainly, the public psychiatric institution was destined to treat the mental pathologies of the individual, but we cannot fail to observe that in many cases it served foremost in the temporary separation and protection of one of the partners in a public or private conflict. This role was well recognized by August Forel, the eminent psychiatrist and fervent supporter of mental hygiene: 'Our asylums have . . . become kinds of public dispensaries of great urgency.'[32]

Conclusion

The quantitative and qualitative evidence that we have been able to collect from the casebooks of the hospitalized patients in the two Swiss psychiatric institutions, in conjunction with the evolution of legislation concerning individuals suffering from mental troubles, allows us to understand a little better the evolution of the demand for admission to a psychiatric institution during the first seventy years of the twentieth century. In summary, and to generalize, we can try to define the 'average' type of person who was admitted to the asylum during our period of study. It was a single man with a psychiatric pathology in which alcohol played an important role or, if it was a woman, she was also single, coming from a humble background with little education and with a poorly defined psychiatric condition. These persons were hospitalized only once or twice and for a relatively short period of time; their admission was involuntary and requested by a doctor. The motivations behind their committal included a complex mixture of reasons, from the strictly psychiatric to social problems in a larger sense. Of course, these 'average' situations do not really exist in reality, but they are confirmed by the analysis of individual cases where the complexity of situations is characterized best by case study. Central to our understanding of this process is the notion of 'dangerousness', which is constantly evoked as a reason for committal and which remains a very poorly defined concept.

We have also argued that, if the asylum played an important role in the maintenance of public order, this role was above all that of diffusing a certain number of conflictual social situations (public or familial) for a short period of time, and less for confining individuals for a long period of time. We have already noted that two thirds of those who were hospitalized remained for fewer than three months, and if long hospitalizations (more than a year) existed, they were rare. However, these long-stay patients played a very important role in the stigmatization of the asylum.

We have also been struck by the relative continuity of our data over the decades under study. The changes in personnel and the type of care for persons suffering from psychiatric problems between 1900 and 1970, the development of an effective psychopharmacology in the 1950s, and even the parallel development of out-patient psychiatry, had little impact on our results: it is as if the

[32] Auguste Forel, 'Pourquoi, quand, comment doit-on interner des personnes dans les asiles d'aliénés?' (Lausanne, 1904), 11 (translation of an excerpt from the ninth annual report of the *Société zurichoise de patronage des aliénés*, 1885).

mental hospital had an internal logic which was little influenced by external factors. On the other hand, certain social factors undeniably played a role in the context of involuntary hospitalization. For example, in the 1920s and 1930s, a period during which social control was strong, the demand for hospitalization from 'non-medical' individuals was higher than average. One can also see the profound changes that occurred after the great discoveries of the 1950s (the increase in the number of admissions, the emergence of voluntary admissions, the shortening of length of stay, and the rehospitalizations). It would probably have been necessary to lengthen our study to the 1980s, the decade during which these changes became most evident, to explore these issues fully. Today, more than half of the patients enter psychiatric hospitals as voluntary patients, their stays are very short (an average of twenty-five days in 1998), and re-hospitalization is more common than ever. The factors that have changed in the last twenty years necessitate, of course, an entirely new research project. It would be profitable to study on the one hand, the modification of the doctor-patient relationship (from a paternalistic to a contractual relationship): one of the examples most characteristic of recent years is the development of patients' groups; another is the demystification of the image of mental illness more generally. On the other hand, economic constraints and, more specifically, the political determination to find a way to control the growth of health spending, have also become central to new models of care.

3 Family strategies and medical power: 'voluntary' committal in a Parisian asylum, 1876–1914

Patricia E. Prestwich

Recent research in the history of nineteenth-century psychiatry has explored the expanding powers of the medical profession and the proliferation of the asylum, that 'magic machine'[1] for curing insanity. This medicalization of madness has usually been portrayed as a 'top-down' process: 'social control imposed from above with greater or lesser success on a population now the unwitting object of medical encadrement'.[2] But as historians have begun to study individual asylums and the complexities of committal, more emphasis is being placed on the role played by families in the process. Asylum doctors, it has been suggested, merely confirmed a diagnosis of insanity already made by families, by neighbours, or by non-medical authorities.[3] Consequently, as the American historian Nancy Tomes has argued, 'the composition of a nineteenth-century asylum population tells more about the family's response to insanity than the incidence or definition of the condition itself'.[4] Such arguments imply

Research for this project was funded in part by the Hannah Institute for the History of Medicine, the Medical Research Council of Canada, the Social Sciences and Humanities Research Council of Canada, and the University of Alberta. I am grateful for their support. I also wish to thank Chuck Humphrey for all his help with the computer analysis of the data. This chapter was originally published in the *Journal of Social History* and has been reprinted by permission.

[1] 'Introduction' in W. F. Bynum, R. Porter and M. Shepherd (eds.), *The Anatomy of Madness: Essays in the History of Psychiatry*, 3 vols. (London, 1988), vol. III, 3. For an introduction to recent work in the history of psychiatry, see the three volumes of *The Anatomy of Madness* (London, 1985–8). The most recent works on French psychiatry are: J. Goldstein, *Console and Classify: The French Psychiatric Profession in the Nineteenth Century* (Cambridge, 1987), I. Dowbiggin, *Inheriting Madness: Professionalization and Psychiatric Knowledge in Nineteenth-Century France* (Berkeley, 1991) and Y. Ripa, *Women and Madness: The Incarceration of Women in Nineteenth-Century France* (Minnesota, 1990).

[2] C. Jones, 'Montpellier Medical Students and Medicalisation in Eighteenth Century France', in R. Porter and A. Wear (eds.), *Problems and Methods in the History of Medicine* (London, 1987), 58.

[3] M. MacDonald, 'Madness, Suicide and the Computer', in Porter and Wear (eds.), *Problems and Methods*, 210.

[4] N. Tomes, 'The Anglo-American Asylum in Historical Perspective', in C. J. Smith and J. A Giggs (eds.), *Location and Stigma: Contemporary Perspectives on Mental Health and Mental Health Care* (Boston, Mass., 1988), 14.

a more 'dynamic and dialectical'[5] interpretation of the process of medicalization, one that requires a careful assessment of family demands for medical services and the degree to which these demands were met, willingly or unwillingly, by the emerging psychiatric profession.[6] Asylum records, although difficult to interpret, can shed some light on family decisions to 'take the road to the asylum';[7] this is particularly true in the case of so-called voluntary committal, where families could avoid the involvement of the police or judicial authorities.

In France, the law of 1838 established *placement d'office*, or official committal, as the normal means of confinement; it was, in the words of the law, restricted to those who 'endanger public order or the safety of individuals'.[8] Implementation of the law of 1838 was the responsibility of departmental officials, and conditions varied widely. In Paris, the procedures for official committal were complicated and a source of continued controversy. Local police sent those who appeared to be both unbalanced and dangerous to the *Infirmerie spéciale*, situated alongside the central police detention cells (the depot).[9] At the *Infirmerie*, potential patients (referred to as *présumés*) underwent a psychiatric examination and, if certified as insane (*aliéné*), were taken by a police van to the Admissions Office of the department of the Seine. This was located in the grounds of the Sainte-Anne asylum, the most important of the public asylums of the Seine, and the only one situated within Paris itself. At the Admissions Office, patients were again examined by a psychiatrist, most likely by the eminent Valentin Magnan, were again certified as insane, and finally, were transported to one of the five asylums in the department of the Seine.[10]

5 Jones, 'Montpellier Medical Students', 58.

6 The term psychiatrist only began to be used after 1860, and for most of the nineteenth century French psychiatrists referred to themselves as alienists (*Grand dictionnaire de la psychologie* (Paris, 1991), 610). In this chapter, I have used the term psychiatrist and asylum doctor interchangeably.

7 Préfecture de la Seine, Direction des affaires départementales, *Rapport sur le service des aliénés de la Seine* (1890), 205.

8 This was article 18 of the law of 1838. On admission certificates the usual phrase was 'the patient is a danger to him/herself or to others'.

9 Both the central detention cells and the psychiatric service were located at 3 quai de l'Horloge, just near the Palais de Justice or central courts. In 1871, the General Council of the Seine voted to separate the police psychiatric service from the prison and name it the *Infirmerie spéciale du dépôt* in order, as the Prefect said, 'that it not be confused with the depot, where prisoners are kept'. In 1970 the service was finally transferred to the grounds of Sainte-Anne. J. P. Soubrier and M. Gourevitch, 'Recherches aux archives de la préfecture sur les origines de l'Infirmerie spéciale', *Perspectives psychiatriques* 96 (1984), 133–4; Archives of the Préfecture de Police, DB 218, letter of the Prefect, 28 February 1872.

10 Valentin Magnan served as the chief examining psychiatrist of the admissions office from 1867 until his retirement in 1912 and also gave weekly clinical lessons there. After patients had been certified by Magnan, the fortunate ones were taken across the grounds to Sainte-Anne itself. Others had to endure a longer trip to the more distant asylums of Vaucluse, Ville-Evrard, Villejuif and, after 1896, Maison Blanche, all of which were outside the city limits of Paris.

For nearly a century, Parisian psychiatrists denounced the continued existence of the *Infirmerie spéciale* and sought to have patients in police custody transferred directly to the Admissions Office at Sainte-Anne. With considerable justification, and a certain self-interest, they maintained that police control of official admissions implied a punitive, rather than therapeutic, approach to the insane. The prison-like atmosphere of the *Infirmerie spéciale* and the claustrophobic cages of the police van imposed, they argued, an unnecessary hardship on both families and patients. As one doctor described the transfer of patients: 'The lunatic is confined to a narrow cage, with the only light coming from a barred opening overhead; he hears the bolt slam shut and the van starts off without his having any idea, in his confused state, where he is going or why he has been locked up.'[11] Patients, noted Magnan, arrived at the Admissions Office 'feeling that they have been imprisoned'[12] and, consequently, were more difficult to treat.

In 1876, Parisian psychiatrists were influential in having a second, more 'humanitarian', type of admission procedure, *placement volontaire*, re-established in the department of the Seine.[13] This form of committal was intended for the mentally disturbed who did not threaten public order but who required treatment. Relatives (or even friends) could place such a person in any public asylum of their choice without police intervention; they could also withdraw the patient at any time, even if the asylum doctor objected. (This so-called voluntary committal did not, of course, involve the consent of the patient: under the law of 1838 the patient was an *aliéné* and therefore incapable of informed consent.)

When this form of admission was re-introduced in 1876, psychiatrists portrayed it as an enlightened reform that circumvented the horrors of the *Infirmerie spéciale* and assured the speedy treatment (i.e. committal) of mentally disturbed but harmless patients. It also, as one doctor observed, 'avoided any formality that could suggest the idea of coercion'.[14] From a medical point of view, this was a distinct advantage at a time when a strong 'anti-psychiatry' movement in the press and among politicians accused doctors of *séquestration arbitraire*, the illegal confinement of people with no signs of mental illness.[15] At first,

[11] *Annales médico-psychologiques* 1 (1877), 428.

[12] 'Procès-verbaux de la Commission de surveillance des asiles publics d'aliénés de la Seine', 19 March 1907, 105.

[13] This type of committal was also authorized under the law of 1838 but had been forbidden in the 1840s because of overcrowded conditions in Parisian asylums. Members of the *Société médico-psychologique* began to study the question of 'voluntary' admissions in 1868. In 1875, a commission of the society recommended the re-establishment of the process as a humanitarian reform that would allow families to avoid police procedures. *Annales médico-psychologiques* 1 (1877), 99ff.

[14] *Annales médico-psychologiques* 1 (1877), 428.

[15] For details of this anti-psychiatric movement, see Dowbiggin, *Inheriting Madness*, chapter five.

asylum doctors did not foresee any great demand for this new type of admission and paid little attention to it.[16] Nevertheless the demand for this service grew steadily. By the mid-1890s, 'voluntary' patients accounted for 21 per cent of all admissions; by 1905 they had risen to nearly 30 per cent and in 1912 they reached 32 per cent.[17] Sainte-Anne, the most prestigious of the asylums of the Seine[18] and the only one situated within Paris itself, attracted the largest number. By the 1890s, nearly half of its annual admissions, male and female, were 'voluntary'. Sainte-Anne continued to receive a disproportionately high number of such patients right up to the First World War.[19]

By 1890, this unanticipated growth in 'voluntary' admissions had attracted the attention and enthusiasm of asylum doctors. For the next twenty-five years, in their annual administrative reports and in their congresses, they welcomed every increase in 'voluntary' patients, complained of their unequal distribution among the five departmental asylums, and urged that more effort be made to attract such admissions. They argued, for example, that because families were offended by the 'promiscuity' of the asylum, that is, the intermingling of classes and medical conditions, separate rooms and better food should be provided for 'voluntary' patients.[20] At a time when asylum budgets were increasingly restricted, there were, as one doctor admitted, 'both economic and humanitarian reasons to facilitate such admissions':[21] at least half of all 'voluntary' committals were paying patients, at a rate of nearly 3 francs per day. But doctors were equally eager to attract non-paying 'voluntary' patients. They complained that neither families nor their physicians were aware of this form of medical assistance and that the procedures for free 'voluntary' admission were too bureaucratic. Marandon de Montyel, the senior psychiatrist at the Ville-Evrard asylum, always informed his patients of these procedures when they were being released, so that in case of a relapse, they could be speedily readmitted.[22]

This enthusiasm for one particular form of admission to public asylums is best understood within the context of the professional ambitions and insecurities of

[16] *Annales médico-psychologiques* 1 (1877), 106.

[17] *Rapport sur le service des aliénés* (1905), 4; (1912), 49.

[18] Sainte-Anne housed both the Admissions Office, where Valentin Magnan worked and taught, and the teaching clinic of the Faculty of Medicine, established in 1877. For psychiatrists of the period, a position at Sainte-Anne was the pinnacle of their career.

[19] No overall statistics for Sainte-Anne were reported in administrative documents. These estimations have been calculated from the annual reports of asylum doctors. In 1895 for example 58 per cent of male patients admitted to Sainte-Anne were 'voluntary'. By 1903 the figure was 53.6 per cent. Female voluntary admissions for the same years were 64.6 per cent and 60 per cent. *Rapport sur le service des aliénés* (1895), 193, 197; (1903), 146, 151. Since 'voluntary' placement did not imply the patient's consent, I have used the term in single quotations throughout.

[20] *Rapport sur le service des aliénés* (1890), 93 and (1896), 172. In fact, 'voluntary' patients received no special treatment in terms of accommodation or diet.

[21] *Rapport sur le service des aliénés* (1889), 234.

[22] *Rapport sur le service des aliénés* (1891), 139.

belle-époque psychiatrists. By the 1890s, the overcrowding in Parisian asylums had become a constant problem. On the one hand, this ever-increasing asylum population and the scant evidence of successful treatment encouraged the anti-psychiatric movement to brand asylums as modern Bastilles and psychiatrists as 'panalienists', self-appointed experts who 'saw madness everywhere except in their own mirrors'.[23] On the other hand, psychiatrists now found themselves responsible for up to 500 patients each, with the result that they could treat only the most urgent or 'interesting' cases. Moreover, there were few such 'interesting' cases because asylums had also become 'silted up' with chronic and incurable patients, such as the aged and the mentally handicapped, who, psychiatrists charged, were responsible for the profession's embarrassingly low rate of cure.[24] Increasingly, asylum doctors denounced the large Parisian asylums as 'barracks', 'depots', and 'lock-ups', places of confinement, not of treatment.[25] Rather than physicians, psychiatrists had become, in the words of one frustrated practitioner, 'the head custodian' of a population that was, from the medical point of view, uninteresting.[26]

The unforeseen growth of 'voluntary' admissions provided psychiatrists with the opportunity to defend their medical expertise and to escape a professional cul-de-sac.[27] As early as 1877, an inspector-general of asylums, L. Lunier, greeted the increase in the number of 'voluntary' patients as evidence that 'families had more and more confidence in public asylums'.[28] In 1890, a young Parisian psychiatrist, Marcel Briand, suggested that the popularity of 'voluntary' admissions was a reflection of medical progress: medical science was now able to diagnose and treat madness at an earlier stage. The result, he argued, was a 'change in attitudes, thanks to which families are more familiar with the road to the asylum, which no longer horrifies them'.[29] In 1912, as Magnan's successor at the Admissions Office, Briand repeated this claim, welcoming the 'more and more numerous' 'voluntary' admissions as an indication that Parisians,

[23] Dr Paul Garnier, *L'Internement des aliénés (Thérapeutique et législation)* (Paris, 1898), 118.

[24] Dowbiggin, *Inheriting Madness*, 141. The cure rate was calculated either on the number of releases in a year or the number of released patients certified as cured. With the most generous calculations, the cure rate in the asylums of the Seine was about 18 per cent (22 per cent for men and 14.5 per cent for women) in 1889 and 16 per cent (20 per cent for men, 12 per cent for women) in 1907. *Rapport sur le service des aliénés* (1889), 25 and (1907), 18. Sainte-Anne had by far the highest cure rate of the asylums of the Seine.

[25] See, for example, Garnier, *Internement des aliénés*, 67.

[26] In French, 'agent surveillant des surveillants'. *Rapport sur le service des aliénés* (1897), 172. Falret, at the Salpêtrière, complained that his service got no acute and curable patients, 'the only ones of medical interest'. *Rapport sur le service des aliénés* (1894), 301.

[27] In a perceptive analysis of recent trends in the history of psychiatry, Nancy Tomes has referred to psychiatry as 'a discipline trapped by its own medical "entrepreneurship" in the therapeutic and administrative cul-de-sac of the mental hospital'. N. Tomes, 'The Anatomy of Madness: New Directions in the History of Psychiatry', *Social Studies of Science* 17 (1987), 360.

[28] *Annales médico-psychologiques* 1 (1877), 87.

[29] *Rapport sur le service des aliénés* (1890), 205.

'better informed about the benevolent aspects of the law of 1838, increasingly appreciate the nature of the care in our asylums'.[30]

But to psychiatrists this growth of 'voluntary' committals symbolized more than the triumph of science and reason over ignorance and fear. It represented a new definition of the asylum and, by implication, a new role for its doctors. When referring to 'voluntary' admissions, psychiatrists increasingly used the term 'hospital', or occasionally 'hospice' and *maison de santé*, rather than asylum. For example, in 1890 Briand praised the 'comfort' and 'luxury' of Parisian public asylums and concluded that 'The Parisian people today consider them almost like ordinary hospitals, where you can enter or leave whenever you want.'[31] In contrast, by the 1890s, asylum doctors frequently associated official or police admissions with imprisonment. As Rouillard, the head of the teaching clinic of the Faculty of Medicine at Sainte-Anne, argued, more 'voluntary' patients were needed in order to show that the clinic was 'a true hospital and not a prison'.[32]

Asylum doctors argued that the two admission procedures resulted in two different types of patients. Those who passed through the *Infirmerie spéciale* were, they believed, inevitably chronic or incurable patients. Although earlier psychiatrists had praised official committal because it 'forced' families to treat the mentally ill,[33] their successors now argued that the horrors of the *Infirmerie* so appalled families that they delayed admission 'until they had lost all hope'.[34] Since it was psychiatric dogma that any delay in admission resulted in the 'manufacture of chronic cases',[35] it was not difficult to associate chronic or incurable patients with the *Infirmerie spéciale*, which psychiatrists increasingly referred to as the depot, the term for the police detention cells. On the other hand, because 'voluntary' patients could be taken directly to the asylum, psychiatrists maintained that they were admitted at the acute, rather than chronic, stages of their illnesses and were, consequently, curable. Only the lack of information about free care, it was assumed, could account for any delay in the admission of 'voluntary' patients.

For asylum doctors, frustrated with the state of their institutions and subject to continued public hostility, the growth of 'voluntary' admissions readily became a symbol of their hopes and a refutation of their critics: asylums could be transformed into hospitals where an influx of acute patients would result in reputable cure rates, expanded medical knowledge, and greater professional status.

[30] *Rapport sur le service des aliénés* (1912), 49.
[31] *Rapport sur le service des aliénés* (1890), 205–6.
[32] *Rapport sur le service des aliénés* (1891), 75. He also argued that 'voluntary' admission avoided the depot (*Infirmerie spéciale*) and the police van.
[33] *Rapport sur le service des aliénés* (1879), 7.
[34] *Rapport sur le service des aliénés* (1898), 92.
[35] Congrès des médecins aliénistes et neurologistes de France, *Rapports* 1 (1897), 298.

But this medical interpretation of the increasing popularity of 'voluntary' procedures is too self-serving to explain why families chose to 'take the road to the asylum'. It assumes that family and psychiatric interests were identical and that families accepted the medical definition of the asylum as a place to treat mental illness through total isolation. It also ignores the fact that Parisian families had already demonstrated their reluctance to submit to psychiatric regulation. 'Voluntary' admissions were reintroduced in the department of the Seine in 1876 not simply because psychiatrists argued for them, but because families were circumventing the law. Either they tried to bypass the *Infirmerie spéciale* and take their relatives directly to the asylum, or they used official admissions for non-dangerous family members, to the detriment of the departmental budget.[36]

Moreover, comments by psychiatrists themselves suggest that it was not administrative procedures but psychiatric power that most disturbed families. Paul Garnier, the examining physician at the *Infirmerie*, admitted that it was difficult to persuade families to commit a relative because they feared that 'the patient will be mistreated, that it will make him hate his family and will drive him completely mad'.[37] Febvré, a psychiatrist at Ville-Evrard, noted that official procedures 'offended' families and 'made them fear the impossibility of ever getting their patients back'.[38] The regulations for 'voluntary' committal, which allowed families to reclaim their relatives without the consent of the asylum doctor, altered the balance of power between the family and the asylum, a fact that psychiatrists rarely acknowledged and never explored in their medical writings.

The tangled motives that led Parisian families to 'knock on the door'[39] of the asylum will continue to elude historians, just as they eluded nineteenth-century psychiatrists. Family decisions were complicated, never fully articulated, and often unrecorded. But asylum records provide some useful indications of family attitudes, particularly when the asylum is Sainte-Anne, the most convenient of the Parisian asylums and, in terms of 'voluntary' admissions, the most popular. An analysis of the records of 'voluntary' patients at Sainte-Anne from 1876 to 1913[40] suggests that families had their own definitions of the asylum and of medical treatment, definitions that did not always accord with the hopes or interests of the psychiatric profession.

[36] Archives de la Préfecture de police, DB 218, letters of 27 September 1871 and 14 July 1877 from the Prefect.

[37] Garnier, *Internement des aliénés*, 64.

[38] *Rapport sur le service des aliénés* (1893), 177. [39] *Ibid.*, 219.

[40] The statistics in this chapter are based on an analysis of every second year of the admissions records (*registres de la loi*) of Sainte-Anne from 1873 to 1913, a total of 7,100 cases (4,488 men and 2,616 women). The admissions records normally give information on the age, marital status, occupation, type of admission, and, for 'voluntary' admissions, the person who committed the patient. These records also give three diagnostic certificates and the result of admission, i.e. release, transfer or death. Not all information was available for all patients, so the number of cases for each statistical table varies slightly.

Although asylum doctors sought to make sharp distinctions between 'voluntary' patients and those admitted under police procedures, asylum records indicate that it was not the method of admission that most clearly distinguished patients, but rather their gender. Men and women had a different experience of the asylum. Both at Sainte-Anne and generally in the department of the Seine, men were most often admitted for alcoholism or for general paralysis, the tertiary and fatal stage of syphilis. At Sainte-Anne, for example, 25 per cent of all male patients were diagnosed as alcoholic and another 20 per cent as suffering from general paralysis.[41] Male patients, if alcoholic, were quickly released after a brief period of detoxification; if paralytic, they died, usually within a year. Women were more likely to be admitted for depression: at Sainte-Anne, depression accounted for 32 per cent of all diagnoses among female patients.[42] Depression was, in medical terms, a less serious condition, but one that did not respond quickly to treatment. Consequently, women stayed longer in the asylum and ran a higher risk of being transferred to other asylums. Therefore, although admission rates for men were higher, at any given moment there were more women than men in Parisian asylums. This pattern was familiar to psychiatrists of the period: as one explained, 'If you exclude general paralysis and alcoholism, women are more predisposed to madness than men.'[43]

Although gender remained the determining factor in patterns of admission, 'voluntary' patients did differ in certain respects from their fellow patients admitted under official procedures. 'Voluntary' committal was, not surprisingly, a family matter and these patients were more likely to be married, although there was little difference in the number of widowed.[44] Nearly 40 per cent of 'voluntary' patients were admitted by their spouses and, in a rare instance of gender equality, wives and husbands committed each other in roughly equal proportions. If not a spouse, it was usually a parent or a child who made the formal committal, although occasionally it was a more distant relative, a friend,

41 Seven per cent of women were diagnosed as alcoholic and 6 per cent as suffering from general paralysis.
42 Among male patients, the rate of depression was 13 per cent.
43 *Rapport sur le service des aliénés* (1896), 68.
44 The statistics for marital status are:

	Male		Female	
	Official %	Voluntary %	Official %	Voluntary %
Single	38.0	27.4	38.0	27.7
Married	53.8	66.1	42.0	53.9
Widowed	7.4	5.9	18.6	17.6

Table 3.1 *Occupations of patients*

	Male		Female	
	Official %	Voluntary %	Official %	Voluntary %
Unskilled	13.9	8.8	18.7	7.3
Skilled	45.3	37.1	39.3	29.2
Petty bourgeois	22.0	32.8	5.6	5.6
Bourgeois	5.6	6.8	2.6	2.6
No occupation	5.6	9.4	31.1	52.6
Vagabonds	3.8	–	1.2	–
Wine trade	2.9	3.9	0.3	1.6
Concierge	0.9	1.2	1.2	1.1

Number of cases: men 4,478, women 2,408. (In all tables, percentages are rounded to one decimal place.)

or even an employer.[45] The decision was usually taken by the immediate family either because, as doctors alleged, families sought to keep mental illness a secret or because, in the late nineteenth century, many Parisians were recent arrivals with no other relatives in the city. The immediate family also took responsibility for patients who were released, although there was a slightly greater tendency for mothers to assume the responsibility for such care.

'Voluntary' patients were also drawn from a somewhat higher social group (see Table 3.1). Among such male patients, 32.8 per cent came from the petty bourgeois or artisan class of shopkeepers, clerks, civil servants and teachers, whereas among men committed by the police, the proportion was only 22 per cent.[46] The number of unskilled and skilled workers among 'voluntary' male patients was also lower: 8.8 per cent and 37.1 per cent respectively, compared with 13.9 per cent and 45.3 per cent for official admissions. Among women 'voluntary' patients, the pattern is even more pronounced. There were significantly fewer unskilled and skilled workers in this group[47] and 52.6 per cent of these women were recorded as having no occupation, compared with 31.1 per cent of official patients. While some of the 'voluntary' women patients with

[45] Of 2,430 'voluntary' patients admitted in the period 1876–1913, 39.6 per cent were admitted by a spouse, 19.5 per cent by an immediate relative (a parent or child) and 23.6 per cent by another family member. Although 17.3 per cent were admitted by a friend, often this was a common-law spouse. Mothers committed 5.7 per cent of patients but assumed responsibility for the release of 8.8 per cent of patients.

[46] Male 'voluntary' admission had a slightly higher proportion of bourgeois patients – 6.8 per cent compared with 5.6 per cent among men committed by the police.

[47] Among female 'voluntary' patients, 7.3 per cent were unskilled and 29.2 per cent were skilled workers. Among official women patients, 18.7 per cent were unskilled women workers and 39.3 per cent were skilled women workers.

no occupation were elderly, the admission registers indicate that most were the wives of skilled workers or of petty bourgeois and did not work outside the home.[48]

The higher social status of the families who used the procedures for 'voluntary' admission suggests that either they were affluent enough to afford the approximately 100 francs a month required of paying patients, or that they were sufficiently informed about the Parisian welfare system to arrange for free 'voluntary' care.[49] Given the complexities of the French class system, where a skilled worker might earn more than a shop owner or clerk, it was not always the petty bourgeois who paid for treatment. Although the records for paying and free 'voluntary' admissions are not extensive, they do suggest the inadequacies of occupational labels. In some cases, the differences between paying and non-paying patients are clear: the head clerk at the tax office, whose son was an employee at the Treasury, was a paying patient as were most patients identified as proprietors, while the day labourer (*homme de peine*) whose wife was also a day labourer (*journalière*) was a non-paying patient. Butchers, grocers and bakers usually paid, as did almost all café owners, but patients identified as investors (*rentiers*) could be found in both categories, as could teachers, tailors, soldiers, seamstresses and clerks.

In some cases, non-paying 'voluntary' patients had received information about the regulations from doctors, as is probably the case with the teaching assistant (*maître-répétiteur*) at the prestigious Lycée Louis-le-Grand, who was committed by his friend, a doctor. Others may have been advised by relatives who worked in the public service, as is likely in the case of two female patients whose husbands were municipal policemen. The head shoemaker at the Vaucluse asylum, who was admitted by his wife, a nurse at Vaucluse, had the advantage of already knowing the asylum system and he received free treatment. But how can one explain the paying patient who was a domestic servant, committed by her sister, a laundrywoman? Perhaps her employer paid, but if so, s/he was a generous one, for the patient was being admitted, at her own request, for the fifth time. At Sainte-Anne, at least, patients seemed to be fully aware of the regulations for free 'voluntary' treatment, perhaps because the asylum also operated a free clinic where Parisians could come for consultations or for baths.[50]

[48] A sampling of registers for the 1880s indicates that over 50 per cent of married women listed as having no occupation were the wives of petty bourgeois and another 25 per cent were married to skilled workers.

[49] The Seine was the only department in France to provide free 'voluntary' committal. It required a police investigation into the financial status of the family, a condition that was not welcomed by families. Officials were allotted 160 non-paying 'voluntary' places in Parisian asylums per year.

[50] The cases cited in the two preceding paragraphs are taken from the admissions records of 1885 and 1887, where paying and non-paying patients are clearly identified. The cases, in order of citation are: *Registre de la loi*, 1887 (men) no. 45; 1885 (men) no. 153, no. 77; 1887 (women) no. 80, no. 145; 1887 (men) no. 206; 1885 (women) no. 123.

Table 3.2 *Admission diagnoses*

	Male		Female	
	Official %	Voluntary %	Official %	Voluntary %
Alcoholism	28.3	13.7	5.2	5.3
Depression	11.3	16.3	27.8	35.2
General paralysis	17.4	30.3	4.6	7.3
Mania	6.4	2.4	13.4	12.3
Persecution	10.6	9.2	18.1	13.3
Senile dementia	3.3	3.1	5.2	5.6
Other	23.0	24.9	25.5	20.8

Number of cases: men 4,456, women 2,505.

At the heart of asylum doctors' enthusiasm for 'voluntary' admissions lay the assumption that such patients suffered from acute rather than chronic conditions. Yet the records of Sainte-Anne indicate that families made the decision to commit not in the early stages of illness but when the situation had become desperate. An analysis of admission diagnoses, for example, indicates a higher rate of chronic or incurable illnesses among 'voluntary' patients (Table 3.2).[51] The rate of general paralysis among male 'voluntary' patients was 30.3 per cent, compared with 17.4 per cent for their official counterparts. These men, usually in their 40s or 50s, often exhibited such violent or bizarre behaviour that nursing homes and hospitals refused to keep them. One such patient, who had been under a doctor's care for a year and was not considered dangerous, was committed when he suddenly attacked his wife with a knife.[52] Patients with general paralysis also tended to have delusions of grandeur, spending hard-earned family resources or incurring large debts. Typically, they would hire a cab for a day-long tour of Paris and then refuse to pay the fare. In one case, the family of a café owner committed him when he began giving free drinks to all his customers.[53] Although general paralysis was not a major reason for the committal of women, the rate among female 'voluntary' patients was also higher.[54]

The incidence of depression was also significantly higher for all 'voluntary' patients. Depression was not always considered a chronic condition, but patients with this diagnosis stayed longer before release, thereby contributing to the continual overcrowding of Parisian asylums. They were also more likely to suffer

[51] The admission diagnoses are taken from the first two certificates in asylum registers, the certificate issued immediately and the certificate issued after twenty-four hours. Both certificates were required by law.

[52] *Registre de la loi* 1879 (men) no. 175.

[53] *Registre de la loi* 1903 (men) no. 26.

[54] 'Voluntary' women patients had a 7.3 per cent rate of admission for general paralysis, compared with 4.6 per cent for other women patients.

Table 3.3 *Death rates*

	Male		Female	
	Official %	Voluntary %	Official %	Voluntary %
Overall	32.1	42.4	18.8	28.0
Alcoholism	13.9	16.5	14.5	17.8
Depression	22.0	19.0	13.9	14.5
General paralysis	72.8	75.5	48.7	84.0
Mania	18.8	36.8	9.4	14.7
Persecution	15.0	30.6	10.0	19.0
Senile dementia	85.0	79.6	58.0	75.0

Number of cases: men 3,425, women 1,977.

a relapse and to be readmitted. Among female 'voluntary' patients, 35.2 per cent were diagnosed as suffering from depression, whereas for other women patients the rate was 27.8 per cent. Although depression was not a major source of admissions for men, it accounted for 16.3 per cent of all diagnoses among 'voluntary' patients, compared to 11.3 per cent for men admitted by the police. Alcoholism, which was responsible for the short stay and high release rate of many male official patients, was conspicuously underrepresented among male 'voluntary' patients. It comprised only 13.7 per cent of these admissions, a lower proportion than diagnoses of depression.[55]

There was also a higher death rate among all 'voluntary' patients, another indication that they were admitted at a later, more chronic, stage of their illnesses (Table 3.3). Male 'voluntary' patients, with a greater incidence of general paralysis, had the highest death rate of all: 42.4 per cent, compared to 32.1 per cent for other men in the asylum. The death rate for women patients was always lower, but 28 per cent of 'voluntary' women patients died, in comparison with 18.8 per cent of other women patients. In some cases, death was the result of old age, as there was a slightly larger proportion of elderly patients, particularly women, among 'voluntary' admissions.[56] Even when patients were diagnosed as suffering from conditions not usually considered fatal, such as

[55] Sainte-Anne may have been atypical of the region's asylums in that few of its patients were mentally handicapped. Such patients may have been placed directly in more distant and cheaper asylums because relatives anticipated long-term care, or they may have been sent to the Bicêtre hospital where D. M. Bourneville had organized special facilities for mentally handicapped children.

[56] Among 'voluntary' women patients, 17.3 per cent were over the age of sixty, compared with 14.9 per cent among official women patients. Among male patients, the proportion of those over sixty was the same (9 per cent), whether official or 'voluntary', but there were slightly more men in the fifty to fifty-nine age group among 'voluntary' patients: 17.6 per cent compared with 15.2 per cent for official male patients.

mania or persecution, the mortality rates for 'voluntary' patients were noticeably higher.

The circumstances surrounding many 'voluntary' admissions reinforce the impression that families turned to the asylum only in periods of crisis or when alternatives had failed. Violence, or the threat of violence, is a recurring theme in the medical certificates and patient files. Of course, families may have exaggerated the violence, either from fear or from an emotional need to justify their actions, but they did not need to do so in order to have the patient committed. 'Voluntary' admission, unlike official procedures, did not require that the patient be dangerous, only in need of treatment. Asylum records suggest that in many cases the fear of violence was valid. One patient, for example, had threatened to kill his mother and sister and had begun digging a grave in the backyard. His mother brought him to Sainte-Anne after he tried to burn his sister's eyes out with a red-hot iron.[57] Wives appear to have used 'voluntary' admission as a temporary means of defence from domestic violence, particularly in the case of alcoholic husbands. In one typical case of alcoholism, the husband had broken the furniture, threatened to strangle his wife, and had provoked numerous complaints from neighbours.[58] But abused wives were not alone in seeking temporary relief from violence through committal. An eighty-year-old man committed his son, suffering from general paralysis, when the son became 'rude and violent', while the family of a female alcoholic, her doctors noted, lived 'under a reign of terror'.[59] A suicide attempt, or even the threat of suicide, was also sufficient reason for admission and was noted as such in 12.5 per cent of cases.

Complaints from neighbours could also provoke a family's decision to commit a relative and, judging from the certificates, asylum doctors found this a reasonable motive. A deaf-mute, who had stayed in her room for fifteen months, was brought to Sainte-Anne by her seventy-two-year-old father because she had become violent and 'her screams disturb the neighbours'.[60] In another case, a family reported that it was impossible to keep the patient in her room and that 'she seeks fights with all the neighbours'.[61] One elderly man 'threatened the neighbours constantly' and was in a 'lamentable state of filthiness' while an elderly woman was committed because she hit the concierge 'without any reason and without any previous disputes'.[62] Doctors may have maintained that families found mental illness so shameful that they kept it a secret, but privacy was a rare luxury in working-class dwellings and the interests of one relative might well be sacrificed to the larger interests of the family. In some cases,

[57] Patient file, no. 100796.
[58] *Registre de la loi* 1895 (men) no. 21.
[59] *Registre de la loi* 1915 (voluntary) nos. 84 and 86.
[60] *Registre de la loi* 1891 (women) no. 258. [61] *Registre de la loi* 1877 (women) no. 32.
[62] *Registre de la loi* 1887 (men) no. 200 and 1887 (women) no. 199.

the influence of neighbours could be more positive: one man claimed that he did not realize that his wife, who was diagnosed as suffering from depression, had been selling their household goods and mistreating their children until his neighbours told him.[63]

Usually, families had already tried other means of treatment before they turned to the asylum. In some instances, patients had been in a hospital, a *maison de santé*, or under the care of a local doctor who finally advised committal. In other cases, families had relied on their own resources, either locking up patients at home or sending them to stay with relatives in the countryside. One patient, a woman who lived alone with her young son, was taken first to a pharmacist and then to a doctor when she began to behave strangely. Neighbours were looking after her until she tried to commit suicide by jumping out of a window. Then she was brought to Sainte-Anne. When she was released, her mother took her to the countryside to recuperate.[64]

Families who turned to the asylum only when the situation had become chronic or unmanageable were equally reluctant to abandon their relatives to the exclusive care of the reputed medical experts. Patients in Parisian asylums were, as doctors remarked, 'much visited'[65] and, given the overcrowded conditions, families could be as well informed about the medical and living conditions of their relatives as was the asylum doctor.[66] Asylum records of 'voluntary' admissions indicate that families were not willing to defer to medical opinion; rather they were able to use their knowledge and legal powers to intervene on behalf of relatives in the two areas of most importance to patients, namely, release and transfer.

Patients committed by their families had a significantly higher rate of release, whether looked at in terms of overall statistics or in terms of specific diagnoses (Table 3.4). Women 'voluntary' patients had the highest general release rate of all patients in the asylum, even though, as a group, they were somewhat older. Their release rate was 48.3 per cent, compared with 40.5 per cent for official women patients. The release rate for 'voluntary' male patients (46 per cent) was slightly higher than the rate for official male patients (45.1 per cent), despite a greater number of terminally ill patients and a markedly lower proportion of alcoholics among male 'voluntary' patients. 'Voluntary' patients, whether men or women, who were diagnosed as suffering from depression had a 64 per cent rate of release, compared with 48 per cent for depressed official patients. There were similar patterns of a higher release rate for 'voluntary' patients suffering from alcoholism and persecution. However, the pattern did not hold

[63] Patient file, no. 88488. [64] *Registre de la loi* 1895 (women) no. 16.

[65] *Rapport sur le service des aliénés* (1894), 89.

[66] *Rapport sur le service des aliénés* (1894), 89. At Sainte-Anne, there was one doctor for approximately 400 patients and doctors regularly complained that they were not able to see most patients, let alone treat them.

Table 3.4 *Release rate*

	Male		Female	
	Official %	Voluntary %	Official %	Voluntary %
Overall	45.1	46.0	40.5	48.3
Alcoholism	69.9	80.1	59.0	68.9
Depression	48.0	64.0	47.6	63.9
General paralysis	14.3	18.2	13.2	4.3
Mania	53.8	55.3	53.8	60.3
Persecution	43.0	50.0	39.0	46.0
Senile dementia	9.5	16.3	9.3	7.5

Number of cases: men 3,425, women 1,977.

true in diagnoses of general paralysis and senile dementia. In these cases, only male 'voluntary' patients were more likely to be released than their official counterparts; female 'voluntary' patients, perhaps because older, were not so fortunate.

In part, these higher release rates among 'voluntary' patients are an indication that many families did not lose touch with their relatives in the asylum. Families were thus available either to accept trial leaves or to assure doctors that they could look after convalescing patients at home. On rare occasions, doctors acknowledged the presence of families as a positive factor that could lead to innovative treatment. At the Villejuif asylum, for example, Dr Marie arranged with the families of paying 'voluntary' patients to lodge them with local families as a preparation for full release.[67] But usually the relationship between families and psychiatrists was less amicable. Families frequently used their powers under the law of 1838 to force the release of a patient in the face of opposition from the asylum doctor. Psychiatrists resented this challenge to their authority. They grumbled about 'the ill-considered impatience of families who, with the text of the law in hand, demand the release of their patient'[68] and blamed the premature release of patients on families who mistook a remission for a recovery.[69] Asylum doctors were careful to note on certificates that they did not agree with the family decision and they took a lugubrious pleasure in pointing out when such releases resulted in readmission or misfortune.

Psychiatrists dismissed this reluctance to defer to their medical expertise as simply the result of families' inclination to yield to pressure from patients or, as one doctor at Sainte-Anne explained, 'families . . . often share the emotions and

[67] *Rapport sur le service des aliénés* (1911), 300.
[68] *Annales médico-psychologiques* 2 (1888), 393.
[69] *Rapport sur le service des aliénés* (1897), 67.

delirious ideas of their patients'.[70] In some cases, families were undoubtedly motivated by optimism, affection, or by insistent pleading. In many cases, however, relatives had formed their own judgement about the feasibility of release in the same manner as did the asylum doctors, that is, from visits, information from asylum attendants, and trial leaves. One husband, for example, reported that his wife had asked him to take her out of the asylum 'with a sweetness that disarms me'. He consented to a trial leave but brought her back as he judged she was not fully recovered.[71] The advice of the asylum physician could be a part of family calculations, particularly when committal had been the result of violence, but relatives were also willing to offer their own diagnoses and to act accordingly. For example, a man who was seeking the release of his sister, whom he and other relatives had visited regularly, wrote to her doctor: 'her condition appears to be much improved and I think that with my supervision she can try to re-establish her own home'.[72] In some cases, the decision was more informal: relatives simply did not return patients who had been released on a trial leave. Doctors usually accepted this as a *de facto* release and issued the appropriate certificate. Of one such patient the asylum doctor wrote, 'he is far from cured, but is not dangerous and since it is a voluntary admission, I am concluding that he has been released on the responsibility of his family'.[73]

The family's decision to have a patient released was also based on less stringent definitions of what constituted a cure. Families often seemed content with some improvement in their relative's condition and were willing to tolerate behaviour that doctors might label abnormal. One husband, asking for the release of his wife because he feared that the asylum was driving her mad, admitted that she still heard voices, 'but I do not think that is a reason for committing her'.[74] This attitude may suggest a tolerance and affection on the part of immediate relatives, or even a different definition of mental illness. But in many cases, it was the result of economic necessity. Women who relied on a husband's or son's income often used economic need as a reason to request the release of a patient. One mother, who offered the opinion that her son appeared to be cured, argued that an immediate release would enable him to find work. If not, the season would be over and 'I will suffer. He helps me out when he works and I am elderly.'[75] Husbands of 'voluntary' patients were less dependent on the income of their wives, who often did not work outside the home, but they did need their domestic work. The prolonged absence of a wife could result in the children being sent to relatives and the husband being left to fend for himself. One woman patient, in asking her doctor for a release that she herself seemed to consider premature, argued that 'my husband is at the end of his patience trying to cope... with the details of housekeeping, which he is not used to'.[76]

[70] *Rapport sur le service des aliénés* (1889), 151.
[71] Patient file, no. 126429.
[72] Patient file, no. 135329, 1908.
[73] *Registre de la loi* 1891 (men) no. 123.
[74] Patient file, no. 129978, 1906.
[75] Clinic, patient file, no. 140432.
[76] Patient file, no. 118305, 1903.

Table 3.5 *Transfer rates*

	Male		Female	
	Official %	Voluntary %	Official %	Voluntary %
Overall	20.8	9.1	38.6	20.0
Alcoholism	14.4	2.9	26.5	11.1
Depression	28.0	13.7	35.7	19.3
General paralysis	11.0	4.0	36.8	10.1
Mania	24.7	7.9	35.0	22.4
Persecution	39.0	16.0	48.0	27.8
Senile dementia	4.2	2.0	31.4	17.0

Number of cases: men 3,425, women 1,977.

The other important area of family intervention was in the decision to transfer patients to more distant asylums. In the latter half of the nineteenth century, asylums in the department of the Seine were so overcrowded that nearly half their patients – usually chronic and often female – were transferred to provincial asylums, where it was almost impossible for families to visit and where the standards of care were suspect. Patients and families alike dreaded a transfer. Families were so upset that asylum officials kept the departure date of convoys a secret from them in order to avoid 'painful scenes' at the railway station.[77] Under the terms of 'voluntary' admission, the transfer of paying patients required consent from the family, and the statistics clearly demonstrate that families were unwilling to give this consent, even for patients who were chronically or terminally ill.

The overall transfer rate for both male and female 'voluntary' patients was nearly half that of official patients (Table 3.5). The difference is most noticeable in the case of those female patients diagnosed as suffering from depression or senile dementia. These patients ran a high risk of being transferred, yet for those admitted by their families, the risk was cut almost in half. Even when 'voluntary' patients were eventually transferred, families were very effective in delaying this decision as long as possible. The contrast in the length of stay before transfer for official and 'voluntary' patients is striking. On average, male official patients stayed 176 days and female official patients 233 days before being transferred; their 'voluntary' counterparts were transferred after 883 days (for men) and 648 days (for women).[78]

[77] *Commission de surveillance*, 12 June 1900, 199.

[78] This is the median length of stay. If the interquartile ratio, i.e. the middle 50 per cent of the data, is calculated, the differences are even greater: official male and female patients were transferred after 618 and 619 days respectively; for 'voluntary' patients the figures are 2,315 days for men and 1,611 days for women.

Families clearly wanted their patients nearby, whether for visits or even for eventual release. (There were cases where the family sought the release of a relative after many years of confinement.) In some instances, families preferred to take their relative home rather than consent to a transfer. Certain families proved particularly adept at manipulating the system. In 1903, one family removed a male relative from Sainte-Anne because the asylum doctor wanted to transfer him to an agricultural colony for non-violent chronic patients. After several months, they returned the patient to Sainte-Anne, claiming that he was very dangerous and potentially violent, particularly when under the influence of alcohol. The patient stayed at Sainte-Anne until his death in 1919.[79]

If the family consented to a transfer, this did not necessarily mean that they were abandoning their relative. Middle-class patients, particularly if chronically ill, were frequently transferred to the *pensionnat* at Ville-Evrard, where the fees were higher but the care more comparable to that of a private clinic. Some patients were transferred to asylums in their native provinces where, presumably, nearby family could visit. (The system also worked in reverse: Sainte-Anne was the preferred destination for patients in provincial asylums whose families wanted them returned to Paris.) Families of paying patients might reconsider their choice of asylum if treatment at Sainte-Anne, the most expensive of the Parisian asylums, proved lengthy. One father, whose daughter had been under treatment at Sainte-Anne for a year, wrote to her doctor: 'I was prepared to make the sacrifice for such lengthy treatment in the hope that, under your good care, she would completely regain her reason. But, despite some improvement, it will still be a long time before I can take her back without danger so I am transferring her to Vaucluse, where it is cheaper.'[80]

In their correspondence and statements, families and patients often referred to Sainte-Anne as a *maison de santé* rather than as an asylum. Given the widespread prejudice against the insane, this may have been simply a form of denial, but in many instances it was also an accurate description of how many families viewed the asylum, namely as a place that was also for the chronic, the elderly, and those who were not, strictly speaking, mentally ill. In this sense families were not, as some historians have suggested,[81] merely enlarging the definition of mental illness; rather they were changing the definition of the asylum from a place for the treatment of the mentally ill to a source of care, whether temporary or long-term, for a wide variety of conditions. Contrary to 'supply' theories about the development of asylums, which postulate that once the asylums were constructed along medical lines 'the working-class family and neighborhood obediently present[ed] its crazy cases',[82] Parisian families were not simply

[79] *Registre de la loi* 1903 (men) no. 227.
[80] Patient file, no. 118305. [81] Tomes, 'The Anglo-American Asylum', 13.
[82] M. Ignatieff, 'Total Institutions and Working Classes: A Review Essay', *History Workshop* 15 (1983), 172.

passive users of asylum services. In reality, their demand for certain types of care shaped the structure of institutions that were still developing.

In all likelihood, this varied use of the asylum was not restricted to those more privileged groups who took advantage of the provisions for 'voluntary' committal. By the 1890s, psychiatrists were complaining that asylums were so overcrowded with other patients that 'true madness has become rarer and rarer in the asylums of the Seine'.[83] In order to return the asylum to its original purpose, as they defined it, asylum doctors began to urge separate care facilities for the elderly, the mentally handicapped, and the 'incurable'. Several such facilities, particularly agricultural colonies, were established in the department of the Seine before 1914. The results of these reforms are evident in the records of Sainte-Anne where, in the last decade before the war, the length of stay and mortality rates for all patients declined, indicating the presence of fewer chronic or incurable patients. Despite a psychiatric ideal of isolation from the community, the development of the asylum must be understood within the context of a complex system of services and institutions for the working classes.[84]

Although families increasingly turned to the asylum for assistance with relatives whose behaviour had become intolerable, they did so – despite the confident rhetoric of psychiatrists – as a last resort and with a clear unwillingness to accept the psychiatric dogma that treatment could only be achieved through isolation. Instead of abandoning their relatives to asylum doctors, they demanded a voice in the release, transfer, or longterm care of 'their' patients.[85] This ambivalent behaviour is not unique, nor is it surprising. As Michael Ignatieff has argued in a discussion of prisons, workhouses, and asylums in Britain and France, 'the poor were suspicious of institutions, but nevertheless supported them'.[86]

Even middle-class families who placed relatives in well-appointed private asylums expressed great anxiety about their decision and a need to be consulted about treatment.[87] Placing a relative in one of the public asylums of the Seine could not have been an easy decision, even when families were armed with legal powers under the terms of 'voluntary' admissions. French psychiatrists,

[83] *Rapport sur le service des aliénés* (1892), 211.

[84] An excellent example of such interconnections is found in Richard Fox's study of the California committal process where he argues that the high percentage of aged women in asylums was in part because almshouses became 'specialized institutions for the care of aged and destitute men'. R. W. Fox, *So Far Disordered in Mind: Insanity in California, 1870–1930* (London and Berkeley, 1978), 131.

[85] Families often referred to patients in a possessive fashion as 'ma malade', 'mon malade' or 'notre malade'.

[86] Ignatieff, 'Total Institutions', 172. Ignatieff credits this insight to M. A. Crowther, *The Workhouse System, 1834–1929: The History of an English Social Institution* (London, 1981).

[87] See, for example, N. Tomes, *A Generous Confidence: Thomas Story Kirkbride and the Art of Asylum-Keeping, 1840–1883* (Cambridge, 1984), 113, 122–23; C. MacKenzie, 'A Family Asylum: A History of the Private Madhouse at Ticehurst in Sussex, 1792–1917', PhD thesis, University College, London (1986), 278–9, 319.

unlike some of their North American counterparts,[88] did not welcome family intervention. Moreover, families did not need the press to conjure up images of Bastilles: they needed only to look at the high walls and locked gates of Sainte-Anne. This reluctant and selective, but increasing, use of 'voluntary' admissions should not be interpreted simply as a form of resistance to medical or state power. Rather, it suggests that families had integrated the asylum into their own well-established systems of treatment for the mentally disturbed or chronically ill, systems that made skilful use of various formal and informal resources available in the family, neighbourhood, and the larger community. When these resources failed, they turned to the asylum, but not necessarily as a permanent or long-term alternative.

French asylum doctors, recognizing the competition, denounced familial treatment as a source of abuse and chronicity. Families, they maintained, could not treat the unbalanced, they could only enchain them.[89] Yet in daily practice, psychiatrists not only had to cope with the needs and views of the family, they were also forced to use the family's resources. Faced with overcrowded conditions, asylum doctors relied on the willingness of families to care for convalescent patients, whether by trial leaves or permanent release. Families often left with specific instructions for the treatment of recovering patients and on occasion they consulted asylum doctors about continued home care. In late-nineteenth-century Paris, the exclusion of families from the treatment of the mentally ill was not, therefore, a realistic model, no matter how much it remained an article of faith for the psychiatric profession. Today, when a 'decarceration' movement has rejected the asylum and aspires to treat the mentally disturbed within the community, it is necessary to understand how community demand helped to shape the nineteenth-century asylum and, consequently, to recognize that new methods of treatment or care must be based on a realistic assessment of family resources and needs.

Finally, as a 'post-Foucaultian' generation of historians focuses on the dialectical relationship between families and the asylum,[90] it is important not to lose sight of the individual patient, that ever-shadowy figure of asylum records. Under law, the patient remained the object of family, administrative, or medical decisions. A truly voluntary admission procedure (*placement libre* in

[88] While psychiatrists in private asylums had an obvious motive for keeping the families of patients informed and pleased, there is some evidence from research on North American asylums for the poor that doctors encouraged a more cooperative relationship with families. See, for example, Constance McGovern's study of the Pennsylvania State Lunatic Asylum where, she argues, working-class families 'interacted with it quite comfortably'. C. M. McGovern, 'The Community, the Hospital, and the Working-Class Patient: The Multiple Uses of Asylum in Nineteenth-Century America', *Pennsylvania History* 54 (1987), 27.

[89] When accused of unlawful confinement, psychiatrists frequently retorted that the only such proven cases were to be found within families. *Rapport sur le service des aliénés* (1905), 187.

[90] The term 'post-Foucaultian' is from Tomes, 'The Anatomy of Madness', 358.

France), with its challenges to medical and family power, was slow to emerge. To what degree did family intervention ameliorate or complicate the patient's situation? Were patients able to use these conflicting systems of treatment to their own advantage? Although current research suggests that patients were not simply 'dumped' in asylums,[91] this says little about the feelings of imprisonment, isolation, and abandonment that patients experienced. Families may have viewed asylums as 'regrettable but indispensable necessities'.[92] It is unlikely that patients agreed.

[91] For example, Tomes, *A Generous Confidence*, 123 and MacKenzie, 'A Family Asylum', 317; E. Dwyer, *Homes for the Mad: Life Inside Two Nineteenth-Century Asylums* (New Brunswick, 1987), 3. Much, but not all, of the evidence for a family involvement in medical decisions comes from studies of private asylums. In contrast, a recent study of a public asylum in Alabama suggests that doctors in that asylum had much more control. J. Starrett Hughes, 'The Madness of Separate Spheres: Insanity and Masculinity in Victorian Alabama', in M. C. Carnes and C. Griffen (eds.), *Meanings for Manhood: Constructions of Masculinity in Victorian America* (Chicago, 1990), 53–66.

[92] Fox, *So Far Disordered in Mind*, 10.

4 The confinement of the insane in Victorian Canada: the Hamilton and Toronto asylums, *c.* 1861–1891

David Wright, James Moran and Sean Gouglas

Introduction

The changing approaches to the history of madness and the lunatic asylum in Canada have broadly reflected the ebb and flow of Anglo-American psychiatric historiography. The first generation of Canadian medical historians privileged the role of individual politicians, the evolution of the medical profession, and the difficulties of state formation in the gradual establishment of state-run mental hospitals. T. J. W. Burgess, for example, thought that Canada had 'shown a gradual process of evolution' in the care of the insane, from 'an era of neglect; then, one of simple custodial care with more or less mechanical restraint; and finally, the present epoch of progress'.[1] Burgess's account was followed by more sobering interpretations of success of the lunatic asylum, such as that written by Harvey Stalwick.[2] In his formulation, social conditions and political priorities beyond the control of pioneering alienists (psychiatrists) undermined the potentially beneficial aspects of institutional psychiatry. His sympathetic assessment of psychiatry, and its shortcomings, was thematically consistent with works produced contemporaneously in Britain and the United States, by Kathleen Jones and Gerald Grob respectively.[3] Their work continues to attract considerable sympathy from researchers in Canada. Endorsing their 'meliorist' interpretation, Peter Keating has argued that the 'moral treatment' of insanity that inspired the first generation of asylums is best understood as a hopeful

The authors would like to acknowledge the following agencies for their funding support: Canadian Institutes for Health Research; Arts Research Board of McMaster University; Associated Medical Services (Hannah Institute), Toronto; the Wellcome Trust, London, UK. An earlier version of this paper was presented to the 2001 Annual Meeting of the Canadian Psychiatric Association, Montreal, Canada.

1 T. J. W. Burgess, 'A Historical Sketch of Our Canadian Institutions for the Insane', *Transactions of the Royal Society of Canada* 18 (1898), 4.

2 H. Stalwick, 'A History of Asylum Administration in Pre-Confederation Canada', unpublished PhD thesis, University of London (1969).

3 K. Jones, *A History of the Mental Health Services* (London, 1972) and G. N. Grob, *Mental Institutions in America: Social Policy to 1875* (New York, 1973). For revisions of these influential monographs, see K. Jones, *Asylums and After: A Revised History of the Mental Health Services* (London, 1993) and G. N. Grob, *The Mad Among Us: A History of the Care of America's Mentally Ill* (Cambridge, 1994).

new breakthrough in medical practice. Keating explains the failure of asylum treatment and the rise of late nineteenth-century degeneration theory in Quebec as the result of social and political factors outside the power and purview of the early asylum promoters.[4]

As historians of madness know too well, the 1970s witnessed an unprecedented scholarly revision of the meliorist view of the rise of mental hospitals and the consolidation of Anglo-American psychiatry. Historians such as David Rothman and Andrew Scull offered a more cynical view of the rise of the asylum, situating the mental hospital within a new range of institutional solutions designed to incarcerate the 'deviant'. Following closely on the work of Michel Foucault, whose *Histoire de la Folie* was translated into English in 1972, these scholars rejected the views of Grob and Jones that placed the establishment of the asylums within the perspective of enlightened humanitarianism. Rather, revisionist scholars suggested that the asylum, and the psychiatric profession produced by it, represented a thinly veiled attempt to control the marginal, antisocial and violent. They differed, however, on their explanations for this sudden need for social control. Rothman emphasized the power and influence of new urban elites eager to find a replacement for the diminishing moral authority of the church; Scull, by contrast, highlighted the major disruptions that the expanding capitalist market brought to older paternalist social relations. Scull also directed his scathing critique at the 'entrepreneurial' ambitions of asylum doctors who, he suggests, had no real claim to expertise over the treatment of the insane.[5]

Rather than moving to broad thematic national histories of the asylum and the psychiatric profession, the colonial configuration of British North America in the nineteenth century, and the decentralized nature of the Canadian confederation in which health care became a provincial jurisdiction, has led historians in this country to examine mental health services at the provincial or institutional level.[6] Thus, although some regional studies incorporated aspects of Rothman's and Scull's critique of the emergence of modern psychiatry and the social control uses of the asylum, there was no overarching revisionist monograph to appear in Canada during this important period in psychiatric historiography. Arguably, the revisionist perspective was received most sympathetically in Quebec where André Cellard, in his *Histoire de la folie au Québec*, placed asylum development into the wider volatile social, cultural and political

4 P. Keating, *La science du mal: L'institution de la psychiatrie au Québec, 1800–1914* (Quebec, 1993).

5 D. Rothman, *The Discovery of the Asylum: Social Order and Disorder in the New Republic* (Boston, Mass., 1971); A. Scull, *Museums of Madness: The Social Organization of Insanity in Nineteenth-Century England* (London, 1979).

6 See, for example, T. Brown, '"Living with God's Afflicted": A History of the Provincial Lunatic Asylum at Toronto, 1830–1911', unpublished PhD thesis, Queen's University, Kingston (1980), 43–51.

contexts of early- and mid-nineteenth-century French Canadian society. Cellard identifies a broadening of the definition of, and a narrowing of tolerance for, insanity among eighteenth- and early-nineteenth-century francophones leading up to the creation of the Montreal and Beauport asylums. In the tradition of Foucault, Rothman and Scull, Cellard problematizes the social constitution and professional classification of behaviours labelled 'insane'.[7]

During the last two decades, Canadian medical historiography has been hugely influenced by the emergence of a new approach to the history of medicine, one that favours doing medical history 'from below'.[8] An explosion of publications reflected new methodologies and primary sources being explored in the field more generally, particularly an interest in the social history of patient populations. On the whole, these works shared with their British and American counterparts a scepticism of the earlier Whig approaches, and a critical engagement with (if not whole-hearted endorsement of) revisionist accounts. The seminal Canadian publications of the last two decades remain S. E. D. Shortt's book on the London Asylum and Cheryl Warsh's monograph on the private Homewood Retreat. Shortt situated alienist Maurice Bucke in the social and intellectual milieu of nineteenth-century psychiatry, while at the same time employing statistical analyses to evaluate the 'human ecology' of the London Asylum patient population.[9] It is not coincidental that his important book was published almost simultaneously with the two other outstanding English-language institutional studies of the period – Anne Digby's book on the Tukes and the Quaker York Retreat, and Nancy Tomes's study of Thomas Kirkbride and the Pennsylvania state asylum.[10] The tradition of excellent, detailed studies of institutions continued with Cheryl Warsh's monograph of the private Homewood Retreat. Warsh examined how the changing nature of the middle-class Victorian family affected the definition of insanity, asylum committal practices, and the diagnostic and treatment strategies of Homewood's medical directors. The trend in institutional histories towards those that incorporate the patient's world has been taken to a provocative new level with Geoffrey Reaume's history of the Toronto Asylum, a book directed entirely to patients' perspectives.[11]

[7] A. Cellard, *Histoire de la folie au Québec, de 1600 à 1850: le désordre* (Quebec, 1991).

[8] R. Porter, 'The Patient's View: Doing Medical History from Below', *Theory and Society* 14 (1985), 175–98.

[9] S. E. D. Shortt, *Victorian Lunacy: Richard M. Bucke and the Practice of Late Nineteenth-Century Psychiatry* (Cambridge, 1986).

[10] A. Digby, *Madness, Morality and Medicine: A Study of the York Retreat, 1792–1914* (Cambridge, 1985); N. Tomes, *A Generous Confidence: Thomas Story Kirkbride and the Art of Asylum Keeping, 1840–1883* (Cambridge, 1985).

[11] C. Warsh, *Moments of Unreason: The Practice of Canadian Psychiatry and the Homewood Retreat* (Montreal-Kingston, 1989); G. Reaume, *Remembrance of Patients Past: Patient Life at the Toronto Hospital for the Insane, 1870–1940* (Toronto, 2000).

Other themes in the Anglo-American historiography of madness in the last two decades can clearly be seen in the scholarship emanating from Canadian medical historians. Wendy Mitchinson's feminist analysis of admissions to the Toronto and London asylums echoes the famous work of Elaine Showalter[12] by emphasizing 'the way the medical profession treated women as patients' when they 'suffered from problems that were or could be related to their being female'. According to Mitchinson, at issue was the use of the male body as the norm against which women's health – including mental health – was gauged.[13] Rainer Baehre has contributed to the growing interest in the interconnection between legal history and the history of madness by putting forth a complex meta-analysis of penal and lunacy reform in early nineteenth-century Canada. Baehre argues that lunacy reform in early Victorian Upper Canada preceded the political unrest of the late 1830s, reflecting more a consensus of political culture between Tory and Reform politicians than the outcome of any classic form of class conflict.[14] More recently Peter Bartlett has analysed mental health legislation in England and Canada in comparative perspective, warning historians against making facile parallels between the two systems.[15] Finally, Alison Kirk-Montgomery and Robert Menzies have examined the relationship between criminality and lunacy through their studies, respectively, of the insanity plea in Victorian Ontario, and asylums for the criminally insane in early twentieth-century British Columbia.[16]

The intense concentration on the asylum, and on patient populations, has led inevitably to a questioning of the centrality of the asylum itself within social responses to mad behaviour. Reflecting the growing interest in the history of care and control of the insane outside of the asylum on both sides of the Atlantic, James Moran has highlighted the importance in understanding traditional responses to madness as practised by local familial, legal, medical and social authorities in New Jersey, Quebec and Ontario.[17] His work reflects a new

[12] E. Showalter, *The Female Malady: Women, Madness and English Culture, 1830–1980* (New York, 1985).

[13] W. Mitchinson, *The Nature of their Bodies: Women and their Doctors in Victorian Canada* (Toronto, 1991).

[14] R. Baehre, 'Imperial Authority and Colonial Officialdom of Upper Canada in the 1830s: the State, Crime, Lunacy and Everyday Social Order', in L. Knafla and S. Binnie (eds.), *Law, Society and the State: Essays in Modern Legal History* (Toronto, 1995).

[15] P. Bartlett, 'Structures of Confinement in Nineteenth-Century Asylums: A Comparative Study Using England and Ontario', *International Journal of Law and Psychiatry* 23 (2000), 1–13.

[16] A. Kirk-Montgomery, 'Courting Madness: Insanity and Testimony in the Criminal Justice System in Victorian Ontario', unpublished PhD thesis, University of Toronto (2001); R. Menzies, '"I Do Not Care for a Lunatic's Role": Modes of Regulation and Resistance Inside the Colquitz Mental Home, British Columbia, 1919–33', *Canadian Bulletin of Medical History* 16 (1999), 181–214.

[17] J. E. Moran, *Committed to the State Asylum: Insanity and Society in Nineteenth-Century Quebec and Ontario* (Montreal-Kingston, 2000); Moran, 'Asylum in the Community: Managing the Insane in Antebellum America', *History of Psychiatry* 9 (1998), 1–24.

interest in care outside the walls of the asylum and a slow decentring of the asylum as the locus of historical enquiry. Along with scholars such as Mark Jackson, he challenges the broader and overused notion of medicalization of madness as a top-down process. Recent and forthcoming research by Thierry Nootens and André Cellard investigates similar patterns of care and control in the community in the context of nineteenth-century Quebec.[18]

The work on the care and control of insane individuals outside of lunatic asylums has refocused an older debate over the reasons for asylum committal. Clearly just as not every person confined in an asylum was 'insane', historians also now appreciate the many thousands of individuals who, although they were recognized as insane by family members and local community members, were never institutionalized. So what combination of medical, behavioural and socio-economic factors combined to create a situation whereby a family would seek an institutional situation? And did these factors change over time?

The historical interest in the context of asylum committal, and in the strategic use of mental hospitals by families and communities, dates back to the late 1970s. At that time, Richard Fox, John Walton and Mark Finnane examined the background of patients admitted to the California state (United States), Lancaster county (England), and Omagh county (Ireland) asylums, respectively, in order to determine the social and economic factors influencing the arrival of patients to the mental hospital. All three agreed that the process of confinement was more complex than a situation of the state (or middle classes) using the asylum as a means of social control. Each pointed to migration as an important factor, though in very different ways. Fox identified isolated unmarried and newly arrived immigrants in California as particularly vulnerable to confinement.[19] John Walton, by contrast, argued that the dislocating process of urbanization placed strains on family and kin resources of those who had left the countryside for work in the developing industrial city of Lancaster.[20] For his part, Finnane suggested that the emigration of fit young men and women from Ireland (following the Famine) robbed households of caring resources, thus making affected families vulnerable to seeking institutional solutions.[21] Building on this work, Cheryl Warsh identified the isolation of young immigrants as she directed her attention at the characteristics of patients admitted to the London Asylum.[22] These studies of sample populations looked for factors that

[18] T. Nootens, 'Famille, communauté et folie au tournant du siècle', *Revue d'histoire de l'amerique français* 53 (1999); A. Cellard, 'Folie, internment et érosion des solidarités familiales au Québec: un analyse quantitative', unpublished conference paper, Folie et société au Québec: 19e–20e siécles, Centre d'histoire des régulations sociales, l'université de Québec à Montréal, 10 March 1999.

[19] R. Fox, *So Far Disordered in Mind: Insanity in California, 1870–1930* (Berkeley, 1978).

[20] J. Walton, 'Lunacy in the Industrial Revolution: A Study of Asylum Admissions in Lancashire, 1848–1850', *Journal of Social History* 13 (1979), 1–22.

[21] M. Finnane, *Insanity and the Insane in Post-Famine Ireland* (London, 1981), chapter four.

[22] C. Warsh, ' "In Charge of the Loons": A Portrait of the London, Ontario Asylum for the Insane in the Nineteenth Century', *Ontario History* 74 (1982), 138–84.

may have made certain individuals vulnerable to being confined. In addition to migration (both emigration, immigration and in-migration), they also analysed the behaviour of patients, gender, marital status and the availability of beds. Many of these factors will be discussed below.

A central theme to emerge from these studies was the active and determining role of the family in the confinement of their insane kin. In her analysis of admissions to the Toronto Asylum, Wendy Mitchinson argued that the Toronto Asylum was used by families to deal with those social excesses of insanity that they thought warranted institutionalization.[23] Similarly, Cheryl Warsh highlighted the social determinants of committal, and the experience of asylum life. Warsh analysed the changing gender dynamics of the Victorian family and their influence on the committal process.[24] Mary-Ellen Kelm's work on the British Columbia Provincial Hospital for the Insane pushed the historiography further by detailing the influence of families *after* the patient had been admitted.[25] Other researchers have focused on particular patient groups. Edward-André Montigny, for instance, has explored the confinement of the elderly to the Kingston Asylum in the latter part of the nineteenth century. He argues that asylum superintendents exaggerated the numbers of aged in their institutions as a means of exculpating themselves for low cure rates. Far from being flooded with elderly admissions, the Kingston Asylum, according to this author, saw a moderate, if slightly increasing, representation of admissions over the age of sixty. Montigny, like many of the other historians just cited, was struck by the distance families went to try and accommodate insane family members within the community, before turning to the asylum as a measure of last resort.[26]

The confinement of the insane in purpose-built asylums, and the social history of madness more generally, have constituted rich fields in the history of Canadian health and medicine, featuring prominently in monographs and in articles in Canadian history journals. This chapter seeks to contribute to this growing body of scholarship by analysing admissions to the Hamilton and Toronto asylums, and placing them in the context of the literature cited above. Before turning to admissions to these two Ontario institutions, it is necessary to survey the establishment of legislation regarding the insane in the province before and after the establishment of Canada through the British North America

[23] W. Mitchinson, 'Reasons for Committal to a Mid-Nineteenth-Century Ontario Insane Asylum: The Case of Toronto', in W. Mitchinson and J. Dickin McGinnis (eds.), *Essays in the History of Canadian Medicine* (Toronto, 1988).

[24] C. Warsh, *Moments of Unreason: The Practice of Canadian Psychiatry and the Homewood Retreat* (Montreal-Kingston, 1989).

[25] M Kelm, 'Women, Families and the Provincial Hospital for the Insane, British Columbia, 1905–1915', *Journal of Family History* 19 (1994), 177–93.

[26] E. Montigny, '"Foisted Upon the Government": Institutions and the Impact of Public Policy upon the Aged. The Elderly Patients of Rockwood Asylum, 1866–1906', *Journal of Social History* 28 (1995), 819–36.

(Confederation) Act of 1867. For the confinement of the insane in any jurisdiction was not only a social and familial phenomenon, it was also a legal and public action, requiring sanction by the state, certification by members of the state-regulated medical profession, and finally admission to state-run public institutions. An understanding of the growth of the state in Victorian Canada is thus critical for understanding the constraints within which families and authorities made choices over institutional confinement.

The construction of psychiatric institutions in nineteenth-century Ontario

Provision for the insane in early Upper Canada (Ontario) among European populations followed the traditions of settler families and communities and included boarding out,[27] warning out,[28] family care and management, and treatment by local medical practitioners.[29] Because Upper Canada did not incorporate the English poor-law system when it was founded in 1791, there was neither organized outdoor relief nor workhouses for the maintenance of the poor insane.[30] This partially explains why, in 1810, legislation was passed making the local jail 'to be for certain purposes a house of correction . . . and that all and every idle and disorderly persons, or rogues and vagabonds, and incorrigible rogues, or any other person who also may by law be subject to be committed to a house of correction, shall be committed to the said common gaols'.[31] The insane were easily subsumed under this list, and the practice of confining them to the York (Toronto) local gaol was eventually made legally explicit in an 1830 Act supporting the maintenance of 'insane destitute persons either in the jail or some other place'. Three years later this Act was extended to include Upper Canada's other district gaols.[32]

Dissatisfaction with this state of affairs emanated from many quarters. Local magistrates and gaol wardens who complained about the costs and chaos resulting from the incarceration of the insane issued Quarter Session petitions for more appropriate asylum provision. Debtors and criminals in Toronto's Home

[27] Boarding out was the practice of placing insane individuals with non-co-residing kin or with strangers for a fee. For an example of its use in Scotland at this time, see R. Houston, 'Not simple boarding': Care of the Mentally Incapacitated in Scotland During the Long Eighteenth Century', in P. Bartlett and D. Wright (eds.), *Outside the Walls of the Asylum: The History of Care in the Community, 1750–1900* (London, 1999), 19–44.

[28] Warning out referred to the banishment of the unwanted outside municipal or county boundaries. For examples see Thomas Brown, 'The Origins of the Asylum in Upper Canada, 1830–1839: Towards an Interpretation', *Canadian Bulletin of Medical History*, 1 (1984), 27–58; Moran, *Committed to the State Asylum*, 99–112.

[29] See Moran, *Committed to the State Asylum*, chapters three and four.

[30] R. Splane, *Social Welfare in Ontario, 1791–1893* (Toronto, 1965), 65–8.

[31] *Statutes of Upper Canada*, 1810, C.5.

[32] Splane, *Social Welfare*, 69; Brown, 'Origins of the Asylum', 29–30.

District Gaol made their own protest over the 'smell' and 'noise' of their mad co-habitants. Individual families also on occasion pleaded with the government for more adequate facilities of safe keeping for their insane relatives. Finally, reform-minded notables both inside and outside of Upper Canada pushed for the establishment of a publicly funded lunatic asylum that would approximate the emerging institutional ideal in other North American colonies, the United States and Britain.[33]

These calls to institutional action stirred activity in the provincial House of Assembly throughout the 1830s. But a combination of political inertia, political fighting between Reform and Tory MPs, and political unrest in the province at large, thwarted successful legislative action.[34] A commission was formed in 1835 to investigate the 'best method of managing and establishing a Lunatic Asylum' in Upper Canada, but the fruits of these labours were not realized in legal form until 1839 with the passing of An Act to Authorize the Erection of an Asylum within this Province for the Reception of Insane and Lunatic Persons.[35] Although this law allowed for the issue of debentures towards the construction of an institution for the insane, a permanent asylum was not ready for patients until 1850. In the interim, a temporary asylum was fashioned out of the Toronto Gaol, which was vacated in 1841.

The temporary asylum was run in accordance with accepted organizational wisdom with a superintendent, Board of Commissioners, and a set of rules and regulations for the management of the new institution.[36] Disputes over authority between the board and superintendent, combined with ongoing political struggles around asylum appointments, and the awkwardness of the converted prison structure, seriously hampered treatment strategies at the temporary asylum. Nevertheless, a form of moral treatment was introduced by Dr William Rees, the first superintendent, and modified by Drs Telfer, Park and Primrose.[37] Admissions during this period were made by petition to the asylum commissioners and to the superintendent himself.

[33] *National Archives of Canada* (NA), RG4 B65, loose documents, Toronto Sheriff to Provincial Secretary, 16 September 1840; P. Oliver, *'Terror to Evil-Doers': Prisons and Punishments in Nineteenth-Century Ontario* (Toronto, 1998), 44–5; Brown, 'Origins of the Asylum', 35–6; Splane, *Social Welfare*, 203. Outside reform pressure was brought to bear by internationally renowned asylum advocates Dorothea Dix and Daniel Hack Tuke. See, NA, RG4 C.1, file 2204, 'Memorial of Dorothea Dix to the Provincial Parliament of Canada East and West'; Dix to Charles Metcalfe, Governor in Chief of the United Provinces, 12 October 1843; Tuke's indictment of the temporary asylum at Toronto and call for better provision is found in D. H. Tuke, *The Insane in the United States and Canada* (London, 1885), 215.

[34] Brown, 'Living with God's Afflicted', *passim*.

[35] *Journals of the Legislative Assembly of Upper Canada*, 2nd Session, 12th Parliament, 1836, 196; *Statutes of Upper Canada*, 2 Vic., C.11,1839.

[36] NA, RG5 C1, File 2883, Report of the Commissioner and Proposed Regulations, 17 February 1841.

[37] Moran, *Committed to the State Asylum*, 50–62.

The opening of the permanent Toronto Provincial Asylum in 1850 soon altered the social organization of treatment and care of the insane in Ontario. This was partly achieved by the passing of the 1853 Act for the Better Management of the Provincial Lunatic Asylum, legislation that significantly increased the power of the superintendent largely at the expense of the Board of Commissioners.[38] Superintendent Joseph Workman vigorously incorporated the principles of this new law, and took advantage of the larger 'purpose built' space of the new asylum during his twenty-two year career, in his efforts to implement moral treatment in Ontario.

The Toronto Provincial Asylum was overcrowded with patients soon after its doors were opened. This led to the establishment of 'branch asylums' at the University of Toronto in 1856, at Fort Malden (near Amhertsberg) in 1859, and at Orillia in 1861. The branch asylums were populated with patients whom Workman considered to be chronically or incurably ill in order that the Toronto Asylum might be maintained as a curative institution, and were closed in 1870 with the construction of the London Asylum.[39] The asylum infrastructure was further expanded in Ontario when a criminal lunatic asylum was opened in Kingston in 1855,[40] when the London Asylum was opened in 1870, and when a new regional institution was built in Hamilton in 1875. The Orillia Asylum reopened as a specialist institution for 'idiot and imbecile' children in 1876.[41] To regulate this expanding public[42] asylum system (and other government institutional infrastructures) a Board of Inspectors of Prisons and Asylums was created in 1857.[43] Thus, for the period under study (*c.* 1861–91) there were four principal lunatic[44] asylums for the Province of Ontario: Toronto, Kingston, London and Hamilton. After Confederation these institutions were considered the hubs of four health 'regions' of the province. Patients were supposed to be admitted to the asylum in the region in which they resided (though, due to overcrowding, and various other social factors, this procedure was not always followed).[45]

Laws governing individual insanity and asylum committal were not well developed in Ontario until the mid nineteenth century. Two early Acts relating

[38] *Statutes of Canada*, 16 Vic., C. 188.

[39] Warsh, 'In Charge of the Loons', 141. This practice of building chronic facilities in the hopes of keeping others as curative institutions was also implemented south of the border. See for example the use of the Willard and Utica asylums in Ellen Dwyer's, *Homes for the Mad: Life Inside Two Nineteenth-Century Asylums* (New Brunswick, 1987).

[40] The Kingston Asylum, also known as the Rockwood Asylum, began to accept 'general patients', in addition to criminal lunatics, in 1862. See Montigny, 'Foisted Upon the Government', 821.

[41] The nineteenth century ended with the establishment of four more psychiatric institutions in Ontario, respectively at Mimico (1894), Brockville (1894), Cobourg (1902) and Penetangueshine (1904).

[42] There was one large private institution – the Homewood Retreat in Guelph – which was established in 1884.

[43] *Statutes of Canada*, 20 Vic. C. 58.

[44] This chapter will not discuss the asylum for idiots in Orillia.

[45] Warsh, 'In Charge of the Loons', 140.

to the estates of specific individuals sanctioned commissions of lunacy whose authority was ultimately vested in the Court of Chancery. The powers of the Court of Chancery were solidified in an 1846 Act. Commitment law was not legally defined until An Act to Authorize the Confinement of Lunatics in Cases where their Being at Large may be Dangerous to the Public was passed in 1851. This act stipulated that the insane could be placed into custody if convicted of a crime, or if considered dangerous either to themselves or to the public.[46] Two certificates of insanity were required to admit such lunatics to an asylum (this was later changed to three in 1853) and the assent of the local mayor or reeve.[47] The lunatic could then be placed in an asylum and remain there until certified sane again by two medical practitioners along with the sanction of the Governor of the Province. This law also allowed for the confinement of suspected lunatics to local gaols, a practice that, in the absence of sufficient asylum provision at mid century, was frequently exercised.[48] Indeed the cumbersome nature of the (regular) civil confinement, where a bed was often not available, and the cost of securing three separate medical certificates, have led some historians to suggest that desperate families resorted to declaring non-violent family members as 'dangerous lunatics', and thus subject to the 1851 statute. Under a dangerous lunatic warrant, local authorities were forced to gaol the individual immediately and then find them an asylum bed.[49] Although precise figures are difficult to tabulate, approximately one-third of admissions to the Toronto and London asylums in the 1860s occurred through the gaol system.[50]

The confinement of the insane in Victorian Ontario thus occurred within the context of an expanding provincial asylum system subject to provincial laws and an implicit desire to locate institutional care close to the family. The tables that follow constitute the first results of a major research project analysing the socio-demographic and medical characteristics of the psychiatric patient population in nineteenth-century Ontario. By the time of completion, the project will examine the admissions of all patients to the four principal lunatic asylums in the province of Ontario, involving the examination of approximately 8,000 patients.[51] For

[46] R. Baehre, 'From Pauper Lunatics to Bucke: Studies in the Management of Lunacy in 19th Century Ontario', unpublished PhD Thesis, University of Waterloo (1976), 118–24.

[47] Bartlett, 'Structures of Confinement in Nineteenth-Century Asylums', 10.

[48] See J. E. Moran, 'Dangerous to be at Large? Folie et criminalité au Québec et en Ontario au 19th siècle', *Bulletin d'histoire politique* (in press).

[49] Montigny, 'Foisted upon the Government', 827. There seems to be some disagreement about the requirement to find a 'dangerous lunatic' a place in the asylum. For a dissenting view to Montigny, see Bartlett, 'Structures of Confinement', 10–11.

[50] Bartlett, 'Structures of Confinement', 10.

[51] The data include all patients admitted to the Provincial Lunatic Asylum in Toronto in the decennial census years of 1861, 1871, 1881, 1891 and all patients admitted to Hamilton Asylum from its construction in 1875 until the end of the year 1885. The authors would like to thank Nanci Delayen of McMaster University for assistance in the input of patient records admitted to the Hamilton Asylum.

the purposes of this chapter, a sample of 1,682 individuals from the Toronto and Hamilton asylums will be analysed for socio-demographic characteristics. These results will be compared with the research conducted by historians on the demographic, social and economic composition of Victorian Ontario.

Characteristics of patients

Ontario[52] was the largest English-speaking province in Canada by population, bordered to the south by the Great Lakes, and stretching from the predominantly French-speaking province of Quebec on the east to the Prairie Provinces on the west. In terms of landmass, the size of the province at Confederation (1867) approximated that of France, though only a small strip of population along the American border and the Great Lakes has ever received sustained non-Aboriginal settlement. Thus, with a population distribution not unlike colonial Australia, the European settlements in Victorian Ontario consisted of a small clutch of cities growing quickly in size and stature, and surrounding agricultural settlements to the southwest and east of the province. To the north lay a vast, largely unpopulated northern interior. The principal cities were, and remain, Toronto (the provincial capital, formerly known as York), Ottawa (which became the nation's capital in 1867), Kingston to the east of the province, London in the heart of the agricultural southwest, and Hamilton (fifty miles around Lake Ontario to the south of Toronto) which became the industrial and steel-producing centre of the province. The major cities were connected by the construction of a rail network in the 1850s running along the northern shores of the St Lawrence and Great Lakes, from Montreal to the border with Detroit, Michigan. As the asylum system expanded, the lunatic asylums were placed in the principal cities, in Toronto, Kingston, London and then Hamilton. Ottawa never received an asylum, but hosted instead Canada's parliament.

The results of the analysis of a sample of 1,682 patients[53] admitted to the Toronto and Hamilton asylums reveal the remarkably diverse backgrounds of asylum patients that have been illustrated in other chapters in this book. To begin with, it has now become common for medical historians to uncover a sex ratio of

[52] During the period under study, what was called the Province of Ontario (after 1867) was variously known as Upper Canada (1791–1841) and Canada West (1841–67). Since 1867, following the Confederation of the provinces of Canada, the province has been known as Ontario. We shall use Ontario in this chapter for matters of consistency and to avoid confusion.

[53] Patients who had been admitted to Toronto and then transferred to Hamilton after its construction in 1875, were included as Toronto first admissions. In order not to bias the results, only first admissions for both institutions were included in the data, thus preventing the duplicated recording of socio-demographic characteristics of the identical individual. Of the patients sampled (1864) 1,684 or 90.3 per cent were first admissions. Excluding second and subsequent admissions tended to increase slightly the *average* length of stay in the tables presented and to increase slightly the age (by excluding multiple readmissions of a few young patients).

patients reflective of the proportion of men and women in the adult population from whence the patients came. In countries with a (relatively) balanced sex ratio, such as western Europe and Britain, the number of men and women at the time of asylum admission was evenly matched.[54] In jurisdictions where there was a surplus of men – such as certain colonial or frontier societies of California, South Africa, Argentina and New Zealand – men were more numerous in the asylums, reflective of their larger numbers in those populations.[55] Moreover, in his examination of the confinement of the insane in colonial New South Wales, Stephen Garton has argued that the preponderance of male admissions in 'frontier society' asylums may also reflect the centrality of the police in the committal process and the desire to reinforce a sense of social stability and conformity in communities that had no tradition of welfare institutions.[56] Certainly his study of New South Wales, and Bronwyn Labrum's examination of New Zealand, suggest that men may have been incarcerated at a rate in excess of their already-disproportionate representation in colonial Australia and New Zealand.

In the first two decades of the provisional and permanent Provincial Asylum in Toronto, Ontario conformed to the colonial pattern of a significantly high male : female ratio of admissions – as much as 3:1. By the 1860s, however, there seems to have been a 'rebalancing' of admissions by gender, a trend seen in other industrializing countries. By the 1870s and 1880s, the gender ratio began to conform, more or less, to the representation of men and women in the adult population of the time.[57] For the sample under study (1861–91), men constituted 53 per cent of the sample population admitted to the Toronto Asylum, and 52 per cent of the sample population admitted to Hamilton (see Table 4.1). Although considerations of gender must have played some role in the perception of insanity and the familial decision to seek (or not to seek) confinement, these preliminary results suggest that gender considerations worked towards the confinement of certain groups of women *and* men, not women instead of men.[58] Indeed, when one brings in quantitative work previously published on the

[54] See Gasser, Malcolm and Dörries in this volume.

[55] See Fox, *So Far Disordered in Mind*; B. Labrum, 'Looking Beyond the Asylum: Gender and the Process of Committal to Auckland, 1870–1910', *New Zealand Journal of History* 26 (1992), 553–74.

[56] S. Garton, *Medicine and Madness: A Social History of Insanity in New South Wales, 1880–1940* (Kensington, Australia, 1988).

[57] W. Mitchinson, 'Reasons for Committal to a Mid-Nineteenth-Century Ontario Insane Asylum: The Case of Toronto', in Mitchinson and Dickin McGinnis (eds.), *Essays in the History of Canadian Medicine*.

[58] For a similar discussion of the role of gender, and the lack of a disproportionate confinement of women in Victorian England and Wales, see J. Busfield, 'The Female Malady?: Men, Women and Madness in Nineteenth-Century Britain', *Sociology* 28 (1994), 259–77; K. Davies, ' "Sexing the Mind?": Women, Gender and Madness in Nineteenth Century Welsh Asylums', *Llafur – Journal of Welsh Labour History/Cylchgrawn Hanes Llafur Cymru* 7 (1996), 29–40.

Table 4.1 *First or subsequent admissions to the institution, admissions to the Toronto and Hamilton asylums, selected years, 1851–1891 (n = 1862)*

	Toronto				Hamilton			
	Men		Women		Men		Women	
	no.	%	no.	%	no.	%	no.	%
First	272	83	238	83	612	94	560	94
Second	40	12	37	13	38	6	31	5
Third or more	15	5	13	5	1	0	4	1
Sub-total	327		288		651		595	
n/a	0		0		0		1	
Total	327		288		651		596	

London Asylum,[59] the Toronto Asylum (from 1841 to 1874),[60] and the British Columbia asylums[61] – work that also confirms that women were admitted in numbers either in proportion to their representation in the adult population (or less than their proportion) – the evidence supporting Showalter's famous claim for the disproportionately high rate of incarceration of women in Victorian asylums becomes very thin, if non-existent.[62] The reasons behind the gradual process of balancing female and male admissions remain difficult to determine from an aggregate study of admission records. It may well be, as Nancy Tomes has argued for nineteenth-century Pennsylvania,[63] that families became less reluctant to send female patients to state institutions, as the asylum slowly distanced itself from its institutional cousin – the poorhouse. Much more research is needed, however, to elucidate the motivations of families in their decision to seek institutional confinement.

The age-distribution of patients at the time of admission also conforms to the general demographic distribution of Ontario *adult* society at the time (Table 4.2). In terms of age and life cycle, these two Ontario asylums under study accepted patients from across the adult age-spectrum. The inter-quartile age range (the middle 50 per cent of all patients) was between the ages of twenty-seven and forty-nine for men and women in both institutions. The median fell between

59 Warsh, 'In Charge of the Loons', 155.

60 Mitchinson, 'Reasons for Committal to a Mid-Nineteenth-Century Ontario Insane Asylum', 92.

61 M. Kelm, 'Women, Families and the Provincial Hospital for the Insane, British Columbia, 1905–1915', *Journal of Family History* 19 (1994), 177–93.

62 For a more substantial exploration of the relationship between gender, class and madness, see chapters in J. Andrews and A. Digby (eds.), *Sex and Seclusion, Class and Custody: Gender, Class and the History of Psychiatry in Britain and Ireland* (Amsterdam, forthcoming).

63 Tomes, *A Generous Confidence*, chapters three and five.

Table 4.2 *Age of patients at time of admission, first admissions to the Toronto and Hamilton asylums, selected years, 1851–1891 (n = 1682)*

		Toronto				Hamilton			
		Men		Women		Men		Women	
	Age	no.	%	no.	%	no.	%	no.	%
	<20	11	4	5	2	18	3	14	3
	20–9	67	25	78	34	167	28	150	28
	30–9	76	29	57	25	152	26	134	25
	40–9	52	20	42	18	112	19	107	20
	50–9	31	12	21	9	80	14	81	15
	60–9	19	7	25	11	42	7	37	7
	>69	9	3	4	2	15	3	12	2
Sub-total		265		232		586		535	
	n/a	7		6		26		25	
Total		272		238		612		560	
Median		35		35		36		36	
IQ Range		28–48		26–47		27–48		28–49	

the ages of thirty-five and thirty-six. There were no significant inter-sex or inter-institutional differences in the ages of patients at time of admission. Rather than representing the aged and demographically marginal, as has been found in case studies of Houses of Industry that were gradually being created in the province in the late Victorian period,[64] the majority of admissions to both institutions were in their so-called prime of life – in their twenties, thirties and forties.

The distribution of ages illustrates that it was extremely rare for children and adolescents to be admitted to a general lunatic asylum in Ontario before the twentieth century. Only one patient out of the sample was under the age of sixteen at the time of admission. Children suffering from ‘idiocy’ were sent to a specialist institution in Orillia, 100 miles northwest of the city of Toronto.[65] It is unclear, however, where adolescents, not identified as ‘idiots’, but exhibiting strange behaviour, would have been sent. It is quite possible that many may well have found their way into the juvenile reformatories also being built at this time.[66] But without a comparison of the behavioural characteristics of

[64] See R. Baehre, ‘Paupers and Poor Relief in Upper Canada’, *Canadian Historical Association: Historical Papers* (1981), 57–80; S. A. Cook, ‘“A Quiet Place . . . To Die”: Ottawa’s First Protestant Old Age Homes for Women and Men’, *Ontario History* 1 (1989), 25–40.

[65] For a history of the Orillia Asylum and ‘mental retardation’ in the province of Ontario, see H. Simmons, *From Asylum to Welfare* (Downsview, Ont., 1982).

[66] R. Nielsen notes that, at the Penetanquishene Reformatory Prison for boys, opened in 1859, ‘the most common charge was larceny; others had been sent to Reformatory for obstructing a railway,

adolescents being admitted to reformatories and adults sent to lunatic asylums, this remains a question demanding further comparative research.

Medical superintendents of Ontario asylums complained that their institutions were filling up with the elderly who were senile, but did not constitute the 'curable insane'. The aged, however, did not constitute a flood of admissions during this time, though they may have been slightly overrepresented. From a low of 3.1 per cent of admissions, the proportion of elderly[67] admissions rose to 14 per cent by 1903, a period in which the elderly increased as a percentage of the general population[68] from 3.1 per cent to 8.4 per cent. In total, the aged constituted only 7.4 per cent of the over 4,000 admissions to the Kingston (Rockwood) Asylum between 1866 and 1906.[69] Comparable figures for Hamilton between 1875 and 1885 show 10 per cent of admissions were sixty years or older. Toronto admissions reflect a similar proportion (10 per cent – 13 per cent). The elderly constituted between 8.4 per cent and 10.5 per cent of the adult population of Ontario society at the time, depending upon whether one uses everyone over the age of fifteen or twenty as a comparable group for asylum admissions.[70] Thus the proportion of elderly, at the time of admission, was hardly remarkable.

The balance of elderly women and elderly men merits further comment. Demographic reconstructions of nineteenth-century Ontario society have suggested that elderly women were commonly kept in extended households, especially if they could contribute to the household economies by childminding. Indeed, they were three times more likely than men to appear in household enumerators' schedules of families living in their early or mid-life cycle.[71] Elderly men, by contrast were more likely to be boarders and to find themselves in Houses of Refuge or Houses of Industry as they were built from the 1860s

keeping found money, breaking windows, *and even lunacy*'. Nielsen, *Total Encounters: The Life and Times of the Mental Health Centre, Penetanguishene* (Hamilton, 2000), 8–9 (our italics); Paul Bennett, 'Taming "Bad Boys" of the 'Dangerous Class': Child Rescue and Restraint at the Victoria School 1887–1935', *Histoire Sociale/Social History* 21 (1988), 71–96; S. Houston, 'The "Waifs" and "Strays" of a Late Victorian City: Juvenile Delinquents in Toronto', in J. Parr (ed.), *Childhood and Family in Canadian History* (Toronto, 1982), 129–42.

67 For consistency, we have defined the elderly as all those sixty years of age or older.

68 Montigny, 'Foisted on the Government', 821. It is unclear, from Montigny's research, whether his figures of comparison are for the *general* population of the time (including children), or the adult population of the time. This is of particular interest inasmuch as children were not, as noted above, admitted to the general lunatic asylums. If he chose the former (the population including those under the age of sixteen years) as a point of comparison, then it is entirely possible that, even at the level of 14 per cent of adult admissions, the elderly were *not* overrepresented compared to the *adult* population of the general population.

69 Montigny, 'Foisted on the Government', 821.

70 Calculated from 'Table VII – Ages of the People'. *Census of Canada, 1870–71*, vol. 2 (Ottawa, 1873), 58–61.

71 M. Katz, *The People of Hamilton, Canada West: Family and Class in a Mid-Nineteenth-Century City* (Cambridge, 1975), 231.

and 1870s. According to one study, elderly men outnumbered elderly women in these welfare institutions by a ratio of 4 to 1,[72] a number more pronounced when one considers the lower life expectancy of men during this period. Clearly over this period, as with the Kingston Asylum, even if the admissions of the elderly to the Toronto and Hamilton asylums were slightly higher than their representation in the population would merit, the Victorian Asylum was 'hardly a refuge for the elderly or the senile'.[73]

The regions surrounding the cities of Toronto and Hamilton were undergoing rapid economic, social and demographic change during the period under study (1861–91).[74] The background of the patients admitted to the Toronto and Hamilton asylums reflects this diverse and changing society. Judging from their occupational backgrounds, they represented what can best be described as the 'lower two-thirds' of Ontario society. However, with fewer than 10 per cent of patients recording 'no occupation' or having no entry in the occupation column of the admission registers, these male admissions clearly reflected the working (rather than marginal) poor of Ontario society. Male admissions to the asylums may have been marginalized by their disorder, but the vast majority seem, by all evidence, to have been gainfully employed in the period before the onset of their 'attack'. Indeed, an analysis of the occupations of Toronto and Hamilton asylum patients reflects a staggering variety of employments performed by Ontarians who were eventually committed to one of these two asylums under study (see Table 4.3a). As in Ontario society at the time, by far the largest male occupational group were listed as 'farmers' in the asylum admission registers (30 per cent of Toronto Asylum male admissions; 33 per cent of Hamilton Asylum male admissions). These figures can be compared with Ian Drummond's statistical table of occupations of Ontario's gainfully employed. We can calculate from Drummond's figures that approximately 49.2 per cent of males were employed in agriculture (excluding labourers) from 1871 to 1901.[75] This suggests that the percentage of the 'farmers' committed to Ontario's asylums (Toronto in particular) was noticeably lower than that of the Ontario population as a whole. This difference might be explained by the relatively stable family structure in place among settled (even recently settled[76]) farmsteads. At the

[72] See S. Stewart, 'The Elderly Poor in Rural Ontario: Inmates of the Wellington County House of Industry, 1877–1907', *Journal of the Canadian Historical Association* (1992), 217–33.

[73] C. Warsh, 'In Charge of the Loons', 154.

[74] Katz, *The People of Hamilton*; G. S. Kealey, *Toronto Workers Respond to Industrial Capitalism, 1867–1892* (Toronto, 1980); D. McCalla, *Planting the Province: The Economic History of Upper Canada, 1784–1870* (Toronto, 1993).

[75] See I. Drummond, *Progress Without Planning: The Economic History of Ontario from Confederation to the Second World War* (Toronto, 1987), table 2.2.

[76] D. Akenson argues that in some Ontario counties, newcomers could form well-established communities rather quickly. See Akenson, *The Irish in Ontario: A Study in Rural History* (Kingston, 1984).

Table 4.3a *Previous occupation of male patients at time of admission, first admissions to the Toronto and Hamilton asylums, selected years, 1851–1891 (n = 884)*

	Toronto		Hamilton	
Occupational group	no.	%	no.	%
No occupation	11	4	22	4
Labourers	51	20	175	31
Servants/porters	5	2	10	2
Soldiers/sailors	6	2	6	1
Skilled artisans	49	19	98	17
Clerks	14	5	18	3
Small shopkeepers	21	8	20	4
Merchants	6	2	11	2
Professionals	9	4	5	1
Farmers	76	30	188	33
Other	8	3	17	3
Sub-total	256		570	
n/a	16		42	
Total	272		612	

very least, farms would have the physical space and some kin resources for the care and management of the insane. In more settled agricultural communities, there would also have been traditions of treatment and care which might have forestalled somewhat asylum committal. The distance of some agricultural communities from the main asylums may also have had an impact, as discussed below.

'Farmers' could include a wide range of individuals, from impoverished tenant farmers to large-scale landowners to farm-labourers.[77] Without further specification as to the size of their land holdings, the precise socio-economic situations of these farming families remain uncertain. One clue to the family's socio-economic circumstances, however, may be gleaned from the negotiation over their contribution (or not) to the cost of asylum treatment. The Toronto and Hamilton asylums accepted patients who entered free of charge (known as 'Provincial' patients), and paying patients who were assessed on a graduated

[77] For a discussion on who could rightly be considered a farmer, please see G. Darroch, 'Scanty Fortunes and Rural Middle-Class Formation in Nineteenth-Century Rural Ontario', *Canadian Historical Review* 79 (1998), 621–59; and R. M. McInnis, *Perspectives on Ontario Agriculture, 1815–1930* (Gananoque, Ont., 1992).

scale of Can$1/week to Can$5/week according to their ability to pay.[78] In just over one-half of the cases of farmer-admissions to the Toronto asylum, the individual or his family paid nothing towards the cost of confinement. This proportion was much higher – close to 80 per cent – in the case of farmers admitted to Hamilton asylum. In the other one-half of admissions to Toronto (and 20 per cent in Hamilton), the person himself, the family or 'friends' were recorded as paying between Can$1 and Can$5 per week. It is difficult to assess accurately the average income of Ontario farmers. Alongside the much-noted poverty and transience of rural existence in Victorian Ontario[79] stood the constancy of many 'middling'[80] and well-established families, suggesting that many could afford the required fees to commit a family member.

If advantages in the domestic care and management of the insane led to the underrepresentation of Ontario's farming population in the province's public lunatic asylums, the reverse argument seems to have held true for the second largest group of male admissions – labourers. Listed as 20 per cent of the patient population in the Toronto asylum and 31 per cent at the Hamilton institution, labourers (including 'farm labourers') constituted roughly 10.6 per cent of the male workers in Ontario between 1871 and 1901. Male labourers often emigrated to Canada alone, working long and physically taxing hours on road, railroad, building and canal construction, hoping to earn enough to support the immigration of other family members or to send needed funds back to the home country. This lack of family as a social resource may in part account for labourers' overrepresentation in the asylums of Hamilton and Toronto. Their poverty and social isolation is reflected in the fact that only one out of fifty-three labourers paid anything towards his upkeep in the asylum (and in this case, it was paid by 'friends').

The third most prevalent recorded group of occupations were those male admissions whose work can be grouped under the general category of skilled, or semi-skilled, artisans. These patients included men who had worked in one of over forty distinct occupational listings, including employment as carpenters, shoemakers, painters, blacksmiths, cabinetmakers and machinists, to the relatively rare marble-cutters, broom makers, turners and saddlers. In comparison to farmers, fewer skilled artisans paid anything at all towards their maintenance (no more than approximately 10 per cent in both institutions). In contrast to

[78] The precise manner in which a family was assessed is not clear from the historical record. Clearly there was some haggling, and no few disputes, over the family's proportion of the upkeep of the patient. Naturally, those who paid the higher amounts received better accommodation and a lower patient : staff ratio. They were also absented from certain chores within the institution. For those who did not pay, either the province, or the municipality in which the lunatic previously resided, paid for the cost of maintenance. See Bartlett, 'Structures of Confinement', 11.

[79] See, for example, Katz, *The People of Hamilton*.

[80] The term comes from Darroch, 'Scanty Fortunes', 623.

farmers and labourers, the percentage of male skilled workers in the Hamilton and Toronto asylums appear consistent with that of the provincial population as a whole.[81]

Although Ontario was predominantly an agricultural province during the second half of the nineteenth century, it was, as mentioned above, experiencing the classic symptoms of industrialization, including accelerated urbanization, the rise of large-scale factory production and the emergence of a larger professional and entrepreneurial middle class. To what extent were these middle-class occupations reflected in the profile of patients admitted to the mental hospitals? The answer: not a great deal. Only 13 per cent of male admissions to Toronto, and 7 per cent of male admissions to Hamilton, were designated as coming from the 'lower middle class', including clerks and schoolteachers, and small shopkeepers, such as grocers, butchers and bakers. Members of the merchant or professional middle class were even rarer: only a handful of physicians and barristers ever made their way into the public institutions, no doubt preferring the privacy of the private Homewood Asylum in Guelph, approximately forty miles west of Toronto, or possibly a private institution for the insane across the border in New York state and Pennsylvania.

Women's occupations, as historians of women and women's work know too well, were poorly and inconsistently recorded in the nineteenth century. Congruent with prevailing property laws that placed legal rights and responsibilities in the hands of the single woman's father or the married woman's spouse, women's 'occupations' in nineteenth-century English and Canadian censuses often listed a woman's position in terms of her relationship to the male head of household. Thus, women, who may well have been receiving wages for casual work, were often lumped under the title 'wife of...' or 'daughter of...'. This practice was less common, however, in instances where women's work was located outside of the domicile, and less usually applied to working daughters than working wives. Keeping in mind, then, the extreme difficulty of using listed occupation as an indicator of women's paid employment, Table 4.3b lists the 'occupations' of women patients admitted to the asylum.

The most frequently cited occupation for female asylum patients was domestic service, which accounted for almost 200 of the 700 female admissions, or 27 per cent of the entire female patient population (the comparable figure for the London Asylum was 32.5 per cent[82]). Of those who listed an actual occupation, domestics represented 70 per cent of all listed waged occupations for women. Borrowing from Carolyn Strange's tabulations of Toronto's female domestics in her book on urban women at the turn of the century, and from Drummond's statistics on the percentage of women in service in his economic

[81] Drummond, *Progress without Planning*, table 2.2.
[82] Warsh, 'In Charge of the Loons', 153.

Table 4.3b *Previous occupation of female patients at time of admission, first admissions to the Toronto and Hamilton asylums, selected years, 1851–1891 (n = 798)*

	Toronto		Hamilton	
Listed 'occupational'	no.	%	no.	%
Wife/housewife, n.o.s.	41	19	16	3
Housekeeper/domestic duties[a]	12	5	71	14
Spinster (n.o.s)	3	1	4	1
Domestic servants/maids/nurses[b]	53	24	144	29
Seamstress/dressmaker/tailoress	6	3	18	4
Teacher	5	2	15	3
Milliner	0	0	5	1
Labourer's w/d/w	9	4	36	7
Artisan's w/d/w	4	2	25	5
Farmer's w/d/w	31	14	89	18
Shopkeeper's or clerk's w/d/w	0	0	6	1
Merchant's w/d/w	2	1	4	1
Professional's w/d/w	1	0	3	1
'Lady'	7	3	1	0
'None'/'no occupation'	32	14	55	11
Other	15	7	13	3
Sub-total	221		505	
n/a	17		55	
Total	238		560	

n.o.s. = not otherwise specified
w/d/w = wife, daughter or widow
[a]including 'housework'
[b]including governesses

history of Ontario, domestic service can be estimated at approximately 45 per cent of the listed female waged labour force in 1881 and 53.5 per cent in 1891.[83] A comparison of the percentages of listed female waged labour in the asylum and in the province at large indicates that, as an occupational group, female domestics were well overrepresented in the asylum. This adds evidence to the argument that female domestic servants, many of whom came to Canada as single emigrants, were more vulnerable to institutionalization.

It has been argued elsewhere that being an unmarried domestic servant was an occupational risk factor that made women more vulnerable, *ceteris paribus*, to being confined during times of emotional or mental crisis. Furthermore,

[83] C. Strange, *Toronto's Girl Problem: The Perils and Pleasures of the City, 1880–1930* (Toronto, 1995), tables A4 and A6; Drummond, *Progress Without Planning*, table 2.2.

unmarried single female workers may have been disproportionately new immigrants, thus suffering in many cases from a distancing of kin that might otherwise have acted as a protective barrier to institutional committal. When compared to the percentage of women listed as female domestic servants in the two asylums (70 per cent of those with an occupation listed), these results seem to reinforce contemporary observations of the extraordinarily high number of female 'domestics' incarcerated during the Victorian era. Two-thirds of them were unmarried, swelling the disproportionate number of unmarried patients mentioned below. Almost 90 per cent were paid by the province, a further proxy to their impoverished state, or their lack of kin support, at the time of admission. One is also struck by the one third of these predominantly unmarried domestic servants who were born in Ireland and Scotland (a proportion higher than the already-elevated higher rate of non Canadian-born patients). The chapter will now turn to this question of the role of migration and the distancing of kin resources that may have had an impact on an individual's vulnerability to being confined in an asylum.

Migration, kin and confinement

The question of migration and kin resources must be considered when assessing the vulnerability of different groups to being confined in the lunatic asylum. Researchers working on the confinement of the insane in the industrializing countries of western Europe have tended to downplay the role of migration (either of patients or their families) as a factor in the likelihood of an individual being confined.[84] By contrast, researchers working on countries experiencing high rates of immigration have tried to account for higher-than-expected numbers of patients who were not born in the country. They have offered two possible explanations for this phenomenon. First, new migrants were more likely, they argue, to be suspected or labelled mentally ill because of their strange appearance, lack of language skills and different cultural habits. An incipient racial approach to incarcerating individuals has often been alluded to in the case of the high incarceration rate of the Irish.[85] Thus, authorities, using confinement in part as a form of social control, may have identified and incarcerated 'undesirable individuals', and have viewed newly arrived single men as part of this potentially troublesome group. Second, other historians have placed greater

[84] See J. Walton, 'Lunacy in the Industrial Revolution: A Study of Asylum Admissions in Lancashire, 1848–1850', *Journal of Social History* 13 (1979), 1–22; R. Adair, J. Melling and B. Forsythe, 'Migration, Family Structure and Pauper Lunacy in Victorian England: Admissions to the Devon County Pauper Lunatic Asylum, 1845–1900', *Continuity and Change*, 12 (1997), 373–401. For an opposing view, see D. Wright, 'Family Strategies and the Institutional Committal of "Idiot" Children in Victorian England', *Journal of Family History* 23 (1998), 190–208.

[85] Mitchinson, 'Reasons for Commital', 93.

Table 4.4 *Place of birth of patients, first admissions to the Toronto and Hamilton asylums, selected years, 1851–1891 (n = 1682)*

	Toronto				Hamilton			
	Men		Women		Men		Women	
	no.	%	no.	%	no.	%	no.	%
Canada[a]	117	44	114	49	292	50	287	53
United States	2	1	7	3	17	3	13	2
England and Wales	57	21	33	14	108	18	66	12
Scotland	22	8	19	8	56	10	40	7
Ireland	61	23	54	23	96	16	116	21
Germany	3	5	3	1	11	2	14	3
Other	5	2	2	1	9	2	9	2
Sub-total	267		232		589		545	
n/a	5		6		23		15	
Total	272		238		612		560	

[a]Including Upper and Lower Canada, New Brunswick, PEI, 'Western Canada'. Not including Newfoundland.

emphasis on the lack of kin resources that may have acted as a protective barrier to institutional committal.[86]

Studying the admissions of Irish immigrants may unpack some of these arguments. Research has detailed the extraordinarily high number of Irish-born admissions to the Toronto asylum in the 1840s, 1850s and 1860s, and to the London Asylum in the 1860s. Over a third of all admissions to these institutions were Irish natives. The 'Great Famine' resulting from massive potato crop failure during the 1840s substantially increased the numbers of Irish immigrants that already existed in Upper Canada. Many of these newcomers worked as labourers either in their geo-social transition to more rural environments, or as more permanent 'urban' dwellers.[87] From 1870s, however, the Irish then diminish steadily in their representation in asylum admissions, to 23 per cent in Toronto (1871, 1881) and 19 per cent in Hamilton Asylum (1875–85) (see Table 4.4).

[86] Yet this argument needs to confront the tradition of chain-migration whereby migrants often located in foreign countries where they had some, albeit limited, kin connections. It is possible that in the case of female domestics, chain migration might have helped to establish some of the familial infrastructure necessary to buffer the fallout from mental trouble. Marylin Barber notes that there are examples 'recorded by government officials [that] illustrate the chain migration process' among female domestics and their families. See M. Barber, 'Immigrant Domestic Servants in Canada', *Canadian Historical Association* 16 (1991), 5. Mitchinson, 'Reasons for Committal', 94, table 2.

[87] Akenson, *The Irish in Ontario*.

Table 4.5 *Marital status of patients at time of admission, first admissions to the Toronto and Hamilton asylums, selected years, 1851–1891 (n = 1682)*

	Toronto				Hamilton			
	Men		Women		Men		Women	
	no.	%	no.	%	no.	%	no.	%
Unmarried	143	53	88	37	333	55	239	43
Married	108	40	120	51	266	44	299	54
Widowed	17	6	27	11	10	2	20	4
Sub-total	268		235		609		558	
n/a	4		3		3		2	
Total	272		238		612		560	

Even at these lower levels, the Irish more than double the representation of the Scottish-born admissions and equal that of the much more numerous (demographically speaking) English-born admissions. Teasing out poverty, prejudice and familial isolation in the social process of confinement is extraordinarily difficult. Just at the time when the Irish were claiming the dubious title of having the highest incarceration rate in their own country, their emigrants were also flooding into colonial mental hospitals.

In the last ten years, analyses of household structure and intra-familial dynamics have also become more of a focus for historians of the asylum. The household is seen both as a place of confinement, control and care before and after institutionalization, but also as a locus for lunacy identification and certification.[88] Certainly, as far as the gross characteristics of patients go, there were several important variables related to marriage and household formation found amongst patients (Table 4.5). First, there was an overrepresentation of the unmarried and widowed (as compared to the population from whence the patients came). Over one-half of all men admitted to both institutions were unmarried, a proportion much higher than the general adult population at the time. The percentage of unmarried women patients was lower than men in both institutions, but still elevated compared to the general population. Historians who have found similar results in patient populations have posited several explanations for the elevated level of the unmarried. First, it may be likely that

[88] P. Prestwich, 'Family Strategies and Medical Power: "Voluntary" Committal in a Parisian Asylum, 1876–1914', *Journal of Social History* 27 (1994), 799–818; M. Kelm, 'Women, Families and the Provincial Hospital for the Insane', 177–93.

Table 4.6 *Length of stay of patients: first admissions to the Toronto and Hamilton asylums, selected years, 1851–1891 (n = 1682)*

	Toronto				Hamilton			
	Men		Women		Men		Women	
In months	no.	%	no.	%	no.	%	no.	%
0 to 6	90	37	86	38	130	22	126	23
6 to 12	32	13	53	23	92	15	85	16
12 to 24	33	14	18	8	69	12	46	8
24 to 60	42	17	31	14	86	14	83	15
60 to 120	19	8	16	7	66	11	66	12
>120	27	11	22	10	155	26	137	25
sub-total	243		226		598		543	
n/a	29		12		14		17	
Total	272		238		612		560	
Median	11.8		8.0		26.5		29.0	
IQR	3.2 to 48.6		3.9 to 32.0		7.2 to 127.7		6.4 to 120.5	

those individuals suffering from recognizable levels of mental disturbance may well have been less likely to have been married in Victorian society (as today). In terms of the social context of committal, those who were married may also have been more integrated into a particular primary household (and thus needed for wages or primary-care giving). Thus, there may have been disincentives to seeking formal institutional committal. Clearly, more work on household structure as a factor in asylum committal needs to be done.[89]

Length of stay and transcarceration

So far these two institutions have revealed remarkably similar socio-demographic characteristics, save for the slightly lower socio-economic background that the Hamilton Asylum patients reflected, *inter alia*, in the statistically significant higher admissions of labourers to the latter institution. Where these two asylums diverge, however, is in the length of stay of patients (see Table 4.6). Toronto, it will be remembered, was the only institution in the province, until

[89] See J. Moran and D. Wright, 'The Lunatic Fringe: Households and the Management of Mad Behaviour in Victorian Ontario', in N. Christie and M. Gavreau (eds.), *On the Margins of the Family: Essays on the History of the Family in Canada* (Montreal-Kingston, in preparation).

Table 4.7 *Outcome of patients: first admissions to the Toronto and Hamilton asylums, selected years, 1851–1891 (n = 1682)*

	Toronto				Hamilton			
	Men		Women		Men		Women	
	no.	%	no.	%	no.	%	no.	%
Discharged	113	45	132	57	225	37	250	46
Died	82	33	49	21	274	45	206	38
Eloped/escaped	8	3	0	0	31	5	1	0
Transferred	46	18	49	21	75	12	83	15
Sub-total	249		230		605		540	
n/a	23		8		7		20	
Total	272		238		612		560	

the construction of the Kingston Asylum in 1855,[90] the London Asylum in 1870 and the Hamilton Asylum in 1875. With these other outlets, Toronto could be more selective in the admission and continued stay of patients. Its medical superintendent, Joseph Workman, by all accounts the most powerful alienist in English-speaking Canada at the time, prioritized Toronto as a curative institution, and seems to have redirected or transferred several score of incurable patients to the Hamilton Asylum. Thus, 50 per cent of men and 60 per cent of women admitted to the Toronto Asylum stayed for twelve months or fewer. Toronto's middle 50 per cent of patients stayed between three months and four years. These figures are directly comparable to major studies of length of stay of general lunatic asylums in Britain and Europe. Hamilton, by contrast, had their middle 50 per cent of patients staying between twelve months and ten *years*. Furthermore, a full one-quarter of all patients stayed for more than ten years. What appears to have happened, then, was a prioritization of Toronto as the primary curative institution of central Ontario.[91]

Differences between sexes emerge in the outcome of patients, their length of stay and their likelihood to be discharged (see Table 4.7). Ten per cent more women than men were eventually discharged from their respective institutions; conversely, 10 per cent more men died in the institution. Here it is very difficult

[90] The Kingston Asylum was built primarily for criminal lunatics, but was transformed into a mixed criminal asylum and a general mental hospital serving the eastern part of the province.

[91] Other North American jurisdictions formalized the idea of asylums for the 'curable' and asylums for the 'incurable'. For an example, see Ellen Dwyer's study of two institutions for the insane in upper New York State – namely the Utica and Willard Asylums. E. Dwyer, *Homes for the Mad*.

to separate cause and effect. It is well known that men had considerably higher mortality rates in the Victorian era, which may have led them to die more quickly in the institution. They may also have had higher rates of fatal diseases (with psychiatric manifestations) at time of admission, such as tertiary syphilis or chronic alcoholism.[92] Further detailed work on medical casebooks will be required in order to answer questions concerning the physical (non-psychiatric) ailments of individuals admitted to the asylums.

The records of discharge from the Hamilton Asylum allow for some general comments about transfer to other institutions. Official transfers to other asylums represented approximately 13 per cent of the eventual outcomes of male admissions and 15 per cent of female admissions to Hamilton Asylum. Transfers reflected not a continuous process, but rather a movement of groups of patients over a relatively discrete period of time. For instance, the Hamilton Asylum transferred forty-five patients to the Kingston Asylum, forty-three in the year 1885/6 alone. A further eighteen patients were removed from Hamilton Asylum to the newly opened Mimico branch of the Toronto Asylum in the year 1892/3, and twenty further patients shortly after the opening of the Penetanguishene institution for the criminally insane in 1909. There were no readily observable characteristics by sex or age for these transferred patients. This reshuffling of patients thus seems to have occurred most often initially after new institutions were established. Indeed a significant proportion of first admissions to the Hamilton Asylum in the first year of its existence were transfers from the Toronto Asylum.

Although transcarceration was an important component of the asylum system in Victorian Ontario, a more significant dimension – one hitherto unexplored – is the movement of discharged asylum patients into the new welfare institutions of the late nineteenth century, be they houses of industry, houses of refuge or old-age homes. Until further research determines the presence of ex-asylum patients in these institutions, the extent to which short asylum stay actually represented the reintegration of thousands of discharged former patients back into their community remains an unanswered question.

The richness of asylum records has inevitably made them a popular resource for social historians of medicine. However, asylums were but one of a range of inter-dependent carceral institutions that emerged in the latter part of the nineteenth century. Future research must focus on tracking patients between different asylums, between asylums and other custodial institutions (such as reformatories and inebriate asylums), and particularly between gaols and

[92] For a discussion of these factors in the characteristics of patients more generally, see E. Shorter, *A History of Psychiatry; From the Era of the Asylum to the Age of Prozac* (New York, 1997), chapter two.

asylums. Almost as many 'insane individuals' were being confined in gaols as were admitted to lunatic asylums in any given year in the province of Ontario. Did these numbers represent the same individuals? What proportion of insane individuals remained in the criminal system? More research also should be conducted on the financing of the asylum. How were decisions made as to who was responsible for the cost of maintenance? And how did these financial responsibilities shape and influence those who were (and were not) confined?

Conclusions

The results of this study have added weight to a new historiography of medicine that fundamentally reconsiders the patient populations of nineteenth-century mental hospitals. As this and other chapters in this book have shown, the Victorian asylum was not populated by the fringe elements of industrial society, at least certainly not from a socio-demographic standpoint. An older body of literature suggesting that women were disproportionately incarcerated in Victorian mental hospitals finds little to no support in empirical research.[93] Nor does there seem to be any convincing research indicating that the aged in Ontario's mental hospitals were overrepresented. Like Gasser and Heller's quantitative study of two asylums in Switzerland in the early twentieth century, this research found that the Toronto and Hamilton asylums received patients from all ages, and equally from both sexes.[94] Indeed, the lack of sex as an important socio-demographic variable when cross-referenced with age, occupation, length of stay, religion and geographical background, is nothing short of remarkable. Moreover, the majority of patients in our sample appeared to be productively employed, and thus functional in one sense prior to their committal. Keeping in mind the mounting evidence of the broad similarity of patient populations in the latter half of the nineteenth-century, it is at least worth posing the question

[93] The major studies suggesting women were disproportionately confined in asylums are: P. Chesler, *Women and Madness* (New York, 1973); E. Showalter, *The Female Malady: Women, Madness and English Culture, 1830–1980* (New York, 1985); Y. Ripa, *Women and Madness: The Incarceration of Women in Nineteenth-Century France* (Minnesota, 1990). For excellent summaries of feminist critiques of psychiatry and the history of psychiatry, see J. Busfield, 'Sexism and Psychiatry', *Sociology* 23 (1989), 343–64 and N. Tomes, 'Feminist Histories of Psychiatry', in M. Micale and R. Porter (eds.), *Discovering the History of Psychiatry* (Oxford, 1994), 348–83. For a more recent discussion of the role gender played in the history of psychiatry, see the collected papers in J. Andrews and A. Digby (eds.), *Sex and Seclusion, Class and Custody: Perspectives on Gender and Class in the History of British and Irish Psychiatry* (Amsterdam, 2002).

[94] The big exception in admissions was the confinement of Native North Americans. This paper could not identify any significant cluster of aboriginal Canadians in the four institutions under study. This absence requires further investigation.

as to whether social historians, in their desire to highlight the importance of social and cultural factors in the construction of ideas about madness, have inadvertently overemphasized the role of these same factors in the confinement of the insane – a bias existing at the expense of a comprehensive analysis of the behavioural and medical characteristics of the patients. For all the work that has been conducted around the world on the socio-demographic characteristics of patients, there has been little attention paid to the symptom profiles of these patients.[95]

This does not undermine the importance of social historical analysis of the asylum. While contributing to a growing literature that challenges some of the long-held assumptions of earlier feminist and social control accounts, this paper has emphasized the many ways that class and gender relations, and the social-familial dislocations resulting from large-scale emigration patterns, have influenced confinement and the representation of individuals in the asylum. But the paper also underscores the need to integrate the concept of madness as mental illness into the social and demographic history of the asylum era. This, as Charles Rosenberg has noted for the history of medicine in general, is no easy task. For Rosenberg, 'disease is at once a biological event, a generation-specific repertoire of verbal constructs reflecting medicine's intellectual and institutional history, an occasion of and potential legitimation for public policy, an aspect of social role and individual – intrapsychic – identity, a sanction for cultural values, and a structuring element in doctor and patient interactions'.[96]

Social historians of the asylum era have been loath to integrate the concept of mental illness as disease (however complicated and variegated a concept that might be when considering a heterogeneous patient population) with the myriad social constructions around mental illness that are the historical inevitability of our response to it. By and large we have found it safer to avoid the first item in Rosenberg's list altogether, seeking academic refuge in the new social history of the asylum. This latter body of literature has, since the late 1970s, vastly expanded our understanding of the intricacies of insanity in the asylum era and it has done so from a range of perspectives. This chapter suggests that the incorporation of analyses of symptom profiles, and, in some instances, of the physiological manifestations of madness, into the new social history of the asylum will lead to further fruitful reconsiderations of nineteenth-century asylum history.

95 A recent exception to this rule is D. Malleck, ' "A State Bordering on Insanity?" Identifying Drug Addiction in Nineteenth-Century Canadian Asylums', *Canadian Bulletin of Medical History* 16 (1999), 247–70.

96 C. Rosenberg, 'Framing Disease: Illness, Society and History', in C. Rosenberg and J. Golden (eds.), *Framing Disease: Studies in Cultural History* (New Brunswick, 1992), xiii.

Appendix

Religion of patients at time of admission, first admissions to the Toronto and Hamilton Asylums, selected years, 1851–1891 (n = 1682)

	Toronto				Hamilton			
	Men		Women		Men		Women	
	no.	%	no.	%	no.	%	no.	%
Baptist	9	3	14	6	25	4	39	7
Bible Christian	2	1	2	1	0	0	2	0
Church of England	74	28	65	28	89	15	52	10
Congregational/Independent	4	2	2	1	3	1	5	1
Disciple	1	0	2	1	3	1	2	0
Episcopalian	0	0	3	1	56	9	42	8
Lutheran	5	2	2	1	8	1	13	2
Mennonite	1	0	1	0	5	1	2	0
Methodist	51	19	49	21	120	20	114	21
Presbyterian	62	23	46	20	117	20	109	20
Protestant	0	0	1	0	10	2	8	1
Quaker	2	1	0	0	6	1	1	0
Roman Catholic	44	17	39	17	113	19	123	23
Salvation Army	2	1	0	0	0	0	1	0
Other	6	2	5	2	32	5	26	5
None	1	0	2	1	4	1	5	1
	264		233		591		544	
n/a	8		5		21		16	
Total	272		238		612		560	

5 Passage to the asylum: the role of the police in committals of the insane in Victoria, Australia, 1848–1900

Catharine Coleborne

Australia's European population arrived in 1788, transported from the British Isles to establish New South Wales as a penal colony. In the first few years of settlement, David Collins, legal advocate on the First Fleet and chronicler of the early history of the colonies, commented on the existence of insanity among convicts.[1] Policies, legislation and practices surrounding insanity in the different Australian colonies developed over the next ten decades. Subsequent histories of the insane in nineteenth-century Australia, following the trajectory of British scholarship on asylumdom, have largely been explored through institutional records.[2] Historians have been interested in exploring the broad concept of asylum committals, and have considered the relationships between agencies of the law, including police and the courts, families and asylums.[3] The custodial character of the colonial asylum meant that 'public disturbances' could result in asylum committals, rather than imprisonment, for women and men. Colonial policing practices of detection and surveillance, and the policing of sex and race, were central to the apprehension of lunatics. Families negotiated with the police in many instances, and the police played roles as intermediaries between the asylum and the families of lunatics. This chapter examines the development of policing practices around lunacy and the asylum in the colony of Victoria in the nineteenth century.

The history of the asylum in colonial Victoria was not unique, and is usefully compared to the histories of other Australian colonies and also the colony of New Zealand.[4] By the mid-nineteenth century, asylums were part of the

[1] See W. D. Neil, *The Lunatic Asylum at Castle Hill: Australia's First Psychiatric Hospital 1811–1826* (Castle Hill, Australia, 1992), 2–3; G. Davison, *et al.*, *The Oxford Companion to Australian History* (Melbourne and Oxford, 1998), 137–8.

[2] On British scholarship, see J. Melling, 'Accommodating Madness: New Research in the Social History of Insanity and Institutions', in J. Melling and B. Forsythe (eds.), *Insanity, Institutions and Society*, 1800–1914 (London and New York, 1999), 1–23.

[3] See especially S. Garton, *Medicine and Madness: A Social History of Insanity in New South Wales, 1880–1940* (Sydney, 1988); M. Finnane, 'Asylums, Families and the State', *History Workshop* 20 (1985), 134–48.

[4] For an account of the potential of historical comparisons between Victoria and New Zealand, see C. Coleborne, 'Making "Mad" Populations in Settler Colonies: The Work of Law and Medicine

Australian colonial welfare landscape. These were large physical structures modelled on the British Victorian asylum, and they housed increasing numbers of the insane. The first official lunatic asylum in New South Wales, smaller and more makeshift than later asylum buildings, was opened at Castle Hill in 1811. By 1866, when the asylum building was demolished, it had already assumed an identity as 'mad house' and as a 'landmark' of colonial history, as reported in the *Illustrated Sydney News*.[5] Before the 'temporary' asylum at Castle Hill, which operated for over fifty years, lunatics, as they were known, were confined in crowded gaol conditions west of the settlement at Parramatta. 'Rules' regarding lunatics' estates had been formed in 1805, and *ad hoc* procedures for the certification and discharge of gaoled lunatics were operating.[6] 'Humanitarian' concepts of authority circulated in the new colony. In 1814 the Governor of New South Wales, Lachlan Macquarie, instructed the asylum superintendent that lunatics were to be treated with 'mildness, kindness, and humanity'.[7]

The evolution of the asylum network was marked by legislative and institutional watersheds, and in the colonial period prior to 1900, was concerned with the distinctions between insanity and criminality. Van Diemen's Land (Tasmania), also a penal colony, had separated the insane from the criminal since 1829. Historians have characterized the lunatics, like the convicts, of Van Diemen's Land as 'desperate characters'.[8] The case of the 'criminal lunatic' raised many questions about the nature of the criminal impulse and the relationship this impulse had to insanity. In 1857 the *Australian Medical Journal* (*AMJ*) reported the case of John Quigley, charged with a violent crime in Tasmania in 1856. The medical testimony produced in this case seemed to be conflicting. Was Quigley a malingerer or a 'dangerous lunatic'? The *AMJ* commented on the phenomena of 'half-mad criminal convicts' and in its editorial concluded that 'in many instances it is difficult to distinguish insanity from crime'.[9]

South Australia, never a penal colony, had an asylum from 1846, and Queensland's Woogaroo Asylum was built in 1864. The insane were confined in the Moreton Bay Hospital before Woogaroo was established. Western Australia was unusual and interesting in that it had a perceptibly more penal approach to the confinement of the insane; there was no asylum, not even a temporary structure, until 1857 and the first permanent building was erected in 1865. After

in the Creation of the Colonial Asylum', in D. Kirkby and C. Coleborne (eds.), *Law, History, Colonialism: The Reach of Empire* (Manchester, 2001).

[5] *Illustrated Sydney News*, 16 July 1866. See Neil, *The Lunatic Asylum at Castle Hill*, 75.

[6] E. Cunningham-Dax, 'Australia and New Zealand', in J. G. Howells (ed.), *World History of Psychiatry* (London, 1975), 706–7; M. Lewis, *Managing Madness: Psychiatry and Society in Australia, 1788–1980* (Canberra, 1988), 4–5.

[7] A. Atkinson, *The Europeans in Australia: A History*, Volume One (Melbourne and Oxford, 1997), 106; Cunningham-Dax, 'Australia and New Zealand', 706.

[8] Cunningham-Dax, 'Australia and New Zealand', 707.

[9] *Australian Medical Journal* 2 (1857), 76–8.

the arrival of convicts in 1850, insanity posed more of a problem. The asylum system in Western Australia developed slowly; legislation was not enacted until 1871 and the first inquiry into the state of the asylum was not held until 1886.[10] Historians have explored the ways in which Western Australia's asylums, including Fremantle Asylum, were 'custodial' in nature. As in Victoria, gendered and racial definitions of inmates as evidenced by readings of patient casebooks at the Fremantle Asylum go some way towards explaining nineteenth-century attitudes to the vagrant, the socially troubling or 'unsettled' populations in Australian colonies.[11] In Victoria, known as Port Philip until 1851, lunatics were sent on a long journey to New South Wales until the first asylum at Yarra Bend was built in 1848.

The Yarra Bend Lunatic Asylum housed all certified insane in the colony until 1867. Two large rural asylums were established that year at Ararat and Beechworth, and by 1871, a new metropolitan asylum at Kew was built to house several hundred inmates. Yarra Bend was located, as its name suggests, on a bend of Melbourne's Yarra River north of the town's centre, and away from centres of population. The Yarra Bend quickly became part of the local imagination. When it first opened, lunatics could be seen by the public as they were transported to the asylum on carts, which drew criticism at an official inquiry of 1852. Wood engravings of the asylum appeared on the front page of the *Illustrated Melbourne Post* in 1862. A few years earlier a visitor from Hobart, Tasmania, commented that the asylum was in a 'sombre, almost gloomy' setting near the river, 'peculiarly lonely'.[12] Inmates sometimes escaped from this 'gloom', like Catherine Canning who made it over the fence and into the yard outside in 1860.[13] In the 1870s Melbourne's residents reportedly hung about the reserve near Kew Asylum for 'sport' and had to be watched by police.[14] Official inquiries into the management of the asylums were held from the 1850s to the 1880s, with the Yarra Bend receiving particular scrutiny in its early years of operation. The custodial character of the Yarra Bend was noted; yet doctors and superintendents brought from England to 'reform' this institution defended the practices at the asylum.

[10] A. S. Ellis, *Eloquent Testimony: The Story of the Mental Health Services in Western Australia 1830–1975* (Perth, 1984).

[11] See B. Harman, 'Women and Insanity: The Fremantle Asylum in Western Australia, 1858–1908', in P. Hetherington and P. Maddern (eds.), *Sexuality and Gender in History: Selected Essays* (Perth, 1993), 167–81; N. Megahey, 'More than a Minor Nuisance: Insanity in Colonial Western Australia', in C. Fox (ed.), *Historical Refractions: Studies in Western Australian History* 14 (1993), 42–59.

[12] See *Illustrated Melbourne Post*, 25 June 1862. R.W. Willson, *A Few Observations Relative to the Yarra Bend Lunatic Asylum* (Melbourne, 1859), 4.

[13] *Argus*, 20 March 1860, 5.

[14] 'Kew Lunatic Asylum – Police Duty At', 12 November 1873, Victorian Public Record Series (VPRS) 937, Unit 60, Inwards Registered Correspondence, Police Department, Bourke District, Bundle 4.

The Yarra Bend, Kew Metropolitan Asylum, and Ararat Asylum, together with other similar institutions in the colony in this period, continued to exist in various forms in the twentieth century and their records, particularly patient casebooks, are extant. These are mostly complete sets of records of patients in nineteenth and early twentieth-century Victorian asylums. Other sources used in this chapter include records of the Victoria police, including police manuals and police regulations, the *Victoria Police Gazette*, and medical journals and official asylum inspectors' reports. Historians in Australia have made useful assessments of the role of police in the matter of lunacy using similar sources for different regions.[15] Yet these historians have not focused on the relationship between police and medicine, nor have historians of lunacy incarceration elaborated on the police role in lunacy committals.[16] Nineteenth-century legislation changed the nature of legal and medical practices regarding lunacy, including police practices, and with it, the nature of the lunatic identity.[17]

The first piece of lunacy legislation enacted in Australia was the Dangerous Lunatics Act of 1843.[18] This Act reflected the fear of 'dangerousness' and of lunacy in its language. It also introduced the medical practitioner to the procedure of confinement, which meant that wrongful confinement was less likely to occur.[19] In the different colonies, specific legislation came to replace this Act. In Victoria, the Lunacy Statute of 1867 implemented an increased surveillance of the lunatic patient, particularly through the requirements for the medical casebook.[20] At the same time the Lunacy Statute made provisions for the building of the two new asylums in rural Victoria at Beechworth and Ararat. With this Statute, police began to play an important role in the process of enforcing laws which attempted to control and describe disorderly behaviour for medicine.

This intervention of police in the matter of lunacy has more significance than it has been given in historical accounts. Increasing scrutiny of settled populations in towns and shifting populations in the mid-nineteenth century

[15] S. Garton, 'Policing the Dangerous Lunatic: Lunacy Incarceration in New South Wales, 1843–1914', in M. Finnane (ed.), *Policing in Australia: Historical Perspectives* (Kensington, Australia, 1987), 77.

[16] See S. Garton in 'Bad or Mad? Developments in Incarceration in New South Wales 1880–1920', in Sydney Labour History Group, *What Rough Beast? The State and Social Order in Australian History* (Sydney, London and Boston, 1982), 89–110; and 'Policing the Dangerous Lunatic', 75–87; see also M. Finnane's *Insanity and the Insane in Post-Famine Ireland* (London and Totowa, 1981). For a history of police in colonial Victoria, see J. McQuilton, 'Police in Rural Victoria: A Regional Example', in Finnane (ed.), *Policing in Australia*, 36–58.

[17] Finnane, *Insanity and the Insane in Post-Famine Ireland*, 121.

[18] Dangerous Lunatics Act, 7 Vic. No. 14, 1843.

[19] C. Coleborne, 'Legislating Lunacy and the Female Lunatic Body in Nineteenth-Century Victoria', in D. Kirkby (ed.), *Sex, Power and Justice: Historical Perspectives on Law in Australia* (Oxford, 1995), 88–9.

[20] Lunacy Statute, 31 Vic., No. 309 (Amended in 1869 and 1878).

was becoming part of the bureaucratic mentality and practice in the colonies.[21] A small number of women in Victoria's asylums were described as prostitutes, and these women were particularly vulnerable to police surveillance. Paula J. Byrne has shown that in New South Wales in the nineteenth century, women were watched differently from men by police.[22] Other groups in Victorian society were similarly watched differently – and closely – by police, and were at times considered dangerous and disruptive: these included the Chinese, the vagrant and destitute and the alcoholic. Lunacy was coming under far greater notice than it had previously, and particularly by official bodies.

The policing practices discussed here evolved over the course of the nineteenth century and were formalized at specific points. The official *Regulations for the Guidance of the Constabulary of Victoria* of 1877 marked one moment of formality for the role of police in regard to lunacy. The increasing concern about asylum management in Victoria suggests that advice to police from asylum authorities which followed was informed by this investigation. This chapter considers both regulatory texts, and policing practices. It was in formal police regulations that evidence of the increasing emphasis on the medical role for police shows that lunacy was itself becoming indisputably a medical matter.

Colonial policing and lunacy in Victoria

William a'Beckett's *Magistrates' Manual for the Colony of Victoria*, published in 1852, had not mentioned lunacy.[23] Yet it is safe to assume that police were involved in the apprehension of alleged lunatics prior to this. From 1867, police were formally involved in the process of lunacy committals. This process was shaped by ideas about sex, race and class as police made discretionary judgements about vagrant, alcoholic, wandering and disorderly women and men. In 1877 the new *Regulations for the Guidance of the Constabulary of Victoria* outlined the particular medical procedure to be followed by police where lunatics were concerned, and the lunatic's passage to the asylum was one increasingly marked by police-enforced medical regulations around her/his body. Police were responsible for the large numbers of committals of lunatics to asylums in New South Wales before 1900.[24] During the same period in Victoria, police

[21] For an indication of what the historian might find within bureaucratic texts, see O. MacDonagh in *A Pattern of Government Growth 1800–1860: The Passenger Acts and their Enforcement* (London, 1961), 17–19.

[22] P. J. Byrne, *Criminal Law and Colonial Subject: NSW 1810–1830* (Cambridge, 1993). See also J. Allen, *Sex and Secrets: Crimes involving Australian Women since 1880* (Melbourne, 1990).

[23] W. a'Beckett, 'The Magistrates' Manual for the Colony of Victoria, containing practical directions to Justices of the Peace, in the performance of their Duties, as Required by the Adoption of Jervis's Acts', *Melbourne Morning Herald*, 1852. See also T. Weber, 'History of the Magistracy', in La Trobe Legal Studies Department, *Guilty, Your Worship: A Study of Victoria's Magistrates' Courts* (Bundoora, 1980), 3–16, esp. 11.

[24] Garton, 'Policing the Dangerous Lunatic', 76.

literally created boundaries between the asylum and the outside world, as their presence was required for the smooth ordering of the division between the insane and curious onlookers, and the 1877 *Regulations* signalled an increasingly medical role for the police in the matter of lunacy.

Fear of disorder and violence in the community, and the increased capability of state intervention into the lives of citizens by this time, made police intervention into everyday life possible and permissible. The so-called 'dangerousness' of the lunatic was less predictable than the violence of assault, murder or robbery; laws regarding criminal offences helped shape lunacy laws, although it was tacitly acknowledged that the lunatic was not 'criminal'. Yet police had the power to suspect the 'mad' person as capable of committing crime, and magistrates could choose gaol or the asylum for the person brought before the court. Where lunacy was concerned, the 'preservation of social order remained paramount in the intentions of the legislators' in early Australia.[25] Harriet Deacon notes that 'dangerousness' was also a key to understanding assessments of the insane in Cape Colony. Deacon comments on police as part of the 'screening mechanism' for asylum admissions in South Africa in the nineteenth century. In Victoria the police were encouraged to use their discretion when apprehending suspected lunatics and other 'idle and disorderly' persons. They were given the power of interpretation of behaviour but asked to be attentive to the processes of the law, and in the case of lunacy were reminded to proceed 'under the provisions of the Lunacy Statute, and not under the provisions of the Police Offences Statute'.[26] This was particularly relevant in arrests of drunk women and men, and police were advised to keep these 'patients' as prisoners until their 'lunacy' had worn off.[27]

In their daily work police had broad categories of 'disorder' to identify, usually based on behaviour and bodily deportment. As early as 1856 police were encouraged to use their discretion when apprehending 'idle and disorderly' persons. Police could arrest without warrants 'all loose, idle and disorderly persons whom [they] shall find disturbing the public peace'. This was extended to include persons found in a state of intoxication or behaving riotously, 'every common prostitute wandering in any street or public highway', persons begging or displaying obscene material and using insulting and threatening words in public.[28] The role played by police was not one exclusively confined to issues of arresting, charging and locking up those who had breached the law but was

[25] *Ibid.*, 76–7. Finnane, *Insanity and the Insane in Post-Famine Ireland*, 122. The situation in England was different, largely because the Poor Law helped to define the lunatic population. Finnane also sees the question of social order in Ireland as central to lunacy legislation.

[26] *Regulations for the Guidance of the Constabulary of Victoria* (Melbourne, 1877), 141.

[27] See for instance the instructions for police in 1873 regarding lunacy brought on by excessive drinking, *Victoria Police Gazette*, September 1873, 236.

[28] *Manual of Regulations for the Guidance of the Constabulary of Victoria* 1856, paragraphs 78–81.

broad and far reaching, suggesting that the policing of social life and the creation of 'order' was important to colonial society.

Evaluating 'order' and 'disorder' in colonial society involves the use of contemporary definitions of crime. Criminal statistics for 1865 in Victoria indicate that 'lunacy' was recorded and considered as an 'offence'. Other offences alongside lunacy included categories such as 'Disorderly characters', 'Disorderly prostitutes', and 'Drunk and disorderly characters', and misdemeanours included 'Drunkenness', 'Nuisance', the use of 'obscene language' and the 'crime' of 'vagrancy'. While large numbers of women and men were taken into custody by police under the charge of lunacy, only ninety-three women were 'convicted' of lunacy in this year, and 214 men were also convicted. Together with lunacy, many of these offences involved vulnerable bodies in public spaces.

People became more or less vulnerable in public spaces for different reasons. As this chapter argues, gender and 'race' were contributing factors. For instance, women figured substantially among the statistics as disorderly, drunk, obscene and homeless, indicating that women were not only disruptive in private but also in public space.[29] Not only did police make choices about what constituted madness, possibly based on 'popular conceptions' of lunacy, but they were also given permission to interpret behaviour.[30] They were involved in the creation of the lunatic asylum population, but also in the construction of the patient identity.

Interestingly, this 'patient' identity was constructed in relation to ideas about colonial laws and order. By 1874 official statistics of police ('Apprehensions, Commitments, Convictions') still listed lunacy as an 'Offence against good order' together with vagrancy, libel, keeping a gambling house, gambling, offences relating to lotteries, nuisances, cruelty to animals and breaches of the Inebriate Act. In this year a total of 294 men and 170 women who were apprehended by police on the charge of lunacy were summarily 'convicted' by magistrates.[31] In 1877, similar police statistics reveal that a range of other offences had been added to the list of 'Offences against good order': breaches of by-laws, offences with horses and vehicles, habitual drunkenness, disorderly conduct, using obscene, threatening or abusive language in public, indecently exposing the person, being an idle or disorderly person or prostitute, and keeping a brothel or disorderly house. The total number of 'convicted' lunatics in 1877 was 434, comprising 285 males and 149 females. In this year the statistical

[29] 'Returns of the Number of Persons taken into Custody by the Victorian Police Force during the Year 1865, with Particulars as to their Disposal, etc.', Criminal Statistics, *Victoria Parliamentary Papers* (*VPP*), 1867, 3.

[30] Garton, 'Policing the Dangerous Lunatic', 80.

[31] 'Return Showing the Number of Males and Females taken into Custody by the Victorian Police during the Year 1874 and the Offences with which they were charged . . .', Statistical Register of the Colony of Victoria for the Year 1874, No. 27, Part IV, 'Law, Crime, etc.', 78–80. *VPP*, 1875–6.

Table 5.1 *Number of lunatics received at Victorian asylums through police, 1875–1890*[a]

Date	1875	1885	1890
Lunatics received through police	407	332	414
Total admissions	659	615	912

[a]The three tables in this chapter were prepared using the Annual Reports of Asylum Inspectors in the Colony of Victoria, published in the Parliamentary Papers in the following year. Researcher Fiona Kean assisted in the preparation of the tables.

register detailed the numbers of people taken into custody by police for each offence in the previous eleven years. Numbers of people arrested under the charge of lunacy remained fairly consistent but there was an actual – if slight – decrease over the period between 1867 and 1877 of police intervention into the matter of lunacy. It is likely that police were arresting large numbers of people in public spaces, and perhaps charging them under the many and increasingly varied categories of 'offences against good order' rather than the charge of lunacy. The fact that there was an increase in the number of people arrested for habitual drunkenness with disorderly conduct and using obscene, threatening, or abusive or insulting language or behaviour in public between 1867 and 1877 suggests that police used a variety of strategies to deal with disorderly people.[32] Yet it was the very medical nature of the problem of lunacy which set it apart from these other offences (as it was still identified) which required more clarification and different practices as far as the police strategies were concerned.

Police scrutiny of a wandering population and of those considered 'disorderly' did not go unnoticed by the wider community in Victoria in the 1870s. In 1873 the *Age* reported that laws surrounding the confinement of lunatics needed some clarification, notwithstanding the efforts of the Victorian legislature in 1867 to refine the Lunacy Statute. The problem seemed to be that it was difficult to 'get out' of the asylum once inside.[33] Furthermore, problems with the policing of the lunatic raised questions about police powers with regard to the definition of insanity. The report raised important questions: the Lunacy Statute of 1867 had not been amended since its enactment; no bills which challenged its operation had been introduced to the Victorian Parliament; and many other lunatics apprehended by police would have been admitted to the asylum

[32] 'Statistical Register . . . 1877', Part VII, 'Law, Crime, etc.', 186–7, 194. 'Persons taken into custody – return for eleven years' (Lunacy), *VPP*, 1878, 3. The return shows these numbers: 1867: 657, 1877: 567 (Habitual drunkenness with disorderly conduct); 1867: 9,351, 1877: 12, 368 (Obscene, threatening or abusive or insulting language or behaviour in public); 1867: 1,473, 1877: 3,487, 194.

[33] *Age*, 3 May 1873, 4.

under the terms of the Act without any controversy or criticism, as far as it is possible to see, by newspapers and the community. The *Police Gazette* in 1873 reported that in September the police were given instructions regarding lunacy brought on by excessive drinking, but this appears to be the only substantial change to the policy of policing lunacy reported that year.[34]

The paragraph of the 1867 Lunacy Statute which was questioned by the *Age* was Section 7.[35] This section detailed the legal provision for police to apprehend people who were possibly at risk from family members or others – being neglected or cruelly treated – and those who were 'wandering at large'. These categories of 'neglected' and 'wandering' raised questions about the legal definition of insanity. The very broad nature of the legal understanding of lunacy reflects a certain willingness to control a large number of possibly disruptive behaviours in public spaces. The very idea of 'dangerousness', the central term used in the Dangerous Lunatics Act of 1843 which shaped the 1867 Act, was also used to create social fears about 'different' members of the community.

By 1874 the *AMJ* was also critical of the role played by police in the committal of lunatics. The *Report of the Acting Inspector of Lunatic Asylums* for 1873 raised serious questions for the medical press.[36] At issue was the ability of police – and the ability of friends or relatives of patients – to understand the 'causes of insanity'. The journal stipulated that 'if a policeman gives the cause, drunkenness, religion, love, or some other very potent cause, is likely to be pitched upon'. Thus statistics which were tabled as representative of the causes of insanity according to judgements made by those committing lunatics to the asylum were considered by the *AMJ* to be 'worse than valueless'.[37] In this way the judgement of the medical man in the asylum, and thus asylum (psychiatric) medicine, was presented as the only valid judgement. Yet the surveillance and arrest of the lunatic by the police in public and private spaces continued, suggesting that somehow the role of the police had to become increasingly 'medical' if the medical fraternity was to accept it as legitimate.

Regulatory texts

The admission of lunatics to the asylum was characterized by bureaucratic processes. When Margaret Tweed was admitted to Yarra Bend Asylum by police in 1873 there were three key documents which accompanied her there. The first was an 'Order for Conveyance to an Asylum, etc., of a lunatic not under Proper Care and Control, etc.' This was signed by two medical practitioners, Thomas Hewlett and Edward Hunt, and two Justices of the Peace, John Falconer and G. B. Heales. The second was a 'certificate that a person is a lunatic, and a

[34] *Victoria Police Gazette*, September 1873, 236. [35] *Lunacy Statute* 1867, s. 7.
[36] See 'Review', *Report of the Acting Inspector of Lunatic Asylums, on the Hospital for the Insane, for the Year 1873. AMJ* 19 (July 1874), 211–17.
[37] *Ibid.*, 214.

proper person to be detained under care and treatment', which was provided for by the Lunacy Statute's seventh schedule under section 8. The third was the 'Police Report of the lunatic Prisoner'.[38] The role of the police was central to the process of making her lunacy both a legal and a medical issue.

In the restraining and treatment of persons apprehended by police as lunatics the police were given medical directives from asylum authorities. By 1877 medical hints for the 'Management of lunatics' en route to the asylum were provided to the police.[39] These were instructions which were about both how to deal with lunacy and also the lunatic's physical condition. Advice about restraint, travel, weather, physical nourishment, posture and pregnancy was offered. The bodies of female lunatics were an oddity here, given that police were always male, so women were to be accompanied by a female attendant. These examples indicate that the police were to 'manage' the bodies of lunatics as much as they were to describe and inscribe them for and in the law.

Police surveillance did not end here: lunatics on trial release from the asylum, and placed with relatives and friends, were to be watched by police in case of relapse. This had been the case since before the 1877 *Regulations*, and directives for the police were issued in the *Police Gazette* in September of 1874.[40] Perhaps following the *AMJ*'s criticism of the police in their role regarding lunatics, in November of 1874 Edward Paley at the Department of Hospitals for the Insane was given the opportunity to amend police regulations regarding lunatics.[41] He wrote to the Chief Commissioner with his thanks, making only a small addition 'marked in the margin of Police note'.[42] But by 1877 it seems that Paley was more implicated in assisting the police with medical information about the management of lunatics en route to the asylum; the police regulations of 1877 were quite detailed in their expectations of police, and further articulated their medical role in relation to lunatics. The *Regulations for the Guidance of the Constabulary of Victoria* of 1877 related to, and were revised and prepared under, section 16 of the Police Regulation Statute of 1873. They were the first new regulations for police since the *Manual* of 1856.[43] Lunacy was not the only 'medical' problem faced by police; the *Regulations* also outlined procedures for the 'Medical Attendance to Prisoners' and rescue procedures for the 'apparently drowned', snakebite victims and the injured.[44]

[38] Margaret Tweed, 14 June 1873, VPRS 7562/P1, Yarra Bend Asylum, Unit 7.
[39] *Regulations for the Guidance of the Constabulary*, 144; see also Finnane, *Insanity*, 108.
[40] *Victoria Police Gazette*, 1 September 1874, 182. [41] See 'Review', 214.
[42] Letter to Captain Standish from Edward Paley, 18 November 1874, VPRS 937, Unit 128, Correspondence to the Chief Commissioner, Department of Hospitals for the Insane, Melbourne, Bundle 3.
[43] As suggested by my research and the research of J. O'Sullivan, *Mounted Police of Victoria and Tasmania* (Adelaide, 1980).
[44] *Regulations for the Guidance of the Constabulary*. See also *Police Regulation Statute*, 1873, no. 476. In the Police *Regulations*, 'lunatics', paragraphs 1037–62, 141–5; 'Medical Attendance for Prisoners', paragraphs 772–8, 102–3 with appendices.

In November 1877 the *Police Gazette* announced that new *Regulations for the Guidance of the Constabulary of Victoria* were available for police. A brief article drew attention to the Chief Commissioner's emphasis on regulation no. 391 'which provides for the form in which police reports are to be prepared'. This signalled attention to the bureaucratic nature of policing, and police responsibilities towards the asylum fell within this ambit. The appendix attached to the police regulations regarding lunacy, to be discussed in some detail below, was entitled a 'Police Report of the lunatic Prisoner arrested', and yet it also required police to identify this 'prisoner' as a 'patient'. The language here was significant: the police were beginning to make distinctions between criminality and lunacy, as they were given more guidelines for this identification.

Regulations for the Guidance of the Constabulary in Victoria provided officially for police practices which had been in effect for some time previously. These regulations took the form of advice to police about lunatics relating to the Lunacy Statute, and included directions regarding the 'Information to be furnished respecting lunatics', the 'Removal of lunatics', particular information regarding female lunatics and their children, and the 'Management of lunatics', a section based on medical advice from the superintendent of Yarra Bend Asylum. Writing about Ireland, Finnane asserts that such regulations can tell us about the 'expectations' surrounding policing 'in the field' and argues that the very 'statement of rules' for police 'embodies some highly symbolic observations on the preferred modes of policing'.[45] The 'Information to be furnished respecting lunatics' is of central interest to me at this point, as this detail was clearly linked to the medical casebook, a technique of surveillance in the asylum. Rule 1043 stated that police must 'procure particulars' of the apprehended and alleged lunatic: her/his history, character and pursuits. This information, for medical and legal purposes, was to be attached to the warrant of committal.[46] The police were being asked to identify *patients* in this situation. They were encouraged to provide extra information which may be 'of assistance to the medical or other authorities in indicating the kind of treatment required by the patient'.

Similar in style to the casebook proforma was the schedule contained in 'Appendix 11', the 'Tenth Schedule Statement', the 'Police Report of the lunatic Prisoner arrested'. The role of police surveillance in this instance was directly informed by medical discourse. The 'Police Report of the lunatic Prisoner arrested' of 1877 was part of the process of identifying the lunatic, and also part of the committal process. Its similarity to the patient casebook proforma which collected details about the medical condition as well as the life situation of the lunatic is important. It suggests that there were in each lunatic's committal layers

[45] Finnane, *Police and Government*, 153.
[46] *Regulations for the Guidance of the Constabulary*, 141.

Table 5.2 *Policing sex of lunatics and passage to the asylum in Victoria, 1885 and 1890*

	Lunatics received through police	
Date	Male	Female
1885	208	124
1890	280	134

of meaning created about her/his condition. Some lunatics were committed on the strength of family or friends' testimony and medical certificates; but the 'lunatic prisoner' was committed with two or three detailed documents already inscribed – even before the asylum began its own process of bodily inscription. The 'Police Report of the lunatic Prisoner' is drawn from the tenth schedule statement of the 1867 Lunacy Statute where the person(s) admitting the lunatic were required to detail the lunatic's illness. In this way other people admitting lunatics were also performing a medical role. The police were given separate instructions which lent them the added responsibility of having some kind of medical 'expertise' in the matter of lunacy. The police were required to ask or ascertain the 'Supposed cause' of the lunatic's insanity, as well as collecting detail about the duration of the attack, the possibility of epilepsy, and 'the form of insanity'. This surely is the most medical of requirements; it appears that police were to have some understanding of the labels in use by the asylum, such as 'mania' or 'melancholia', and to apply these to the arrested lunatic.

Policing sex

The policing of 'sex' in nineteenth-century Victoria involved police observation of women and men which was shaped by ideas about appropriate sex-role behaviour, and the construction of gendered 'mad' identities. Police were aware of local communities and kept a close eye on events. While Table 5.2 indicates that more men than women were admitted to asylums by police in the latter part of the century, significant numbers of women were also apprehended and admitted by police. Mary Ann Cook was taken by police to the asylum in 1870 suffering from melancholia. The casebook recorded that: 'this unfortunate woman was the wife of the man Cook who shot Mrs Moss in Ballarat some time ago, then shot himself. She is very desponding and has attempted to strangle herself.'[47] Police brought Mary Jane Secombe from Geelong to Yarra Bend in

[47] Mary Ann Cook, 9 July 1870, VPRS 7400/P1, Ararat Asylum, Unit 4, 9.

1878. Suffering from melancholia and suicidal, Mary Jane had jumped from an upstairs window in the hospital at Geelong to 'escape the chloroform'.[48] Police were also called to restrain women who posed problems for others. In the 1867 cases of Eleanor Jenkin and Mary Jane Squires, the police were called upon by family and husband to commit women identified as lunatics to Ararat Lunatic Asylum. Eleanor Jenkin was suffering from puerperal mania and in decline. The casebook noted that 'she and her family had been in great poverty'. Mary Jane Squires was suffering from acute mania and became violent.[49] The unfortunate Sarah Stynes, deaf, dumb and blind from birth, was transported by police to Ararat from her previous residence, the Geelong Gaol.[50] There are many other cases similar to these of police playing a very central role in the committal of needy women – women who were neglected, ill, and also troublesome to family.

Other women were clearly disturbing 'good order' and arrested under the charge of lunacy. Alice Rose Patterson, Bridget Callaghan, Louisa Gee, Mary Hegarty, Martha Gorwood and Anne Ah Lou were identified as prostitutes. Women who worked as prostitutes were certainly far more vulnerable before the law and the police than many other women; like vagrant women or women already understood to be 'criminal' they were under surveillance by police for most of the time. Upon her arrival to Ararat Asylum in police custody, Alice Rose Patterson 'stated that she had been drinking' and that she had experienced delusions.[51] Bridget Callaghan was perhaps known to police as she was 'intemperate', described as 'dangerous and destructive' and had been previously imprisoned for vagrancy and in and out of gaol for offences relating to her drinking.[52] Louisa Gee was transferred from Yarra Bend to the asylum at Kew, and the label 'prostitute' followed her there; her child, born at Pentridge Gaol, was taken from her and sent to an Industrial School.[53] Mary Hegarty and Martha Gorwood went to Kew as prostitutes, both 'disordered through drink', with delusions in the case of Mary and 'loss of memory' and violence in the case of Martha.[54] Anne Ah Lou was arrested by police at Sandhurst some years later, diagnosed as suffering from mania and described as a prostitute.[55]

In 1877 several *Police Gazette* reports of escaped male lunatics detailed the clothing and appearance of the escapees for police detection and recapture.

[48] Mary Jane Secombe, 10 April 1878, VPRS 7400/P1, Ararat Asylum, Unit 5, 269.
[49] Eleanor Jenkin, 22 October 1867, VPRS 7401/P1, Ararat Asylum, Unit 1, 4; Mary Jane Squires, 29 October 1867, VPRS 7401/P1, Ararat Asylum, Unit 1, 5.
[50] Sarah Stynes, 6 December 1867, VPRS 7401/P1, Ararat Asylum, Unit 1, 12.
[51] Alice Rose Patterson, 18 December 1870, VPRS 7401/P1, Ararat Asylum, Unit 1, 250.
[52] Bridget Callaghan, 12 April 1873, VPRS 7400/P1, Yarra Bend, Unit 5, 122.
[53] Louisa Gee, 29 July 1872, VPRS 7397/P1, Kew Asylum, Unit 1, 102.
[54] Mary Hegarty, 28 February 1874, VPRS 7397/P1, Kew Asylum, Unit 2, 94; Martha Gorwood, 18 January 1875, VPRS 7397/P1, Kew Asylum, Unit 2, 276.
[55] Anne Ah Lou, 27 May 1887, VPRS 7397/P1, Kew Asylum, Unit 8, 113.

Around twelve reports of escaped lunatics (some of whom were recaptured) resembled criminal profiles and, in their descriptive style, were precursors to later surveillance technologies used by police:

THOMAS MILNE escaped from the Yarra Bend lunatic Asylum about 7.20pm on the 11th instant. Description: – 55 years of age, 5 feet 2 or 3 inches high, small features, grey eyes, long grey hair; wore the asylum clothing; was sent to the asylum from Camperdown. – 0.5082. 14th August 1877.[56]

The escaped lunatics listed throughout 1877 were all men.[57] Wife deserters and male criminals were described in similar ways, as were 'absconders' and vagrants. However, lunatics were mostly identifiable by their 'asylum clothing' which literally branded their bodies as they fled the place (unless they had the wherewithal to dispose of it).[58]

Men who were arrested by police on charges of lunacy and taken to Yarra Bend Asylum in the 1870s included thirty-three-year-old Matthew Larkin of Melbourne, a single tailor who came from Ireland and was Roman Catholic. Larkin suffered from delusions: 'Fancies that he hears everything, and everybody talking, that the Bible is being read all night.'[59] Achilles King was a publican who 'used to keep the "café de Paris" and managed the Athenaeum Club' who suffered a loss in business. He was taken to the asylum by police perhaps with the co-operation of his wife, Lavinia Anne King.[60] Police found George Peters and Thomas Williams, who had been in the bush for ten days without food and 'found wandering about the streets half naked'.[61] Being arrested in the bush or the streets was not uncommon for men who were understood to be insane as a result of loneliness, isolation, sunstroke: the 'casualties' of colonial conditions.[62]

The female lunatic's passage to the asylum was a particularly complex rite as, caught within a medico-legal framework, the potential disorder of her sexual difference created some police anxiety. Prior to the 1877 *Regulations*, police had

[56] *Victoria Police Gazette* 15 August 1877, 221.

[57] See references to James Morrison and Patrick William Daglish (14 February, 39), Patrick Byrne (21 February, 53), Christian Burke (28 February, 53), George Henry Heather (7 March, 59), James Baker (28 March, 81), R. G. Bayldon (23 May, 136), Joseph Guella (30 May, 142), Vincent Mitchell (8 August, 215), Thomas Milne (15 August, 221), Edward Gangaillot (26 September, 257) and William Johnstone Cooke (28 November, 312) in the *Victoria Police Gazette* for 1877.

[58] For a detailed description of asylum clothing printed in the police gazette for the police, see *Victoria Police Gazette*, 22 May 1878, 142.

[59] Matthew Larkin, 30 January 1872, VPRS 7399/P1, Yarra Bend Asylum, Unit 1, 12.

[60] Achilles King, 31 January 1872, VPRS 7399/P1, Yarra Bend Asylum, Unit 1, 13.

[61] George Peters, 6 Feburary 1872, 'fancies that people are running after him', VPRS 7399/P1, Yarra Bend Asylum, Unit 1, 19; Thomas Williams, 8 February 1872, VPRS 7399/P1, Yarra Bend Asylum, Unit 1, 20.

[62] As Garton also suggests in 'Policing the Dangerous Lunatic', 81–2. This phrase is suggested by Kathryn Cronin's *Colonial Casualties: Chinese in Early Victoria* (Melbourne, 1982).

to make decisions about how to transport and police the lunatic female. In 1866 Sergeant King of the Belfast Police Barracks in rural Victoria reported to his superintendent that 'when Mrs Chastele was arrested she was quite incapable of looking after herself'. He arranged for a woman to attend the patient 'cheaper than any person he could find'.[63] In 1869 another police constable offered his wife this role when a female lunatic was transported to Yarra Bend Asylum, but he had to wait for some time before her payment was made.[64] There was still confusion about payments to police in 1873.[65] One historical account of mounted police in colonial Victoria comments that the task of transporting lunatics was 'one of the most unpleasant tasks faced' by police, as unpleasant as transporting corpses from outlying and remote areas.[66]

Policing 'race'

Sex was one characteristic police used in their observations of the population; 'race' was another. There are virtually no records of Aborigines in Victorian asylums, suggesting that 'asylum' was not automatically a destination for the indigenous peoples in the colony. There were however new arrivals, including Chinese, in the colonial population.[67] Anne Ah Lou and Margaret Ah Lee were both arrested by police in 1887. There were not many Chinese women in Victoria, and Margaret Ah Lee was identified by the Kew Asylum as 'A Chinese half-caste girl'. As women with Chinese heritage they were both obvious to police. In fact Margaret was apprehended twice by police, first in March 1887 when she was seventeen, and later in August of 1888 when she was nineteen. Her age, her status – she was a single servant – and her appearance made her particularly vulnerable to arrest.[68] Of the Chinese men found at Kew Asylum in the 1870s, all were taken there by police. They were also subject to more scrutiny inside the asylum. A number of these men were miners, and others were labourers. Ah Lop was 'said to have murdered a man on the diggings'

63 'Belfast – The Hire of attendant on female lunatic prisoner', VPRS 937 Victoria Police Force, Unit 35, Belfast District.

64 'Re Mrs McCraith, attendance on lunatic Eliza Wernmouth en route to asylum', VPRS 937 Victoria Police Force, Unit 105, Castlemaine District, no. 5.

65 See another letter from Kew Station on 21 December 1873 regarding the 'lunatic prisoner' John Kenny and payment to police, VPRS 937 Unit 60, bundle 4. This issue of payment to police involved in the transporting of lunatics seems to have been one common to English and Australian lunacy practices. James Adam, Superintendent of Caterham Asylum in Surrey, recorded in his diary in 1876 that Sergeant Biddlecomb asked 'is there anything allowed for the capture of the female lunatic who was found wandering at Lingfield . . .?' 9 December, 1876, J. Adam, *Diaries 1872–82*, MS 5510–19, London.

66 Victoria Police Management Services Bureau, *Police in Victoria 1836–1980* (Melbourne, 1980), 30–1.

67 See Coleborne, 'Making "Mad" Populations in Settler Colonies'.

68 Margaret Ah Lee, 19 March 1887, VPRS 7397/P1, Kew Asylum, Unit 8, 88; 20 August 1888, VPRS 7397/P1, Kew Asylum, Unit 8, 272.

which even the asylum notes recorded as a possible flight of fancy: 'this is only hearsay'. Nevertheless, Ah Lop was 'very insane', and other Chinese men admitted by police to Kew were similarly constructed as dangerous and excessive in their madness.[69]

Ah Lang was 'rather dangerous'; Ah Gee 'dangerous and destructive'; Ah Shung 'not to be trusted'; Ah Sin 'potentially dangerous' and an opium user; and Ah Him was both 'dangerous' and of 'immoral habits'.[70] Like most of the Irish women who were identified as intemperate, vagrant or prostitutes, these Chinese men were vulnerable to police and had no one to speak on their behalf. They usually remained in the asylum until their deaths. The Chinese were feared to be sufferers of mania in large numbers, as an early article in the *AMJ* asserted, connecting the problem with the use of opium among the Chinese. In the article the police were identified as the proper agents of restraint and conveyance to the asylum for one case of a 'mongolian' with the 'disease' of insanity who had run away into the bush attacking people who crossed his path, and was captured by 'a European'. The police were obliged to bind him 'hand and foot in a sack' as he was 'kicking, hitting out, spitting, and biting with terrific strength and energy and hideous cries'.[71]

The writings of Chinese miner Jong Ah Sing provide another glimpse at the process of negotiation by a patient with authorities like police, magistrates and doctors in the event of committal.[72] The painful memoir by Ah Sing reveals much about his sense of frustration with these authorities, perhaps due to his poor skills in English. The memoir is a record of a possibly wrongful confinement and one that may also have been motivated by a skirmish on the goldfield which led to the arrest of the Chinese man.

Jong Ah Sing was an inmate of Yarra Bend Lunatic Asylum sometime around 1869. His diary or memoir, or possibly his 'plea', is an extraordinary account of his experiences under arrest, at trial and in the asylum. Ah Sing was apprehended by police around rural Dunolly or Maryborough, possibly because of a disturbance on the diggings. He appears to have been quite ill, and found himself in hospital in Dunolly. It is difficult to understand the story completely, but by piecing it together a number of revealing themes emerge. Ah Sing wrote about his feelings during his encounters with colonial law enforcers. 'My freedom

[69] Ah Lop, 16 January 1874, VPRS 7397/P1, Kew Asylum, Unit 2, 53.

[70] Ah Lang, 22 April 1874, VPRS 7397/P1, Kew Asylum, Unit 2, 130; Ah Gee, 10 November 1874, VPRS 7397/P1, Kew Asylum, Unit 2, 287; Ah Shung, 6 March 1875, VPRS 7397/P1, Kew Asylum, Unit 3, 89; Ah Sin, 26 March 1875, VPRS 7397/P1, Kew Asylum, Unit 3, 104, Ah Him, 28 June 1878, VPRS 7397/P1, Kew Asylum, Unit 6, 149.

[71] 'Chinese lunatics', *Australian Medical Journal* 10 (1865), 202.

[72] Jong Ah Sing, MS 12994, Box 1718/10, La Trobe Australian Manuscripts Collection, State Library of Victoria, 91pp. Jong Ah Sing was a Chinese inmate of Yarra Bend Lunatic Asylum, December 1872. No record of his admission to Yarra Bend Asylum can be found, due to the lack of extant male casebook records from the Yarra Bend in the period of Ah Sing's confinement.

my body', he wrote, and 'my no likey englishman tucker'.[73] His observations about the different goldseekers are also very interesting; the conflicts between British, Scottish, Irish and Chinese are clearly as much about race as they are about gold. But his encounters with police were starkly portrayed:

> Police . . . tear my body clothes take my money go, police take my money go . . . police tear my clothes 1 China good coat 1 china cotton thick coat 2 wool warm shirt 1 moleskin breeches undress my pair top boots
>
> Police keep batton stick step my shoulder 2 time police go out police go back my yard keep batton fight my body 3 time.[74]

Jong Ah Sing was evidently taken to Yarra Bend Asylum by the police. He mentions a 'gaoler', Thomas Peacock, who came with a 'cart' to take him to B Ward of the Yarra Bend. One line announces his arrival: 'Dr Paley lunatic Asylum'.[75] That Ah Sing met the asylum through its doctor, Paley, indicates that for him the institution and its medical keeper were closely associated; Paley's presence marked his first encounter with the asylum. The police had transported him from a punitive space to one defined by medicine.

Families

Police made observations of sexual and racial differences within the population which aided them in their work to identify trouble and danger. At times members of the community asked for their assistance as they too identified difficult behaviour, and thus the police were sometimes required by friends and families of alleged lunatics to monitor private space. In 1873 it was with difficulty and some reluctance that William Robertson wrote to the Chief Commissioner of Police in Victoria regarding his 'unfortunate friend' Mr Pettet, requesting police assistance in his confinement. Acting on behalf of Pettet's wife, Robertson asked that 'a member of the Police Force (preferably a Detective)' arrange the committal of Pettet to a 'private asylum'. Pettet's condition, wrote Robertson, 'border[ed] on insanity' and medical advisers recommended restraint.[76] The request was approved by Chief Commissioner of Police, Frederick Standish. The letter, and others like it, suggest that there was an awareness that police could perform such a role in relation to lunatics and troubled families.

Asylum authorities also made contact with the Chief Commissioner of Police when they needed assistance in a lunacy matter. In February of 1878 the Master-in-Lunacy, who watched over all insane asylums in the colony, wrote to the Chief Commissioner about lunatic patient Margaret Houston at Cremorne Asylum, a private establishment. In this instance police were being asked to co-operate

[73] Jong Ah Sing, 29, 31. [74] *Ibid.*, 32, 50. [75] *Ibid.*, 50.
[76] Letter from William Robertson to Chief Commissioner, 13 September 1873, VPRS 937 Unit 128, Chief Commissioner, Correspondence, Bundle 5.

in the continuing surveillance of the patient's family of four children, left to survive under the care of the eldest girl who earned money as a dressmaker. She earned 12*s*. a week, but the other children, aged nine, four and two, needed care. The Master-in-Lunacy asked for the address of the visitors of the Ladies' Benevolent Society in Collingwood, and also suggested that Margaret Houston might be moved to Yarra Bend or Kew Asylums as funds were not available to keep her at Cremorne.[77] Some days later the local Collingwood police wrote to the Chief Commissioner to report that the older sister was caring for the family and wanted to continue this by herself. She had also asked that their mother be transferred to Kew Asylum as it would be a saving for them. The police letter reported that the girl had admitted her mother acting on the advice of a Lay Minister of the Gipps Street Wesleyan Church.[78] This case reveals that family members negotiated with police not only about the problem of lunacy but also about the care of dependent children and welfare.[79] It also reveals that religious figures played a part in the committal of some of those deemed to be lunatics.

Police were sometimes also required by asylum authorities to help trace lunatic patients and their families. A number of police stations were involved in tracing the whereabouts of the wife and daughter of lunatic patient Thomas Birch at the Kew Asylum in 1878. On 25 February 1878 the Chief Commissioner was able to inform the Master-in-Lunacy that Thomas Birch had lived with a prostitute at Black Creek where he was employed butchering sheep, and had been locked up for drunkeness on a number of occasions; his wife and daughter were living in New Zealand.[80] In 1880 police were asked to trace the brother of Georgina Watson, a patient at Yarra Bend, regarding the payment of asylum costs in her 'maintenance' there as a patient. When traced, Hugh Watson refused to make any payments.[81]

Readings of the patient casebooks and asylum inspectors' reports indicate that police were in some instances perhaps performing work which families could not. Table 5.3 indicates that the majority of asylum inmates were brought to asylums by police over the years of this study. Families perhaps counted among the friends who also admitted 'lunatics', but police also assisted. Young women living at home were sometimes placed in the asylum by police, possibly because family members were unable to undertake this task themselves.

[77] Cremorne Asylum (a private institution) 14 February 1878. VPRS 937, Unit 299, Melbourne Police Inward Correspondence, Complaints against, Melbourne District Police Magistrates (hereafter abbreviated as MPIC – MDPM).

[78] 18 Feburary 1878. VPRS 937, Unit 299, MPIC – MDPM.

[79] On welfare policing see C. Twomey, *Deserted and Destitute: Motherhood, Wife Desertion and Colonial Welfare* (Melbourne, 2002).

[80] Correspondence to Chief Commissioner: 15 January 1878, 14 February 1878, 20 February 1878, 25 February 1878. VPRS 937, Unit 299.

[81] Hugh Watson, Master-in-Lunacy, letter of 20 February 1880. VPRS 937, Unit 302, Inwards Registered Correspondence, Police Department, Melbourne District, bundle 5.

Table 5.3 *Passage of lunatics to Victorian asylums, 1875–1890*

Date	1875	1885	1890
Brought by friends	123	150	268
Received through police	407	332	414
Sent from benevolent asylum	26	4	4
Sent from hospital	0	6	9
Sent from gaols	21	23	37
Sent from other institutions	71	83	154
Recaptured	11	17	26
Total	659	615	912

Whether some of these women were so uncontrollable that police were required is not known in all cases, suggesting that at least in some instances families needed more than physical assistance to deal with the problem at hand – they needed the support and validation that the police could provide in the committal of their family member.

Conclusion

In identifying lunatics and lunacy in the colony of Victoria the police operated in both public and private spaces. Their role in the identification and detection of lunacy existed in relation to the asylum. By the early twentieth century criminal investigation was theorized by writers who emphasized the role of scientific investigation in police work, especially in the detection of crime. By 1906 one published work on criminal investigation declared that insanity was one of the many 'practices' adopted by criminals and it was important for police to know the difference between feigned and actual insanity.[82] Thus lunatics were not only branded by their asylum clothing when inside the asylum or trying to escape from it. They were increasingly branded by the official language of bureaucracy which grew up around them during the nineteenth century. By 1893 the *Police Gazette* reminded its constables that:

> Members of the Force are instructed that, when drawing requisitions for the conveyance per rail of lunatics and their escorts, they must write the word 'lunatic' at the top of the requisition conspicuously or in red ink. They are not to include in the same requisition lunatics with other prisoners.[83]

[82] *Criminal Investigation: A Practical Textbook for Magistrates, Police Officers and Lawyers*, adapted by J. Adam and J. Collyer Adam from the *System Der Kriminalistic* of Dr Hans Gross (London [1906], 1962), 195.

[83] 26 July 1893, *Victoria Police Gazette*, 202; for the increasing complexity of police procedures regarding lunacy see also 10 August 1887, 231.

This marking of the lunatic – and marking out of the lunatic from the prisoner – was by 1893 a sign of two things: that the police still struggled in their dealings with lunacy, and that the way to deal with lunacy was to separate it and categorize it as different from criminality, a process underway in the 1870s. Despite the decrease in the role of police where lunacy admissions was concerned by the early 1900s, in part because police discouraged the public from involving them, in the final decades of the nineteenth century the bureaucratic and quasi-medical management of the insane by police was evident.[84] The police were being asked to perform both bureaucratic and medical roles, roles which were created by the asylum and medical bureaucracy as much as they were by the increasing bureaucratization of police practices.

Police work happened in conjunction with the asylum and its inmates and their families. Police liaised with the asylum, they transported lunatics to the asylum, and they watched families left alone when parents spent time away in the institution. Whether this might be understood as 'welfare policing' or surveillance is a tension in the social history of the nineteenth-century city in Victoria. During the nineteenth century police work was becoming professionalized and the expansion of duties and techniques of police, including policing the passage of lunatics to the asylum, may be understood as part of this process. As agents of colonial medicine, law and order they further defined the colonial 'lunatic' for Victoria's asylums.

[84] Dean Wilson, 'On the Beat: Police Work in Melbourne 1853–1923', Unpublished PhD thesis, Monash University, 2000, 197.

6 The Wittenauer Heilstätten in Berlin: a case record study of psychiatric patients in Germany, 1919–1960

Andrea Dörries and Thomas Beddies

Introduction

In 1933, Gustav Blume, a psychiatrist at the Wittenauer Heilstätten Asylum in Berlin, wrote:

> It is no secret to say that reading psychiatric case reports is not an unspoiled pleasure. Often it is a hopeless torture! I am not talking about the content of the reports, but about the technical process of reading them. For example, you have to work out a case history of an old schizophrenic, which covers some 20 to 30 years and more than a dozen stays in different hospitals. You sit worried in front of a chaotic package of more or less faded, damaged, and mostly loose sheets of paper from which stacks of illegible and crumpled letters and papers emerge. You try unsuccessfully to find out where the case history begins, where the most recent entries can be found; you reorganise, sort, and take notes. You dig deep into the scientist's last reserves of courage and dive into the stormy sea of faded or fresh hand-written psychiatrists' notes, and – you finally collapse. You then despair (or become enraged, depending on your temperament) of decoding your colleagues' notes and you are driven over the precipice to complete frustration. To document a case history by handwriting required a slower pace of life compared with today. To then read these old-fashioned entries, however, is impossible for the modern rational man in a time of portable typewriters. He refuses to do this as an enormous waste of time and power.[1]

Despite these remarks, during the last two decades the use and decoding of patient records as a historical source has become increasingly important.[2] The development of computer programs – graphical user interface – offers an expanded range of quantitative and qualitative approaches. Major advantages are an improvement in precision, a 'dynamic view'[3] and 'new insights not easily

[1] G. Blume, 'Über die Einrichtung Psychiatrischer Krankheitsgeschichten', *Allgemeine Zeitschrift für Psychiatrie* 99 (1933), 84–97.

[2] A. Digby, 'Quantitative and Qualitative Perspectives on the Asylum', in R. Porter and A. Wear (eds.), *Problems and Methods in the History of Medicine* (London, 1987), 153–74; Ö. Larsen, 'Case Histories in Nineteenth-Century Hospitals – What do They Tell the Historians?' *Medizin, Gesellschaft und Geschichte*, 10 (1991), 127–48; V. Rödel, 'Möglichkeiten und Grenzen der Archivierung medizinischer Unterlagen', *Der Archivar* 3 (1991), 427–35; G. E. Berrios, 'Historiography of Mental Systems and Diseases', *History of Psychiatry* 5 (1994), 175–90.

[3] Digby, 'Quantitative and Qualitative Perspectives', 153–74.

acquired by more traditional methods of analysis',[4] as well as a change of perspective regarding the persons involved in hospital life.[5] Though case records seem to be informative as historical sources, their limits can be overcome by supplementing case-record information with other historical sources such as statistical data, administrative reports, and proceedings of meetings.[6]

The aim of our study was to focus – within the constraints of the German data protection law[7] – on patients' lives in a psychiatric hospital, to define patient profiles and to reconstruct the activities on the wards and in the hospital's treatment facilities.[8] The intention was to do this on the background of the changing social life of three different political systems.

Psychiatric institutions and the Wittenauer Heilstätten in Berlin

The opening of the City of Berlin's first public lunatic asylum in 1880 coincided with a phase of numerous constructions for new institutions considered necessary due to a strong increase in patients needing the care of an institution.[9] In the German Empire, the number of public institutions for psychiatric patients almost doubled between 1877 and 1904 (from ninety-three to 180); the number of in-patients nearly tripled (from 33,023 to 111,951).[10] Various approaches exist to explain the causes of this increase, ranging from the scientific advancements of psychiatry and its establishment as an academic medical discipline, the consequences of industrialization, urbanization and rural drain, to the impacts of jurisdiction resulting – last but not least – from the bourgeois-bureaucratic desire for security and order.[11] In the case of Prussia, this applies to the Poverty Law (Landarmengesetz) of 11 July 1891,[12] which facilitated an increase in admissions by redistributing expenditure. All in all, a whole bundle of reasons

[4] *Ibid.*,153–74.

[5] M. MacDonald, 'Madness, Suicide and the Computer', in Porter and Wear (eds.), *Problems and Methods*, 207–29; G. B. Risse, 'Hospital History: New Sources and Methods', in Porter and Wear (eds.), *Problems and Methods*, 175–203.

[6] Digby, 'Quantitative and Qualitative Perspectives', 153–74; Risse, 'Hospital History', 175–203; G. B. Risse and J. H. Warner, 'Reconstructing Clinical Activities: Patient Records in Medical History', *Social History of Medicine* 5 (1992), 183–205.

[7] W. Rössler, 'Überlegungen zur Archivierung Psychiatrischer Krankenunterlagen', *Der Archivar* 3 (1991), 435–42.

[8] T. Beddies and A. Dörries (eds.), *Die Patienten der Wittenauer Heilstätten in Berlin, 1919–1960* (Husum, 1999); A. Dörries and T. Beddies, 'Coping with Quantity and Quality: Computer-based Research on Case Records from the "Wittenauer Heilstätten" in Berlin (1919–1960)', *History of Psychiatry* 10 (1999) 59–85; Risse and Warner, 'Reconstructing Clinical Activities', 183–205.

[9] D. Blasius, *Einfache Seelenstörung. Geschichte der deutschen Psychiatrie 1800–1945* (Frankfurt, 1994), 61.

[10] *Ibid.*, 64. [11] *Ibid.*

[12] Gesetz-Sammlung für die Königlichen Preußischen Staaten 1891, Nr. 9471. Where the effects of this law in comparison to Bavaria are concerned, see G. Kolb, 'Die Familienpflege unter besonderer Berücksichtigung der bayrischen Verhältnisse', *Zs. für die ges. Neurologie und Psychiatrie* 6 (1911), 273–304.

determined the increase in asylum patients which was enormous even when considering a strong population increase. The decision in 1868 to allocate the mentally diseased to major integrated hospital and care units outside their communities came at a time when a fundamental argument broke out about the foundations and forms of institutions within psychiatry. The protagonists of this discussion were Walter Griesinger, Professor of Psychiatry at Berlin University, and the asylum psychiatrist Heinrich Laehr. At a vote among the psychiatric section at the Gesellschaft deutscher Naturforscher und Ärzte in Dresden in September 1898, Laehr's position prevailed. The most essential consequences of this argument were the lasting distance of institutional psychiatry to the upcoming academic psychiatry and the creation of large remote hospital and care units.

In the empire's particularly thriving capital, Berlin, the rapid increase in the number of mostly poor people needing institutional care posed a special challenge. This is where the fundamental argument between Laehr and Griesinger became particularly explosive, as both competed for an influence on planning and setting up the first public lunatic asylum.[13] As early as 1853 and corresponding to the lunatic doctors' beliefs at that time, Laehr had pleaded for the erection of a 'relatively integrated hospital and care institution' for Berlin 'in a rural and friendly setting'; however, respective fundamental decisions stipulated by the community council were not followed by actual steps.[14] The project was taken up again and submitted to a group of experts for evaluation as late as the early 1860s, under the City Mayor Karl Theodor Seydel. In the spirit of the ten-year-old plea by Laehr, the overwhelming majority of experts favoured the construction of a lunatic asylum in conjunction with a sick unit for lunatics. In 1866 (meanwhile, Griesinger had joined the evaluation group), the experts were heard once again, and once more the majority of them voted for a centralized, closed unit, thus rejecting Griesinger's suggestion for decentralized accommodation in small colonies or in family care. Griesinger nevertheless regained influence on the planning phase due to a power struggle between the magistrate and the city's delegates by pleading for the site of the Dalldorf domain in North Berlin and the establishment of a rural asylum there, but the defeat he suffered – in his absence – in Dresden as well as his death shortly afterwards, led to the fact that his concept was no longer followed up and that a hospital and care unit was established in the traditional style.

However, the Dalldorf institution, opened in 1880 after three years of construction, could not cope with the demand for accommodation of mentally ill

[13] As to the following details on the argument between Griesinger and Laehr, please see K. Sammet, '*Über Irrenanstalten und Weiterentwicklung in Deutschland. Wilhelm Griesinger im Streit mit der konservativen Anstaltspsychiatrie 1865–1868*' (Hamburger Studien zur Geschichte der Medizin, 2000), 1, 41–72.

[14] *Ibid.*, 45.

patients from the very first day,[15] so that the allocation of patients to private institutions was to some extent inevitable. Facing an average annual increase of 130 to 160 patients, a requirement for 2,800 to 3,000 beds was envisaged until two projected institutions were established. Whereas these figures alone proved the necessity for new buildings, it was also considered urgently necessary that epileptics were separated form the actual mentally ill and admitted to a special institution. In 1893, the first transfers from Dalldorf to the newly established second city lunatic asylum Herzberge became possible.[16] Also in 1893, the third institution, Wuhlgarten, for epileptic patients was opened.[17] In 1906, a further unit finally went into operation in Buch, northeast of Berlin. But even this fourth institution for mentally ill Berliners (three lunatic asylums and the institution for epileptics) would not meet the expected requirements. Towards the end of 1911, barely two-thirds of the registered 8,431 mentally diseased patients in Berlin (5,270 = 62.5 per cent) were allocated to the public units, a considerable proportion of the rest found accommodation in private institutions and in family care. In 1919 (the beginning of the actual period of investigation for this study), far fewer admissions were recorded for the hospital and care units than, for instance, in 1914 due to the enormously increased mortality rate among patients during the First World War.[18] This allowed for an allocation of the remaining 4,255 patients (as per 1 April 1919) to the existing city institutions without any problems; the communal wards in private Berlin institutions, which had accommodated some 2,350 patients just before the war, disappeared completely.[19]

Hardly any generally applicable statements can be made for the reported period regarding the legal grounds of admission and hospitalization in German hospital and care units in general, as these differed considerably from one region to the next according to the federal structure regarding care for the mentally ill. Although a supposedly progressive lunatic law had been passed in the Grand Duchy of Baden as early as 1910,[20] developing respective regulations for Prussia or the German Empire proved unsuccessful despite intensive negotiations in the twenties.[21] Only after the foundation of the Federal Republic of Germany and the passing of the German constitution (*Grundgesetz*), relevant state laws

[15] *Statistisches Jahrbuch der Stadt Berlin* 10 (1884), 229.

[16] C. Moeli, *Die Irrenanstalt der Stadt Berlin in Lichtenberg – mit Bemerkungen über Bau und Einrichtung von Anstalten für Geisteskranke* (Berlin, 1896).

[17] J. Bresler (ed.), *Deutsche Heil- und Pflegeanstalten für psychisch Kranke in Wort und Bild* (Halle, 1912), 2, 315.

[18] H. Faulstich, *Hungersterben in der Psychiatrie 1914–1949. Mit einer Topographie der NS-Psychiatrie* (Freiburg Breisgauim, 1998), 58.

[19] G. A. Waetzold, 'Die Berliner Heil- und Pflegeanstalten in den ersten 10 Jahren des Gesetzes Groß-Berlin', *Zs für das gesamte Krankenhauswesen* 19 (1931), 533.

[20] L. Holdermann, *Das badische Irrenfürsorgegesetz mit Vollzugsordnung* (Karlsruhe, 1927).

[21] A. Beyer, 'Irrengesetzgebung', *Psychiatr. Neurol. Wschr* 26 (1924), 3–7.

were passed based on Article 104 of the constitution (restriction of personal freedom).[22]

Probably owing to the federal structure of the empire, too, hardly any reliable records on either the number of existing beds in the public hospitals and care units or their utilization to capacity are available. In a fundamental study about starvation in German psychiatry from 1914 to 1949, Heinz Faulstich gathered the recorded figures for the years 1928 to 1937 and arrived at an overall figure of 101,470 (1928) or 119,591 (1937) for these years.[23] This is significant inasmuch as the increase in the number of beds did not correspond with the building of new institutions; instead the existing institutions were utilized to a higher concentration of beds in order to save costs.

The psychiatric institution of our investigation was a typical representative of the centralized, closed unit type of psychiatric hospitals mainly for the lower classes. It was founded in 1880 under the name of Irrenanstalten der Stadt Berlin in Dalldorf. As late as 1920, when Greater Berlin was established, it was situated within the city limits of Berlin in the new district of Reinickendorf. In the mid-twenties, its name was changed to Wittenauer Heilstätten of the City of Berlin due to programmatic reasons. As 'Dalldorf' before, to the Berliners the term 'Wittenau' became a common synonym for the psychiatric institution existing there. It was renamed Karl-Bonhoeffer-Heilstätten at the end of the fifties, and was finally given the name Karl-Bonhoeffer-Nervenklinik (KBoN). As the name Wittenauer Heilstätten was most commonly used during most of the investigated period, it shall be the term used throughout the following.

The patients

The patients of the Wittenauer Heilstätten (between 1919 and 1960) lie at the centre of the study. With the aid of supplementary source material, the analysis of 4,000 case histories from a total of forty-two admission years, aimed at making the acquaintance with human beings who were admitted to the institution and who remained there for a relatively long time compared to modern conceptions, very often dying there.

It becomes apparent that the respective concerns and needs of the sick and healthy very often differed considerably. The public interest outside the institution walls was hardly ever the interest of the individuals inside the institution who were sick or considered diseased, especially as the latter had for a long time been considered a risk to public safety and order, their status as patients in need of healing and care taking only second place. By placing the individuals who

[22] F. Berlin, *Gesetz über die Unterbringung von Geisteskranken und Süchtigen*, GvBl. (1952), 636 and GvBl. (1958), 521.

[23] H. Faulstich, *Hungersterben in der Psychiatrie 1914–1949*, 129.

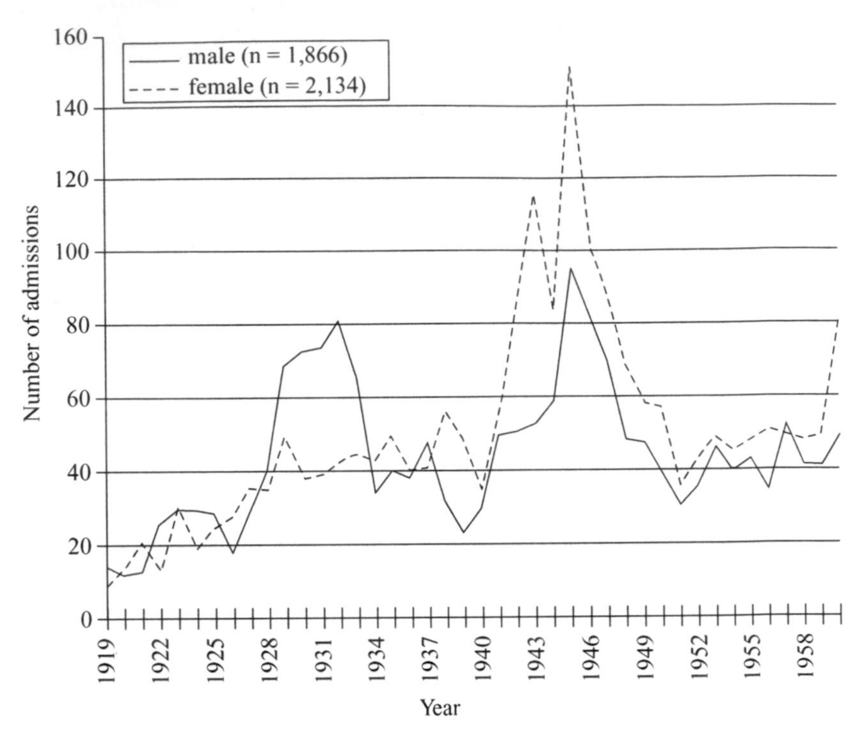

Figure 6.1 Number of male and female admissions per year (1919–1960)

were either sick or declared sick first in our survey, the (psychiatry-relevant) scientific-historic approach was reduced in favour of an approach relating more to social history. It is not only the disease and its therapy which mattered, but also a strong emphasis lay particularly on the patients' social environment, the way they and their families were treated, and the question as to how their elementary rights were upheld or withdrawn.

Furthermore, the relationship between the patients and the medical staff of the institution – physicians and nurses – was examined. The results of the study remain unaffected as we draw attention to the fact that, on one hand, the physicians at the institution particularly acted as the committed expert administrators on behalf of the sick individuals they looked after, especially as they literally found themselves on the same side of the clinic walls as the patients in their charge. On the other hand, however, it is also true that some physicians – and some of the most dedicated at that – were involved in the unprecedented crimes that patients during the Nazi regime fell victims to in their thousands. The Wittenauer Heilstätten doctors, too, were representatives of their faculty 'who were not born as monsters but rather acquired expert knowledge and sought-after positions in our society in a more or less inconspicuous way and with average talent before they took to anaesthetising the acquired skills of humanity',[24] in order to participate actively or by tolerating the atrocities 'with the gift of self-appeasement bordering onto the artistic'.[25]

Period of investigation

When we decided on the period to be investigated, the years between 1930 and 1960 became our first preference for reasons, among others, of the structure of the hospital archives. In 1930, the case-history administration within the clinic was reformed at the suggestion of a physician. This included a change from the traditional stitch files to letter files. Due to the increased need for more space, it became necessary to revise the filing system itself, so that a comprehensive system comprising roughly 40,000 units of case histories (between 1930 and 1960) in alphabetical order arose over the years in the KBoN archives. It became apparent that a representative random selection of this number should be examined.

Thematically, it was recommendable to extend the period investigated, accepting a certain workload when extracting the respective case histories from the old system. It was repeatedly noted that the roots of the excesses and crimes committed in the name of psychiatry during National Socialism could be traced to the Weimar Republic. The end of the World Economic Crisis, beginning in the late twenties and bringing about a general deterioration of living conditions, especially elevated the rise of racial hygiene (or eugenics) to a doctrine of national salvation. National Socialists supplied conditions for this type of science that led to the well-known results such as forced sterilization and the killing of psychiatric patients. However, it is also recorded that towards the end of the Weimar Republic, when the economic factors seemed to be compelling, the living conditions of patients in mental institutions deteriorated drastically. It was believed that the 'inferior' should have fewer liberties than the people outside who were healthy and fit for work. Stretching the period of investigation to the entire post-war years of the First World War enabled us to take a look at the development within the institutions from its lowest level in 1919 over the years of revolution and inflation, the economic boom in the mid-twenties right up to the years of crisis towards the end of the Republic, thus abandoning the focus on the last years of the Weimar Republic as the years of preparing the ground for National Socialism. Widening the scope of time offered an opportunity for looking at the poor treatment of psychiatric care units in the end phase of the

[24] A. Mitscherlich and F. Mielke (eds.), *Medizin ohne Menschlichkeit. Dokumente des Nürnberger Ärzteprozesses* (Frankfurt, 1962), 7.

[25] *Ibid.*, 8.

Weimar Republic as a continued resumption of the type of behaviour already familiar from the First World War.

The selection of the 1960s as the end of the investigated period was partly due to the filing system of the hospital archives. The year 1960 saw a new assessment of files corresponding with the creation of a so-called 'new archive', which was in operation right up to the era of electronic data-processing some years ago. In the 1960s the division of Berlin was consolidated with the building of the Wall. The transfer of patients between east and west was no longer possible. At the same time, a second psychiatric hospital in Berlin was opened in the western part of the city, in Spandau, ending the role of the Wittenauer-Heilstätten as the only major unit to provide for the mentally ill in the western part of Berlin. Another contribution was the gradual change within German psychiatry, which began to shape in the early sixties and lead to the so-called psychiatric reform. This development could not be included as an issue in our study.

Looking at the relatively long period of forty years called for the need for periodization; thus, changes in the political systems – from the Weimar Republic to National Socialism and from National Socialism to the West German Democracy – suggested themselves. Within these three large political eras: from 1919 to 1933, from 1933 to 1945, and from 1945 to 1960, sub-periods could in turn be made: for the Weimar Republic, these are primarily the restless years until the end of inflation in 1923, followed by the phase of relative stability until the beginning of the World Economic Crisis of 1929/30, and finally, the years up to 1933 which were marked by material hardship and political radicalization. The era of National Socialism can be subdivided into the years prior to 1939 (i.e. the era of consolidation and the beginning of internal economic prosperity while the war as well as the 'final solution' and the killing of patients was prepared by the National Socialists), and the Second World War itself, when a still-unknown number of psychiatric patients was murdered. The years from 1945 to 1960 should be subdivided into the immediate post-war era which lasted as long as the end of the blockade of Berlin in 1949, and, finally, the period of beginning normality since 1950, where increasingly easier restructuring, facilitated by the improvement of material preconditions, replaced the administration of shortage.

Linking the periodization of the outside world with the history of the psychiatric unit was of double importance. Firstly, the exterior conditions represented the economic, scientific, ethical, healthcare-relevant and socio-political framework for the institution and the people working within. Secondly, these outward events of course also exerted an influence especially on a psychiatric hospital and its patients, whose scope of experience in turn was determined by the respective events.

Patient records

The patient records, which were analysed, consisted of a representative selection from the total records dating back to the period concerned which were still in existence (8 per cent of approximately 50,000, i.e. 4,000). This does, however, not imply a representative selection of all patients who were admitted. Regarding data protection, we have sought approval in respect of archive legislation as well as the Berlin Hospital Law.

In this context it is worth mentioning a fundamental decision concerning the treatment of evaluated case histories. Case histories, and psychiatric case histories in particular, pose some considerable methodical problems as to their use as historical sources, however, they open up special opportunities. The hardly transparent entanglement of fact and fiction in these records is no doubt problematic. To illustrate this, one may imagine a 'reality scale': the accurately completed and well legible laboratory note recording the result of a blood examination would be at one end of the scale, whereas a patient's detailed social success story which turns out to be the megalomaniac imagination of a paralytic would probably belong to the other end. The most diverse data in terms of formality, contents, origin and time range somewhere between these two extremes. The often multiple refraction of information is significant in this context: the physician remarks about the patient's brother telling him that their father had also been a heavy drinker; furthermore, the mother is said to have been of a nervous disposition. In the end, there was only one way of accommodating the difficulties in recording the patients' non-verifiable social data and life stories: we decided generally to believe the statements made by the patients and their family members. Whenever there were contradictions in the details we made a note of it. Wherever the contradictions were unresolvable, the information was not used.

A special chance resulted from recording the patients' delusions (paranoia, excessive jealousy, megalomania, etc.). Beyond the medically interesting statement that the patient was under the influence of delusions, the opportunity opened up to link the contents of this statement with reality. Corresponding with Reinhart Koselleck who demonstrated the importance of dreams for historic science,[26] it became possible to introduce a source based on the delusions of the Wittenau patients. This becomes especially apparent in the years of the

[26] R. Koselleck, 'Terror und Traum. Methodologische Anmerkungen zu Zeiterfahrungen im Dritten Reich', in R. Koselleck (ed.), *Vergangene Zukunft. Zur Semantik geschichtlicher Zeiten* (Frankfurt, 1979), 278–99. The reference to this comprehensive article was found in D. Blasius, *Der verwaltete Wahnsinn. Eine Sozialgeschichte des Irrenhauses* (Frankfurt, 1980), 108 and 283.

Third Reich where patients in their delusions directly referred to political events taking place around them.

The analysis of the patients' life circumstances, their actions and their treatment in the social environment added up to a picture of many individual destinies, which could, however, be summarized under a diversity of aspects. We have undertaken to do this especially concerning the patients' gender, age and their diagnoses, while taking care to elucidate individual lengths of time in order to take into consideration external and time factors and their influence on the patients.

Diagnostic classification

In 1930, a classification of psychological disorders, later to be called the Würzburger Schlüssel, was elaborated for statistical purposes and tested for a period of two years.[27] It replaced the so-called Reichsirrenstatistik of 1901 and listed new groups of disorders. The pragmatic categorization was mainly descriptive and age-orientated with etiological components. The Würzburger Schlüssel focused largely on addiction (especially alcohol), psychopathies (differentiating between adults and adolescents), and syphilis-related disorders. Hereditary aspects were taken into consideration. Several disorders affecting elderly patients were not listed in individual descriptive categories, but categorized as age-related. Using the discussion about the Würzburger Schlüssel and its application in psychiatric clinics not only facilitates the illustration of scientific opinion of that era, but – due to the pragmatic background – the evaluation of specific syndromes towards the end of the Weimar Republic. A basic, socially compatible agreement on normality and disease is immanent in this classification of psychological diseases – as well as in those classifications still valid today. The psychiatrist George Agich demonstrated this in his analysis of antisocial personality disorder belonging to a contemporary classification system (Diagnostic and Statistical Manual for the American Psychiatric Association, DSM-III-R).[28] The psychiatrist Fulford even demanded that classification systems be used not only for statistical purposes, but to examine, publicize and deliberately utilize the value-orientation of such classification systems today.[29]

[27] See appendix. A. Dörries and J. Vollmann, 'Medizinische und ethische Probleme der Klassifikation psychischer Störungen, dargestellt am Beispiel des "Würzburger Schlüssels" von 1933', *Fortschr Neurol* 65 (1997), 550–4; A. Dörries, 'Der "Würzburger Schlüssel" von 1933 – Diskussionen um die Entwicklung einer Klassifikation psychischer Störungen', in Beddies and Dörries (eds.), *Die Patienten der Wittenauer Heilstätten in Berlin, 1919–1960*, 188–205.

[28] G. J. Agich, 'Evaluative Judgement and Personality Disorder', in J. Z. Sadler, O. P. Wiggins, M. A. Schwartz (eds.), *Philosophical Perspectives on Psychiatric Diagnostic Classification* (Baltimore, Md., 1994), 233–45.

[29] K. W. M. Fulford, 'Closet Logics: Hidden Conceptual Elements in the DSM and ICD Classifications of Mental Disorders', in J. Z. Sadler, O. P. Wiggins, M. A. Schwartz (eds.), *Philosophical Perspectives on Psychiatric Diagnostic Classification*, 211–32.

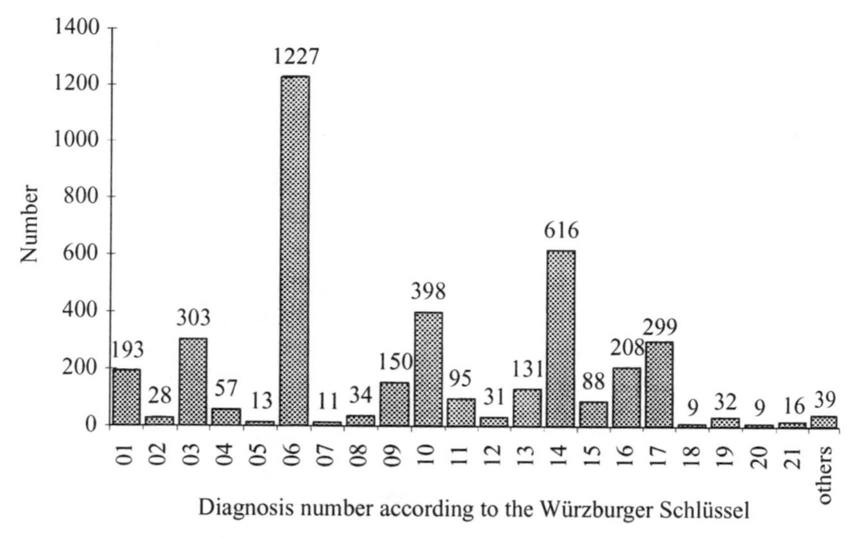

Figure 6.2 Distribution of diagnoses (N = 3983)

In the present study of a description of patients in the Wittenauer Heilstätten in Berlin, a representative selection of 4,000 case files for the period 1919–60 was analysed. The methodical approach facilitated a vertical section to be investigated and a search for annual distributions.[30] The case files at the Wittenauer Heilstätten proved to be a rather informative source, containing entries and complementary medical documentation about various individuals and occupational groups (psychiatrists, nursing staff, administration), thus enabling us to extract patterns.[31]

According to our experience with this study, the quantitative analysis should always be supplemented by a qualitative focus analysis and an analysis of certain groups of files. It makes sense to have research carried out by an interdisciplinary group (e.g., medical history, medicine, or history) with experience in handling large amounts of data as well as a basic knowledge of statistical methods in order to conduct a quantitative investigation of case files. The quantitative analysis required the clear categorization of the various fields. The preliminary diagnosis at the time of admission, however, was taken over word by word, so that moral implications of medical practice could be analysed for certain periods of time. The diagnosis at the time of discharge was based on the Würzburger Schlüssel. For the evaluation, the necessary categories corresponded to the time-relevant categories, not to the present classification of psychological disorders, International classification of Diseases (ICD 10). The qualitative analysis was elaborated by means of key words and supplemented

[30] Risse and Warner, 'Reconstructing Clinical Activities', 183–205. [31] *Ibid.*

by additional research into source material the traditional way. It proved to be essential for the overall picture when evaluating individual life data and helpful in respect of the patient's view, as it facilitated a subjective interpretation of the reasons stated for the admission, of the social context, and of the political behaviour.

Although a large amount of detailed information was theoretically possible, the entire scope could only be exploited in just a few cases due to the varying contents of the individual files.[32] In general, however, it can be said that the social data of psychiatric patients were documented in much more detail than the case files in other areas of medicine. Newer files showed more complete and detailed entries than those dating back to the twenties and thirties. They contained more details on diagnostics and courses of treatment; the precise descriptions of patients of the past which were often signified by personal terminology, however, were missing. In the fifties, the choice of words for describing the patients changed. The kind of documentation (e.g., the summaries in the discharge papers) altered the file profile and subsequently led to a new number of files from 1960.

The greatest challenge in the investigation was the elaboration of constancy and change during the four decades. The renamings of the clinic indicated how internal conditions and self-image changed. By means of the representative data, for instance, it became possible to illustrate and interpret the change in the range of diseases, the conditions for admission and an accurate determination of the conditions inside the hospital. For a more comprehensive evaluation, however, the conditions outside the clinic and the interactions with the political system had to be determined by studying additional sources. Comparisons with other hospitals provide a useful tool with which to judge the importance of local influences.

However, some problems and limitations occur when working with case files of the twentieth century. As an analogy to Daniel Greenstein's conditions that he laid out regarding computer-assisted work with historic patient files, the procedure can be extrapolated to our study. For the computer-assisted analysis of patient files, Greenstein's theoretical concept identified, in particular, an awareness of the limits of information, the inclusion of the context and the necessary self-criticism when analysing the case files; at the same time, however, he dispelled any frequently made reservations as to the respective historical importance of patient files, the risk of taking a purely scientific view, the necessary acquisition of comprehensive computer skills, and the time factor involved.[33] His notes and reservations proved to be justified.

[32] M. W. Dupree, 'Computerising Case Histories: Some Examples from Nineteenth Century Scotland', *Medizin, Gesellschaft und Geschichte* 11 (1992), 145–68.

[33] D. I. Greenstein, *A Historian's Guide to Computing* (Oxford, 1994), 108–13.

Evaluation of medical data

The greatest number of patients by far was admitted for age-related psychological disorders, followed by schizophrenia, alcoholism, progressive paralysis and reactive disorders. Mentally diseased juveniles and children as well as purely neurological disorders were documented in only a few individual cases, as the former were kept in separate files and the latter were not treated primarily at Wittenauer Heilstätten. In addition to the psychiatric diagnoses, further organic diseases were frequently diagnosed, especially in elderly patients, paralytics and patients with somatic psychoses. The sexes were unevenly distributed among the individual diagnostic categories. Whereas alcoholism was almost exclusively diagnosed in men, and while men prevailed among the diagnoses for psychopathy, progressive paralysis, injuries, addictive diseases, and brain disease, women formed the larger part of cases with schizophrenia, reactive disorders, suicide attempts, age-related psychological disorders, epilepsy, the few cases of *encephalitis epidemica* and *chorea huntington* (Huntington's disease), as well as some uncertain diseases.

Based on 630 case files of schizophrenia patients at Wittenauer Heilstätten, some statements as to genesis, development and therapy of this disease could be made for the period under investigation.[34] Some of the results confirmed the existing impression of this group of patients. Part of this picture is the good somatic condition of the patients, the low percentage of married persons, or the relatively high intensity of the new shock and convulsive treatments. However, those observations deviating from the expected which elucidate a gender-specific differentiation are naturally far more interesting. Not only do the files at least show a considerable female majority among the patients. The female schizophrenia patients also received medical treatment later while subsequently staying in the clinic twice as long as their male fellow sufferers, undergoing the various shock and convulsive therapies noticeably more intensely. In this context, the first elucidation of the recorded file material has at least yielded a research desiderate specific to the clinic.

In a timely respect, the increase in admissions of old-age-related mental disorders since the Second World War was especially significant. However, prior to that elderly people had increasingly been admitted as in-patients. At the same time, mentally disabled patients whose institutions had been either destroyed during the war or dissolved were increasingly admitted to the Wittenauer Heilstätten. In the forties, the change of the annual average age for the mentally disabled allowed for some respective conclusions as to the admission of a larger group of patients from outside. High figures could be established for

[34] R. Giel, 'Schizophreniepatienten in den Wittenauer Heilstätten (1920–1960)', in Beddies and Dörries (eds.), *Die Patienten der Wittenauer Heilstätten in Berlin*, 399–434.

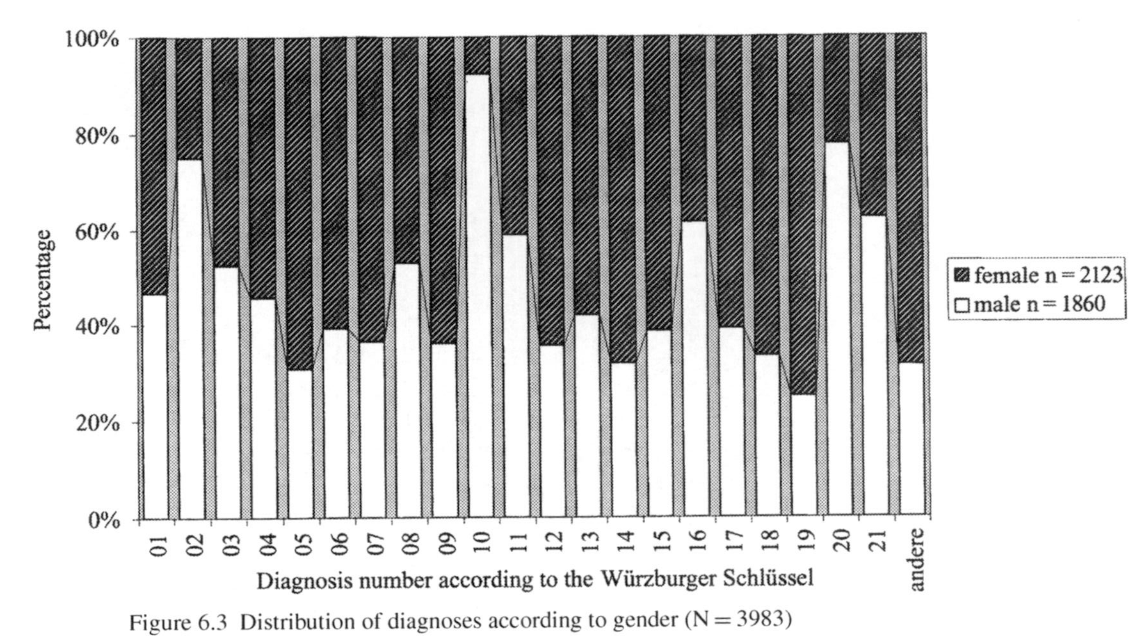

Figure 6.3 Distribution of diagnoses according to gender (N = 3983)

alcoholics towards the end of the twenties and the early thirties. It was not until the end of the forties that the admission figures rose again. The former was due to the establishment of a new special ward for alcoholics (1928–32), the latter may be regarded as a renewed admission of alcoholics after the interruption of the Second World War. In the fifties, the proportion of women admitted with alcoholism increased markedly. The number of patients with reactive disorders was also constantly on the increase during the period of investigation. It is possible, however, that this was due to the increased use of a new diagnosis category in conjunction with the Würzburger Schlüssel; on the other hand, the reason could be a modern therapeutic concept with an in-patient component. Patients with progressive paralysis were admitted mainly from the early twenties up to the forties, which was partly associated with the malaria therapy carried out at Wittenauer Heilstätten. This high-risk therapy was applied in only a few specialized clinics; the Wittenauer Heilstätten took up this therapeutic concept at an early period.

During the entire period of investigation, the average admission age remained remarkably constant for certain diagnostic categories, e.g., for age-related psychological disorders and schizophrenia. Whereas schizophrenia was mainly diagnosed in younger women, the average age for the predominantly male alcoholics was significantly higher, but still under the average age for the group of patients with age-related psychological disorders. A high proportion of suicide attempts could be established, which prevailed for the diagnoses 'psychopathy' (including adolescents), manic-depressive psychosis and reactive disorders. Elderly patients with psychological disorders and schizophrenics made significantly fewer suicide attempts.

On the therapeutic side, any state-of-the-art methods were applied in their respective time: medication, occupational therapy, various forms of shock therapy. Electro-convulsive therapy was employed from the mid-thirties, requiring the purchase of new technical equipment. It was mainly applied for schizophrenics and reactive disorders and partly replaced the expensive and high-risk 'insulin shock therapy'. By far the most common forms of medication were sedatives, followed by analeptics, antibiotics, major tranquillizers and a great number of other drugs. The percentage of patients treated with at least one type of drug rose continually during the period investigated especially at the end of the fifties. Although some uncertainties are assumed in the documentation, this result proved the rapid increase in the number of newly discovered and approved drugs. Some drugs, such as major tranquillizers and antibiotics, were prescribed in increasing quantities soon after their approval. Sedatives were applied to a large part in cases of reactive disorders, schizophrenia and for elderly patients. Analeptics were used for the elderly (e.g., camphor) and for schizophrenics (e.g., cardiazol). The application of major tranquillisers was above-average in schizophrenics and patients with reactive disorders. The malaria blood therapy

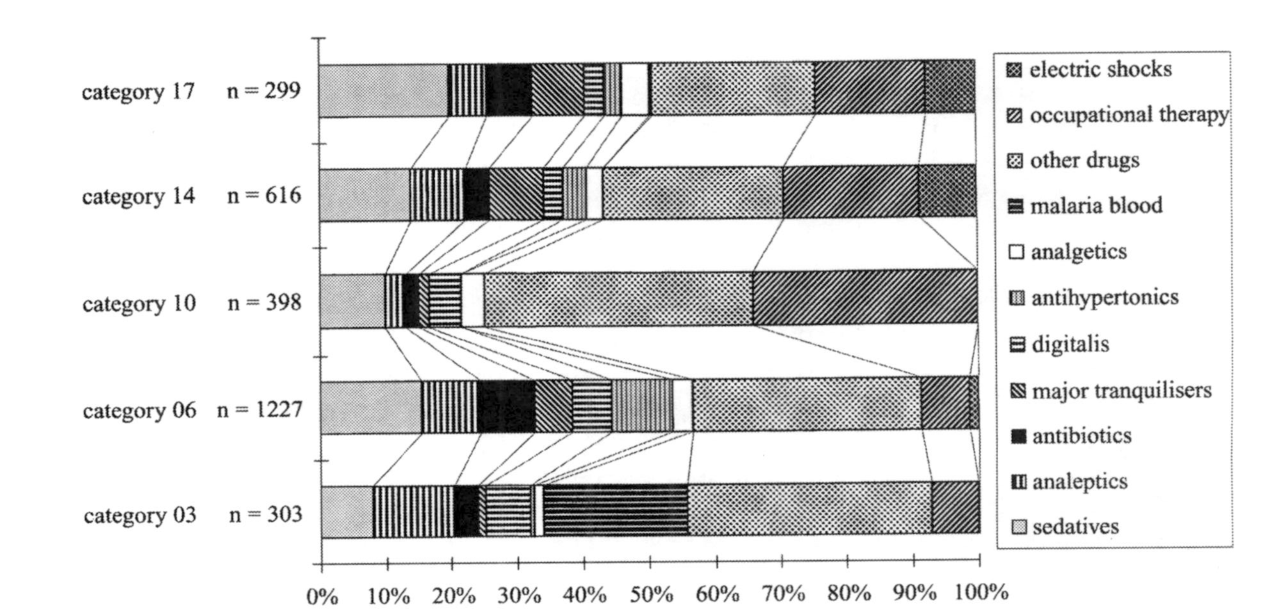

Figure 6.4 Therapies (including occupational therapy and electric shocks) concerning certain diagnoses (numbers according to Würzburger Schlüssel)

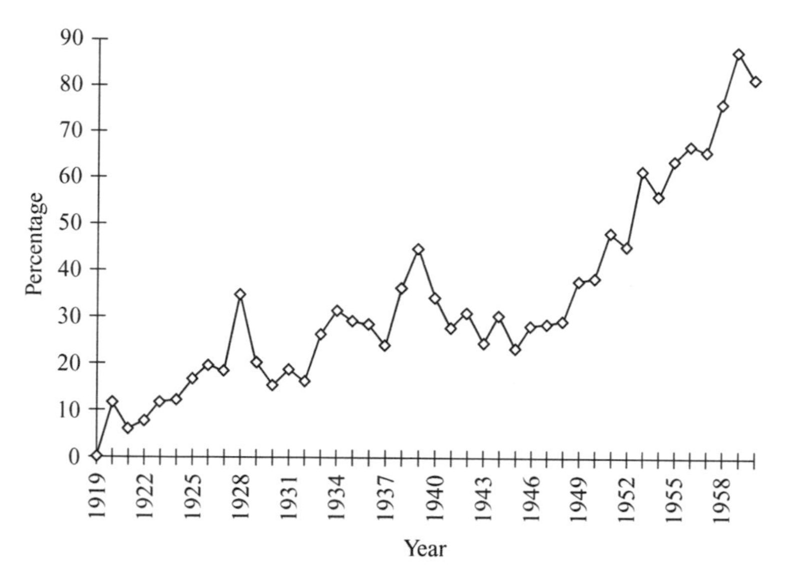

Figure 6.5 Annual percentage of patients treated with at least one drug (N = 4000)

was exclusively prescribed for progressive paralysis and included careful documentation.

Occupational therapy formed an important supplement to drug therapy, especially for alcoholics. It was utilized throughout the entire period of investigation and was specially supported. In this context, the proportions of male and female patients were almost even. Only in the thirties, more men than women were asked to join occupational therapy. According to the case files, an economic advantage could only be supposed in a few individual cases. The patients were mainly occupied in the 'field crew', with gardening or housework; in addition to housework, female patients were also employed in the sewing room. Plucking, which was considered the 'lowliest' activity, was mainly carried out by the elderly, epileptics, paralytics and mentally disabled persons.

Approximately 37 per cent of the admitted patients died in the institution. Discharge as a 'recovered' patient was effected in about half the cases, while the term 'recovered' is only justifiable when compared to the states of agitation, attacks and depressive phases that preceded the time of discharge. Due to the reasons described above, transfers to care units were documented comparatively rarely. The majority of patients was moved to general hospitals (sometimes for sterilization), moved or returned to prison, or – during the Second World War – to one of the 'euthanasia units'. Handing over to the medically supervised 'family care' did not class as a discharge from the institutional care.

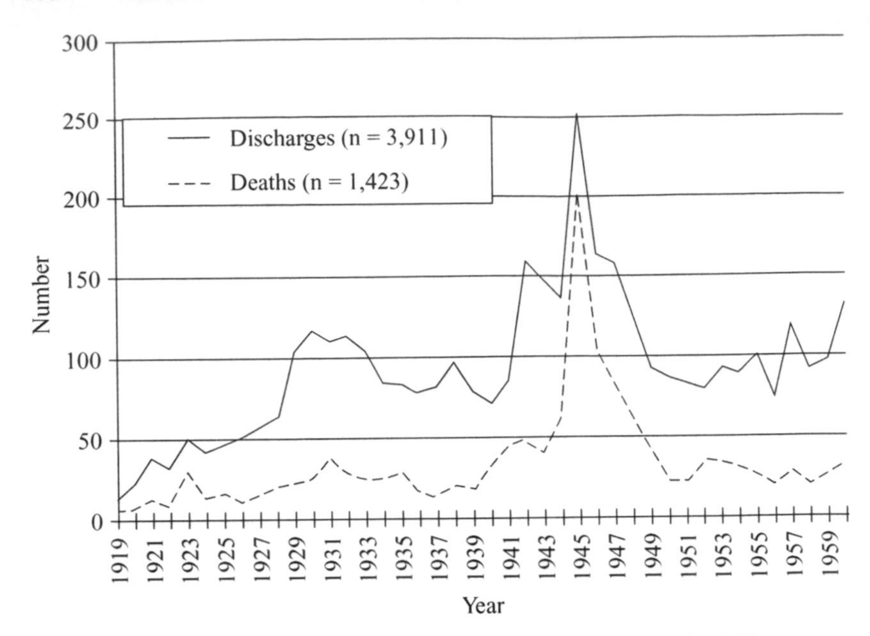

Figure 6.6 Number of discharges and deaths per year (1919–1960)

Roughly a third of the 4,000 patients examined died in the clinic; by far the highest mortality rate occurring in the period immediately after the Second World War. The major cause of death was diagnosed as cardiac arrest, followed by lung infections, other organic diseases and other infections. Suicides – as opposed to suicide attempts – only occurred in a few individual cases. 'Cardiac arrest' as the recorded cause of death increased since the forties, whereas the proportion of missing details after a case of death decreased. In the autopsies, mainly organic diseases previously not recognized were diagnosed. The greater part of the pathological examinations took place between the twenties and early forties. In 1923 and 1924, significantly fewer autopsies were carried out – a result of the economizing measures in the institution. The rate of dissected patients markedly decreased since 1942 for about a decade, only to increase marginally afterwards. The strongly fluctuating annual dissection rates in the early fifties may be explained in the context of problems in co-operation and a pronounced absenteeism of the prosecutor employed.

Social history

During the entire period of our investigation, the Wittenauer Heilstätten was an institution for the initial psychiatric care of patients in Berlin who had shown peculiar behaviour. If it became apparent after a certain phase of observation and possible therapeutic measures that an improvement in the patient's condition was not to be expected, a transfer to another institution – mostly a psychiatric unit in the province of Brandenburg – was aimed at for financial reasons and, above all, to reduce overcrowding.

Probably a large part of the case histories of these transferred long-term patients could not be used for the random check, as the files often remained in possession of the institution the patients had been moved to. Quantitative data from the analysed random check should be seen in the light of this background. In particular, it needs to be pointed out that the average hospitalization period for all patients recorded would probably have been considerably longer than the period of 300 days obtained in the random check. On the other hand, the mortality rate in comparison to the total number of admissions – i.e., including the patients who had been moved – may have been lower than the mortality rate obtained in the random check (slightly higher than a third).

The function of the Wittenauer Heilstätten as a psychiatric hospital for acute cases explains the low rate of patients who were moved to it from other psychiatric units. At the time of their admission, the overwhelming majority of patients came straight from home or from a general hospital where they had been considered intolerable due to the additional care they required or because of their behaviour.

The patients of the Wittenauer Heilstätten were not a homogenous group of 'mentally ill' behind asylum walls. Rather than that, the analysis of the patients' circumstances, their actions and their treatment within their social environment, add up to a picture of a great number of individual fates, which can however be summarized under various aspects. This has primarily been done in respect of gender, age, and the diagnoses for the patients while examining preferably short periods of time in order to be able to consider external and time-relevant factors and their influence on the patients.

Under the aspect of gender, the increasing number of elderly women among the admitted patients was especially striking. Since the forties, it rose steadily and under the aspect of the patient's gender, the increase in elderly women among the admissions was especially significant. Since the forties, it rose continuously and, in the fifties, outnumbered the admission rate for men by far.

The care unit and hospital nature of the institution that had surfaced since the forties is possibly one of the reasons for the increase in the proportion of female patients with psychological disorders in conjunction with old age. The analysis of the female patients' life circumstances showed that they lived alone much more often than was the case for men. It became apparent that the loss of a spouse also had a negative influence on the material living conditions of these women, so that, according to the negotiations during the admission procedure, the female patients often showed symptoms of malnutrition and neglect in addition to their alleged mental disorders, especially during the war and post-war years. In conjunction with supposedly endangering themselves and others, this diagnosis sufficed for an admission to Wittenauer Heilstätten. Since at least 1943, the admission to the hospital had been synonymous with the real danger of being transferred to the 'euthanasia unit' in Obrawalde, about 200 km east of Berlin, for numerous female patients with the diagnosis 'mental disorder of older age' who needed care or were regarded as troublesome and 'incurable'.[35] In this context, the social component of the Nazis' 'euthanasia activities' supposedly based on research on hereditary diseases became evident.

In the collapse situation of 1945, the number of admissions especially of old, helpless and feeble people increased to such an extent that the institution could no longer work appropriately. In 1945/6 the mortality rate for admitted patients was almost 80 per cent.

Under the aspect of gender, the extreme differences in the average period of stay for various diagnoses is striking. On average, schizophrenic women spent more than twice as much time in the institution than schizophrenic men (335 : 748 days). A possible explanation for this phenomenon could be the significantly greater number of therapeutic measures for female schizophrenics.[36] Another aspect – which can, however, hardly be quantified – is the deliberate confinement to the institutions of female patients up to their enforced sterilization. It is well known that in the mid thirties there had been longer waiting lists for the more elaborate sterilization methods for women. For the growing number of female patients with mental disorders associated with older age, the care unit and hospital nature of the institution, which became more and more predominant from the forties, may have been the cause for longer periods of stay.

The educational standard and the details of the patients' occupations initially showed the clear discrepancy between the sexes again; this can, however, not be regarded as a specific feature of patients in psychiatric institutions. Women in general had a poorer school education, often received no occupational training, did less paid work and worked in socially low-grade jobs. This supposition is relative when one looks at the patients' spouses. Here, social categorizing leads to significantly different results from the corresponding categorization of the female patient itself. The proportion of unskilled labourers was much lower, the proportion of skilled workers and craftsmen significantly higher. The female patients therefore must be seen in a context of a social position dependent on their husbands, a social position which could deteriorate rapidly if the husband died – especially as, compared to the women and considering

[35] T. Beddies, 'Die pommersche Heil- und Pflegeanstalt im brandenburgischen Obrawalde bei Meseritz', *Baltische Studien* 84 (1998), 85–114.

[36] Robert Giel, 'Schizophreniepatienten', in Beddies and Dörries (eds.), *Die Patienten der Wittenauer Heilstätten in Berlin*, 399–434.

all due differentiation in respect of age and diagnosis, the markedly higher proportion of married men indicates that the chances for men to marry or remarry despite their conspicuous psychological features or relatively old age was greater than for women.

It could be clearly seen that the social position of the patient strongly depended on the age at the first incidence of the disease. Diseases which had existed since birth ('congenital feeble-mindedness') or which had first appeared in younger years ('schizophrenia') would naturally have a negative effect on a patient's life in view of marriage and occupation. Thus it is not surprising that a large part of the patients with mental disorders of older age were judged in a more favourable way in the social anamnesis than most of the younger patients. Similarly, this applies to alcoholics and drug addicts. Addictions allowed for a relatively long occupational period of paid work as well as a family life functioning at least on the outside. However, the comparatively high proportion of upper-class drug addicts among the patients must not be overestimated. All persons with an academic background were considered to belong to this group. Nevertheless, the economic situations of numerous patients thus classified had deteriorated in such a way at the time of their disease or other unfavourable conditions that they could hardly be considered as belonging to the social upper class. This applied especially to the strongest group of drug-addicted physicians, who in many cases had already been deprived of their qualification.

Under critical source-relevant aspects, it is worth noting that – as a rule – interest, opportunities and the relevance of anamnesis in respect of family background and education would have been greater in the case of younger patients than for the elderly. Considering this, the question arises as to the direction of causality if it can be determined that there would be a higher proportion among younger patients of illegitimately born children or children who did not grow up with their parents than in the case of older patients.

By contrast, the alcoholics and drug addicts as well as the paralytics and patients with mental disorders of older age had more favourable prospects when it came to education, occupation as well as marriage and family bonds; this was, however, due to the large proportion of relatively young male alcoholics. The considerable fraction of war participants, especially among the alcoholics and paralytics, was significant for the social 'functioning' of this group up to the time of their admission. On the other hand, many men particularly blamed the war for their syphilis infection, their tendency to drink alcohol, or their morphine addiction.

Conclusions

The purpose of the analysis of 4,000 case histories from forty-two recording years was to become acquainted with the details of patients who were admitted as a result of their psychological disorders and behaviour and, compared to modern experiences, spent some considerable time and often died there. Detailed information on diagnosis, diagnostic procedures, therapies and causes of death could be obtained. In the eyes of healthy contemporaries and the authorities, the patients had become estranged from their familiar environment, they appeared to be troublesome, in need of care and were regarded as potentially and immediately dangerous. When one looks at the phenomenon of emotional or mental deviation from the perspective of the individuals concerned, however, one discovers human beings who had become estranged from themselves, felt disturbed or threatened, isolated and in need of help. Furthermore, the greater part of the clients in the city's mental care institution originated from the socially disadvantaged population groups of Berlin. In the majority of the cases, the costs of their institutionalization had to be borne by the respective welfare or social security authority. Thus the deviation caused by poverty and its related problems often contributed to the psychological disorder.

A rather favourable evaluation of social status for a large part of the patients must not distract from the fact that this was only derived in a comparison with the significantly more disadvantaged patients with mental disorders acquired in early life. In addition, the social classification only recorded their status before the time of admission. One of the problems for psychiatry in institutions and thus for the institutionalized patients for the whole duration of the investigated period, however, was the stigma the admission or transfer entailed. Misgivings regarding job loss, the partner's intention to end the relationship and general social isolation prevailed particularly in patients who were formally settled when they demanded their immediate discharge. These symptoms were neither changed by frequent name changes for the institution which was founded as a 'lunatic asylum' nor by the psychiatrists' noted repeated efforts to suppress the mental home nature of the unit in favour of an outward appearance as a hospital.

Appendix 'Würzburger Schlüssel' classification system (1933)

(1) Congenital and childhood mental deficiencies (idiocy and imbecility):
[Angeborene und früherworbene Schwachsinnszustände]
- (a) with unknown cause
- (b) due to known brain damage
- (c) cretinism

(2) Mental disorders due to brain damage (concussion or contusion of the brain):
[Psychische Störungen nach Gehirnverletzungen, Gehirnerschütterung und Gehirnquetschung]

(a) acute traumatic psychosis (psychosis due to concussion)
(b) traumatic sequelae (epileptic change of character, etc.)

(3) Progressive paralysis
[Progressive Paralyse]

(4) Mental disorders due to lues cerebri, tabes and lues latens
[Psychische Störungen bei Lues cerebri, Tabes und Lues latens]

(5) Encephalitis epidemica
[Encephalitis epidemica]

(6) Mental disorders of old age:
[Psychische Störungen des höheren Lebensalters]
(a) arteriosclerotic forms (including general hypertonia)
(b) presenile forms (depressive and paranoid forms)
(c) senile forms
(d) other forms (Alzheimer, Pick)

(7) Huntington's disease
[Huntingtonsche Chorea]

(8) Mental disorders associated with other brain diseases (tumours, multiple sclerosis)
[Psychische Störungen bei anderen Hirnkrankheiten]

(9) Mental disorders associated with infections, diseases of internal organs, general diseases and cachexia ('symptomatic psychoses in the narrower sense'):
[*Psychische Störungen bei akuten Infektionen, bei Erkrankungen innerer organe, bei Allgemeinerkrankungen und Kachexien: symptomatische Psychosen im engeren Sinne*]
(a) due to infectious diseases (including chorea minor)
(b) due to diseases of internal organs, general diseases and cachexia (diseases of circulation, the intestine, diabetes, uraemia and eclampsia, anaemia, carcinosis, pellagra)
(c) due to Morbus Basedow, myxedema, tetany and other endocrinological diseases
(d) symptomatic psychosis in puerperium and during lactation

(10) Alcoholism:
[Alkoholismus]
(a) intoxication
[Rauschzustände]
(b) chronic alcoholism (excessive jealousy)
[Chronischer Alkoholismus: Eifersuchtswahn usw.]
(c) delirium tremens and hallucinations
[Delirium tremens und Halluzinose]
(d) Korsakow's psychosis (polioencephalitis hemorrhagica)
[Korsakowsche Psychose: Polioencephalitis haemorrhagica]

(11) Addiction (morphinism, cocainism)
[Suchten]

(12) Mental disorders due to other intoxications:
[Psychische Störungen bei anderen Vergiftungen]
(sedatives, lead, mercury, arsenic, carbon disulphite, carbon monoxide)

(13) Epilepsy:
[Epilepsie]
(a) with unknown cause
(b) symptomatic epilepsy (if not mentioned in another group)

(14) Schizophrenic forms
[Schizophrener Formenkreis]

(15) Manic depressive forms:
[Manisch-depressiver Formenkreis]
(a) manic and depressive phases
(b) hyperthymic, dysthymic and cyclothymic constitution

(16) Psychopathic personality
[Psychopathische Persönlichkeiten]

(17) Abnormal reactions:
[Abnorme Reaktionen]
(a) paranoid reactions and paranoid development (paranoia querulans)
[Paranoide Reaktionen und paranoische Entwicklungen: Querulantenwahn]
(b) depressive reactions not included in 15a
[Depressive Reaktionen, welche nicht unter 15a fallen]
(c) psychotic reactions to imprisonment
[Haftreaktionen]
(d) compensation neurosis
[Rentenneurose]
(e) other psychogenic reactions
[Andere psychogene Reaktionen]
(f) induced mental illness
[Induziertes Irresein]

(18) Psychopathic children and adolescents
(up to eighteen years of age)
[Psychopathische Kinder und Jugendliche]

(19) Unresolved cases
[Ungeklärte Fälle]

(20) Nervous diseases without mental disorders
[Nervenkrankheiten ohne psychische Störungen]

(21) No nervous disease and free of mental deviations
[Nicht nervenkrank und frei von psychischen Abweichungen]

7 Curative asylum, custodial hospital: the South Carolina Lunatic Asylum and State Hospital, 1828–1920

Peter McCandless

The South Carolina State Hospital, formerly the South Carolina Lunatic Asylum, is one of the oldest public mental institutions in the United States. The oldest, in Williamsburg, Virginia, dates from 1773. For several decades after its opening, Virginia's asylum remained an anomaly, the only institution in the country founded specifically to care for the insane. The South Carolina Lunatic Asylum, however, was founded at the beginning of a sustained wave of asylum construction. During the early nineteenth century, increased population density, growth of towns, and expansion of a market economy brought dissatisfaction with existing modes of caring for the insane in private homes and public poorhouses. Enlightenment empiricism encouraged a faith in human ability to solve human problems. European psychiatric innovations, particularly moral treatment, inspired therapeutic optimism, the belief that lunatic asylums could cure large numbers of the insane and restore them to productive labour.

Between 1817 and 1824, philanthropists in several northeastern states, often aided by public subsidies, opened private charitable asylums intended to serve patients of all social ranks. These institutions were influential, but they ended up catering to a small, mainly affluent clientele and did not provide the organizational pattern for American asylums. Neither did private proprietary asylums, which first appeared in the 1820s, and had become common in some parts of the country by the 1870s. The dominant type of mental institution in the United States has been public. But the American model differed from that of most European countries in that the federal (central) government did not pass legislation mandating the construction of public asylums. The initiative in building and maintaining asylums remained with individual states. The public model was adopted by three southern states in the 1820s: South Carolina, Kentucky, which opened an asylum at Lexington in 1824, and Virginia, which opened its second state asylum at Staunton in 1828. In 1833, Massachusetts became the first northern state to open a state asylum, at Worcester. By the late nineteenth century, nearly every state had one or more asylums and most mental patients resided in them. These state asylums were promoted by small groups of reformers and were often designed, like the private charitable asylums, to serve a socially mixed clientele. Funding varied but initially included some combination of revenue from the state, local government, and families of patients.[1]

Although the earliest state asylums were in the south, few historians have focused on southern psychiatric developments.[2] Moreover, historians of insanity in the United States have tended to portray the south as psychiatrically backward. According to this scenario, founders of early southern asylums were ignorant of moral therapy and therapeutic optimism, and created custodial welfare institutions.[3] The evidence for South Carolina does not support this argument. But some of these historians have rightly stressed a more important psychiatric distinction between north and south. The south's large black and (until 1865) slave population made race a much more significant issue for southern than northern asylums, at least before the twentieth century. In South Carolina, blacks constituted more than 50 per cent of the population between 1820 and 1920, and reached a high point of 60 per cent in 1880.[4]

The question of southern distinctiveness that such a comparison arouses has been debated in American historiography since the1920s. Recently, historians have begun to explore how differing regional experiences of disease and its treatment may have contributed to southern distinctiveness.[5] The debate over southern distinctiveness raises a number of important questions for historians of American insanity. How psychiatrically distinct was the south? How did southern responses to insanity and southern mental institutions differ from those in other regions of the country? How did issues of slavery and race influence the care and treatment of the insane? To what extent and in

[1] G. Grob, *The Mad Among Us: A History of the Care of America's Mentally Ill* (Cambridge, Mass., 1994), chapters one and two, G. Grob, *Mental Institutions in America: Social Policy to 1875* (New York, 1973), chapters one to three, five; D. Rothman, *The Discovery of the Asylum: Social Order and Disorder in the New Republic* (Boston, 1971), chapters five and six. On South Carolina, see P. McCandless, *Moonlight, Magnolias, and Madness: Insanity in South Carolina from the Colonial Period to the Progressive Era* (Chapel Hill and London, 1996); P. McCandless, 'A Female Malady? Women at the South Carolina Lunatic Asylum, 1828–1915', *Journal of the History of Medicine* 54 (1999), 543–71.

[2] See S. B. Thielmann, 'Southern Madness: The Shape of Mental Health Care in the Old South', in R. L. Numbers and T. L. Savitt, *Science and Medicine in the Old South* (Baton Rouge, 1989), 256–75; N. Dain, *Disordered Minds: The First Century of Eastern State Hospital in Williamsburg, Virginia, 1766–1866* (Charlottesville, 1971); J. S. Hughes, 'Labeling and Treating Black Mental Illness in Alabama', *Journal of Southern History* 58 (1993), 435–60; R. F. White, 'Custodial Care for the Insane at Eastern State Hospital in Lexington, Kentucky, 1824–44', *Filson Club Quarterly* 62 (1988), 303–35.

[3] A. Deutsch, *The Mentally Ill in America: A History of their Care and Treatment from Colonial Times* (Garden City, NY, 1937), 106; Grob, *Mental Institutions in America*, 95–96, 190, 195, 342–4, 359–68; G. Grob, *Mental Illness and American Society, 1875–1940* (Princeton, NJ, 1983), 25–6, 104, 159–60, 218–20, N. Dain, *Concepts of Insanity in the United States* (New Brunswick, NJ, 1964), 128, 177, 225n.2, 242n.17.

[4] Grob, *Mental Institutions in America*, 243–55; Grob, *Mental Illness and American Society, 1875–1940*, 22–3, 26, 220–21; Dain, *Concepts of Insanity*, 90–1, 104–8.

[5] T. L. Savitt and J. H. Young (eds.), *Disease and Distinctiveness in the American South* (Knoxville, Tenn., 1988).

what ways did the region's economic problems after the Civil War affect the insane?

South Carolina presents an excellent venue for the examination of these questions. In South Carolina, as in the south as a whole, race has always been a powerful historical theme. From the early eighteenth century until the 1920s, the majority of South Carolina's population was African-American. The state produced some of the staunchest defenders of slavery and precipitated secession and Civil War. For more than a century after the war, it, like the south generally, was marked by poverty, racial segregation, and nostalgia for a mythical Old South of cavalier planters, plantation mistresses and happy slaves.

In many ways, the history of the South Carolina Lunatic Asylum resembled that of contemporary public asylums in the northern states and Europe. Its founders were inspired by moral treatment and therapeutic optimism to create a curative institution for patients of all social classes. In 1821, they convinced the legislature to establish an asylum at Columbia, the state capital. The commissioners who superintended its construction incorporated features of recent European and American asylums into its design. The architect was influenced by British innovations in asylum architecture, particularly those advocated by one of the most important advocates of moral treatment, Samuel Tuke of the York Retreat.[6] Although the asylum never fully achieved the founders' therapeutic hopes, its officers came closest to replicating the ideal community of moral treatment during the antebellum decades. This may have been partly because the antebellum asylum cared for a relatively small and homogeneous clientele. Although the patients included both rich and poor, the number of paying patients nearly balanced the number of paupers. The number of patients never exceeded 200 before the Civil War, and most of them were white and native born. The asylum did not formally accept blacks until 1849 and housed only a few black patients when the Civil War began in 1861.

For both South Carolina and its asylum the war was a watershed. Economically and politically, the state changed radically in the post-war decades. In 1800, it had been one of the richest states. By the late nineteenth century, it had become one of the poorest. Economic decline began in the antebellum period. The 1860s and 1870s brought defeat in war, the emancipation of the

[6] Special Committee on Lunatics, 1818, General Assembly Papers (hereafter cited as GAP); South Carolina Department of Archives and History (hereafter cited as SCDAH); W. Crafts, *Oration on the Occasion of Laying the Corner Stone of the Lunatic Asylum at Columbia, July, 1822* (Charleston, SC, 1822), 12–16; Joint Committee of the Senate, Report on the Lunatic Asylum and the School for the Deaf and Dumb, 1822, GAP; *Reports and Resolutions of the General Assembly of South Carolina* (hereafter cited as RR), 1822, 103–4; *The Statutes at Large of South Carolina* (Columbia, SC, 1836–), vol. 6, 168; J. M. Bryan and J. M. Johnson, 'Robert Mills' Sources for the South Carolina Lunatic Asylum, 1822', *Journal of the South Carolina Medical Association* 75 (1979), 264–8; J. Bryan (ed.), *Robert Mills, Architect* (Washington, DC, 1989), 85–8.

slaves, federal reconstruction, severe economic dislocation, and enormous loss of wealth. Despite the emergence of a textile industry after 1880, the state's inhabitants became more dependent than ever upon the cultivation of cotton and rice in a time of falling agricultural prices. For much of the late nineteenth and early twentieth centuries, South Carolina suffered severely from depression.[7]

The South Carolina Lunatic Asylum also changed radically after the war. The patient census grew enormously, from 192 in 1860 to over 2,200 by 1920, and altered both socially and racially. The number of paying patients declined to insignificance and the institution was flooded with publicly supported patients, many of them black. Emancipation fully opened the asylum to the black majority and the number of black patients rose to nearly 1,000 by 1920. As the asylum population grew larger and more heterogeneous, its therapeutic function declined. The proportion of patients discharged as recovered dropped substantially, and it effectively became a custodial welfare institution. These changes coincided with a marked deterioration in patient care and internal conditions, due to grossly inadequate funding. To a large extent this development resulted from the state's steep economic decline and a profound legislative antipathy to debt and taxes. But it also reflected the significant changes in the institution's clientele. Legislators committed to fiscal conservatism and white supremacy showed little sympathy for a patient population that was increasingly poor, black and chronic. Many patients endured horrific conditions as a result, but black patients suffered more than whites.

To be sure, the asylum faced severe obstacles from the beginning. Its promoters failed to achieve a broad consensus of public support for their goals. Building it cost much more and took much longer than they had predicted. Cost overruns and delays angered legislators, and conflicts arose over the nature, functions and even necessity of an asylum. Many legislators were sceptical about the feasibility of a curative institution for all ranks of citizens. Like most early nineteenth-century Americans, they viewed hospitals as disreputable institutions for people too poor to get care at home. In the 1820s, such institutions were few, and they served a predominately lower-class clientele. The only hospital in South Carolina prior to the establishment of the state lunatic asylum was connected with the poorhouse in Charleston. Although the poorhouse hospital admitted some paying patients, many people viewed a public lunatic asylum as a pauper institution no respectable family would wish to patronize.[8] Legislators

[7] The best history of South Carolina is W. Edgar, *South Carolina: A History* (Columbia, SC, 1998).

[8] Journal of the South Carolina House of Representatives, South Carolina Department of Archives and History, 1827, 194–7, GAP; Joint Committee of the Senate on the Lunatic Asylum, 1822, GAP; Governor's Messages, Thomas Bennett, 1822, no. 1318, GAP; Reports from Commissioners of the Poor in Obedience to the Act of 1821, GAP; Charles Rosenberg, *The Care of Strangers: The Rise of America's Hospital System* (New York, 1987), 18–22.

who held these views charged that the asylum commissioners were building a palace for paupers and were able to cut funds for outfitting the building when it was finally completed in 1827. As a result, the asylum's officers were initially unable to purchase items considered essential to effective moral therapy, such as gardens, pleasure grounds, indoor plumbing, cattle, horses and carriages. The absence of these things initially increased the asylum's difficulties in attracting wealthy patients and implementing its therapeutic ideals. These problems were exacerbated by the legislature's refusal to provide annual appropriations for operating expenses. For this outcome, the asylum's advocates were largely to blame. They had repeatedly assured legislators that the institution would support itself from patient fees.[9]

For several years after the asylum opened in 1828, few patients arrived. In 1831, when it nearly closed down for lack of funds, only thirty-five patients were in residence in an institution designed for a hundred.[10] The number of patients did not reach a hundred until 1849, and its financial condition remained precarious until the late 1830s. Critics who had doubted the demand for such an institution found confirmation in its inability to attract enough patients to sustain itself financially.[11] The officers blamed the difficulty in attracting patients largely on unenlightened public attitudes and inadequate laws. Too many families had outmoded, negative stereotypes of asylums; too many local officials valued keeping taxes low over curing the insane poor; and the law gave these officials too much discretion in deciding whether or not to commit pauper lunatics (insane persons local poor-law officials declared unable to afford to pay for treatment). After the asylum's Board of Regents threatened to close the asylum due to a lack of patients and revenue in 1831, the legislature passed an Act which required counties to transfer pauper lunatics to the asylum.[12] The Act

[9] Journal of the House of Representatives, 1826, 240–1, 1827, 195–6; Report of the Trustees to Effect the Operation of the Lunatic Asylum, 1827, GAP; *Columbia Telescope*, 30 January 1829; *Pendleton Messenger*, 6 February 1828; McCandless, *Moonlight, Magnolias, and Madness*, chapter two.

[10] South Carolina State Hospital (hereafter cited as SCSH), Admission Book, 1828–74, Department of Mental Health (hereafter cited as DMH), SCDAH, 1; *Charleston Mercury*, 28 November 1828; *Pendleton Messenger*, 10 December 1828; SCSH, Minutes of the Board of Regents (hereafter cited as MBR), 5 July, 12 December 1828, DMH, SCDAH; SCSH, *AR*, 1831.

[11] SCSH, Admission Book, 1828–1874, *AR*, 1829–1850; W. G. Simms, 'The Morals of Slavery', in *The Proslavery Argument* (Charleston, 1852), 226–27.

[12] *Statutes*, vol. 6, 437; MBR, 29 November, 22 December 1828, 4 February 1832; *AR*, 1828, 1830. In South Carolina, paupers were technically persons unable to support themselves without public assistance. South Carolina had adopted the English Poor Law in 1712, and established a system of relief funded by a tax collected and distributed by overseers of the poor in each parish (after the Revolution, by commissioners of the poor in the districts or counties). Pauperism did not carry the legal disqualifications it did in England, and for many whites the stigma of it was reduced somewhat by the need to maintain a united white community in the face of the black majority. The stigma attached to pauperism was also less when it was the result of disease. For example, a pauper lunatic was often simply a person deemed to lack sufficient personal or

brought a modest stream of admissions. But it did not provide any penalty for non-compliance, and local poor-law commissioners continued to keep many of the insane paupers in poorhouses or board them out.[13]

A flood of pauper patients would not have solved the asylum's financial problems. It needed to attract enough high-paying private patients to subsidize the care of the poor. During the asylum's first decade, few wealthy families sent insane relatives to the asylum, either out of prejudice against public hospitals or fear of exposing the family shame.[14] Their reluctance led the asylum's regents to seek patients in neighbouring states. The commitment law allowed the regents to admit patients from other states and the asylum was well positioned geographically to serve insane southerners of means. Until the 1840s, it was the only asylum south of Virginia. Georgia, North Carolina, Alabama and Florida did not open theirs until 1844, 1856, 1860 and 1877, respectively. In 1829 the first of many Georgia patients was admitted to the asylum. Others soon came from North Carolina, Alabama and Florida. Throughout the antebellum period, the asylum received wealthy patients from these and other southern states.[15]

By the mid-1830s, wealthy South Carolina families were also becoming more willing to patronize the asylum. Positive comments about the asylum from its promoters, physicians and a few prominent families who had sent relatives to the asylum helped make it a respectable alternative to home care, boarding out, or an asylum in the north.[16] As one can see from Table 7.1, paying admissions outnumbered pauper admissions during the years for which such information is available.

To be sure, the asylum never attracted as many private patients as its supporters had hoped; and paupers normally constituted a slight majority of the resident patients, because they tended to remain longer than paying patients. But, as Table 7.2 shows, the number of pauper patients never greatly exceeded that of paying patients before the Civil War.

An obvious way to increase the number of patients would have been to admit blacks. The asylum's regents began to receive applications to admit insane

family income to afford medical treatment. Paupers were necessarily treated generously. As in England and other American states, their treatment varied considerably according to time, place and local attitudes towards poverty. *Statutes*, vol. 2, 593–598; Barbara L. Bellows, *Benevolence Among Slaveholders: Assisting the Poor in Antebellum Charleston, 1670–1860* (Baton Rouge and London, 1993).

13 *AR*, 1842, 23, 1847, 8–9; *Statutes*, vol. 6, 437. For more information on boarding out and other forms of community care, see McCandless, *Moonlight, Magnolias, and Madness*, 149–151, and chapter eight.

14 *Pendleton Messenger*, 11 February, 20 May 1829; *Columbia Telescope*, 30 January 1829; see also, W. G. Simms, 'Lunatic Asylum', *Magnolia; or Southern Apalachian* 1, n.s. (1842), 394.

15 MBR, 7 March, 2 May 1829, 3 April 1830; SCSH, Admission Book, 1828–74.

16 Townes Family Papers, correspondence, May–August 1835, South Carolina Library, Columbia (hereafter cited as SCL); J. C. Calhoun to A. Burt, 18, 27 May 1838, Ms. Dept., Perkins Library, Duke University; R. L. Meriwether (ed.), *The Papers of John C. Calhoun* (Columbia, 1981), vol. 14, 292–3, 305, 311–12, 354, 394, 499, 593.

Table 7.1 *Paying and pauper admissions to South Carolina Lunatic Asylum, for selected periods, 1828–1852*

Period	Paying	Pauper
1828–38	117	110
1837–41	68	59
1848–52[a]	118	65

[a]Statistics for 1842–47, 1851 are unavailable.
Sources: South Carolina State Hospital, Admission Book, 1828–38; *Annual Report*, 1842, 1848–52.

Table 7.2 *Average number of paying and pauper patients resident in South Carolina Lunatic Asylum, for selected periods, 1830–1864*

Period	Paying	Pauper
1830–4[a]	15	18
1840–4	26	39
1845–9	34	49
1850–4	68	78
1855–9	88	97
1860–4	82	94

Note: Averages are rounded to the nearest number.
[a]Statistics for 1832, 1835–39 are unavailable.
Sources: South Carolina State Hospital, *Annual Report*, 1831, 1833, 1842–64; Minutes of the Board of Regents, 1 May 1830; Journal of the South Carolina House of Representatives, South Carolina Department of Archives and History, 1834, 164–5; Charleston *Courier*, 20 December 1864.

blacks soon after it opened, but at first decided not to admit them because the law did not expressly permit blacks' admission and because their presence would increase expenses and complicate the running of the institution. They could not be mixed with the whites, and they would need a separate keeper and exercise yard. Ironically, one of the first patients to be admitted was a fourteen-year-old slave named Jefferson. He was admitted as a favour to his owner, who had committed his brother at the same time. Jefferson was not housed in the main building with the white patients, and his name was not recorded in the asylum's admission book. His owner removed him after a few months. Only one other non-white patient appears in the asylum's record before 1850, a free mulatto who spent several months there in 1839. Several years after the asylum opened, its regents declared themselves willing to admit blacks if the legislature approved it. They periodically advocated the reception of black patients during the 1830s and 1840s.[17]

The legislature voted to legalize the admission of blacks in 1848, largely in an effort to counter abolitionist propaganda. To be sure, the Act's advocates stressed the standard humanitarian, medical and social control arguments for asylum care. They argued that the state had a moral responsibility to care for faithful servants, and to protect slave owners' families from the dangers of living in proximity to lunatic slaves. But they also stressed the political dangers of refusing to admit blacks to the asylum. Charleston writer William Gilmore Simms warned that 'discrimination between the sufferings of [insane] blacks and whites' could help give anti-slavery propaganda credibility. A legislative committee noted that opening the asylum to blacks would enhance South Carolina's reputation for humanity and serve as a rebuke to the 'idle and vicious fanaticism' of the abolitionists.[18]

That the Act admitting blacks to the state asylum was largely a political exercise is supported by the minimal impact it had on their situation. During the 1850s, only around thirty blacks were admitted to the asylum, compared to more than 600 whites. At the end of 1858, seven of 180 patients in the asylum were black.[19] The accommodation the asylum provided for blacks was much inferior to that for the white patients, consisting of a couple of small brick outbuildings placed near the main asylum structure. Soon after the first black patients arrived, the physicians protested that provision for them was unacceptable medically, racially and administratively. The black patients could not get proper exercise because their building was located in the white patients' exercise court. The presence of the blacks distressed the whites and inhibited their recovery. The asylum had to hire a special attendant for the few black patients, since the regular attendants refused to care for them. Because of these problems, the regents decided in 1858 to release the black men patients and admit no more until the state agreed to fund a proper building and grounds for them. The asylum continued to admit a few black women, but turned down numerous applications for the admission of male slaves.[20] South Carolina's reluctance to provide accommodation for insane blacks was not unusual, although the state's

[17] *AR*, 1829, 1832, 1844; H. Martineau, *Society in America*, 2 vols. (New York, 1837), 291–2; MBR, 7, 27 June, 28, 29 November, 1828, 4 April, 6 June, 3 October, 1829, 5 January, 2 February, 2 March, 1839, 2 November, 1844; SCSH, Patients' Treatment Record, I, Jefferson, II, David I. Duncan, DMH, SCDAH.

[18] Simms, 'Lunatic Asylum', 395; RR, 1848, 77.

[19] *AR*, 1858, 12–13; SCSH, Admission Book, 1828–74.

[20] *AR*, 1850, 1851, 8, 1853, 24, 1858, 12–13, 1860, 13; MBR, 6 November 1858, 5 November 1859, 4 August, 17 September 1860.

black majority made its consequences significant. Prior to the Civil War, most asylums in the United States either did not accept black patients or provided them with separate and inferior facilities to those for whites. Southern asylums admitted more blacks than northern ones, but the north had a tiny black population. Only the Eastern Virginia Lunatic Asylum accepted significant numbers of black patients before the 1860s.[21]

The founders of the South Carolina Lunatic Asylum not only aimed to attract large numbers of patients of all classes, but they also expected to cure most of them. James Davis, who became the institution's first physician, argued that an asylum should not be merely 'a place of comfort' but also 'a house of cure'. Davis and the other antebellum physicians claimed that 80 to 90 per cent of the insane could be cured if sent to an asylum in the early stages of the disease. This therapeutic optimism, as in the north and Europe, was largely based on faith in moral treatment. Soon after the asylum opened, its regents advertised that it was conducted on the principles of moral treatment. The physicians repeatedly claimed that they governed their patients by moral methods. They employed mechanical restraints and other forms of coercion only when necessary for medical or safety reasons and applied them as leniently and gently as possible. The asylum's reports frequently compared the patients to a happy family, in which the physician played the role of the wise and benevolent father, and the patients were like children. Nineteenth-century asylum officials often used such a rhetoric of domestic patriarchy, both to define their mission and to allay suspicions of madhouses.[22]

Such descriptions represented an ideal. In reality, lack of money hampered the asylum's efforts to supply the facilities and personal attention moral treatment demanded, especially during the first decade. Coercion in the form of mechanical restraint, seclusion and cold showers was common during the early years. The physicians also relied heavily at first on drastic medical therapies such as purging, bleeding and blistering. Yet, the evidence indicates that they conscientiously tried to employ moral methods within the limits of their resources. During the 1840s and 1850s, they greatly decreased their use of drastic medications and mechanical restraints. Over time, the asylum was able to supply occupation and amusement for many of the patients. The regents gradually purchased or were donated land for farms and gardens, horses and carriages for riding, a library, a bowling alley, billiard tables and other games.[23]

But the South Carolina Lunatic Asylum, like other contemporary asylums, never achieved the high cure ratios its promoters had anticipated. During the

[21] 'Proceedings of the Association', *American Journal of Insanity* 12 (1855), 43; Grob, *Mental Institutions in America*, 243–55; Savitt, *Medicine and Slavery*, 258–79; Dain, *Disordered Minds*, 19, 105, 109–13; Dain, *Concepts of Insanity*, 107–8.

[22] *AR*, 1829, 1830, 1842, 1844; *Pendleton Messenger*, 20 May 1829.

[23] McCandless, *Moonlight, Magnolias, and Madness*, 84–118.

first four years, 19 per cent of the patients were discharged as cured. Between 1835 and 1855, the proportion of recoveries claimed rose to about 45 per cent, still far from the predicted 90 per cent.[24] The officers explained their failure to cure a larger percentage of patients much like asylum authorities did elsewhere: the admission of too many chronic cases, inadequate facilities, and an inability to get and keep a sufficient number of qualified attendants. They also blamed unenlightened social attitudes which led families and local officials to delay sending patients to the asylum and to squander the best chances of obtaining a speedy and lasting cure. The commitment law did not help: it required the asylum to admit idiots and epileptics as well as lunatics.[25]

From the 1840s, the officers often cited inadequate means of classifying patients as a major hindrance to the asylum's therapeutic and financial success. Implementing moral therapy required separating the quiet from the noisy, the peaceful from the violent, the clean from the dirty, the respectable from the indecent, rich from poor, women from men. To separate these classes from one another in one building in which the patients also had to be separated by sex, and after 1850 by race, was no simple matter.[26] On the other hand, the asylum's officers never complained about a classification problem that beset many American asylums in the 1840s and 1850s, the admission of large numbers of immigrants, especially Irish. At some northern asylums immigrants constituted a majority of the patients by mid-century. But South Carolina's immigrant population was small. In 1850 and 1860, less than 1 per cent of the state's inhabitants were foreign born, and most of them lived in Charleston. The foreign born never made up more than a small minority of the total patient population at the South Carolina institution. During the 1850s, when the percentage of foreign-born patients was highest, they never exceeded 15 per cent of the total in residence. As in the north, most of them were Irish.[27]

Yet the officers often complained about inadequate means of classification, not only because it hampered moral treatment, but because it had important financial consequences. The asylum's inability to segregate pauper and paying patients made it harder to attract wealthy patients, whose fees subsidized the paupers. Unless the asylum provided suitable accommodation for paying patients, paupers would predominate and it would have to be supported entirely by public funds. In 1842 and 1848, the officers used these arguments successfully to convince the legislature to fund extensions to the original building. By the

[24] *AR*, 1829–33, RR, 1855, 155–8.
[25] *AR*, 1829, 1842, 11–13, 20, 1852, 4, 1857, 6; *Statutes*, vol. 6, 323, 382, 437.
[26] *AR*, 1842, 7, 16, 22, 1844, 1–2 , 1852, 4.
[27] *AR*, 1853, 27–31, 1859, 16–21, 23; US Bureau of the Census, *A Century of Population Growth* (Baltimore, 1967), 128. US Bureau of the Census, *Statistics of the United States in 1860* (New York, 1976), vol. 20; Grob, *Mental Institutions in America*, 230–6; Dain, *Concepts of Insanity*, 99–104.

early 1850s, they decided that proper classification was impossible within the existing asylum structure, and began a campaign to replace it with a new one. A second building was begun in 1858, but construction was interrupted by the outbreak of the Civil War in 1861 and it was not completed until 1885.[28]

Despite problems in attracting and curing patients, the asylum came closer to achieving its founders' therapeutic and social goals before the Civil War than after it. The war proved a radical disjunction in the history of the asylum as well as the state. Soon after the outbreak of war in 1861, the financial situation of the asylum became desperate. War created an emergency situation only the state government had the resources to deal with, but it also diverted most of the state's revenue to military purposes and greatly weakened its economic position. By preventing the export of cotton and rice, the federal blockade deprived the state of hard currency. Many citizens were caught between eroding incomes and rampant inflation. The end of the war brought no relief. South Carolina had suffered staggering economic losses. Many people were impoverished by the accumulation of debt, loss of land and slaves, and the breakdown of traditional patterns of trade and agriculture.[29]

The asylum shared the misfortunes of the general populace. The officers had to pay highly inflated prices for food, medicine, clothing and other essentials. They found it increasingly difficult to collect payment for patients and debts mounted quickly. The legislature raised the fee for paupers several times during the war, and began making annual appropriations for the asylum. But inflation rendered such assistance inadequate before it was granted. To meet the emergency, the asylum's officers economized, borrowed, begged and pledged their own resources and credit. They also discharged many patients and prohibited admissions from other states. The patient census dropped from 192 in November 1860 to 128 at the end of 1865. Conditions deteriorated badly during the war's later stages. The officers could not get enough money to feed and clothe the patients properly or pay the attendants regularly. Mortality, which had averaged 8 per cent of the patients under treatment in the decade 1853–62, increased to 13 per cent between 1863 and 1865. The officers kept the asylum running only by putting off badly needed repairs and improvements. By the late 1860s various observers declared the asylum's condition a disgrace to the state.[30]

Following the Civil War, the asylum came under closer state supervision and control. Between 1868 and 1877, the victorious federal government imposed

[28] *AR*, 1842, 15–17, 1848, 2, 4–9, 1851, 3–4, 1852, 4, 8–9, 1853, 5; Journal of the South Carolina House of Representatives, 1842, 17; RR, 1842, 99, 1848, 77; MBR, 7 October 1848; McCandless, *Moonlight, Magnolias, and Madness*, chapter six.

[29] W. Roark, *Masters Without Slaves* (New York, 1978), 40–52, 77–8, 88–9, 132–53, 170–80; E. M. Lander, *A History of South Carolina, 1865–1960* (Chapel Hill, NC, 1960), 3–5.

[30] *AR*, 1853–69; MBR, 1861–69, 2 February, 2 March 1867; RR, 1868, 114–16; *Anderson Intelligencer*, 8 December 1870.

Table 7.3 *Number of patients and number of black patients in South Carolina Lunatic Asylum/State Hospital, 1840–1920*

Year	All patients	Black patients	SC population	Percentage of blacks in SC population
1840	51	0	594,398	56.4
1860	192	7(1858)[a]	703,708	58.6
1880	490	187	995,577	60.7
1900	1027	393	1,340,316	58.4
1920	2205	988	1,683,724	51.4

[a]The number of black patients was not reported in 1860 but was in 1858.
Sources: Annual Report, 1840, 1858, 1860, 1880, 1900, 1920; *South Carolina Statistical Abstract, 1998* (Columbia, South Carolina: Budget and Control Board, Division of Research and Statistical Services, 1998), 325, 335.

reconstruction on South Carolina and other former Confederate states. Its goal was to ensure that the civil rights of the newly emancipated blacks would be respected. Protected by federal troops, northern Republicans took control of the state government, supported by newly enfranchised blacks and a minority of white residents. The asylum was soon affected by this political revolution. A new state constitution of 1868 gave the governor the power to appoint the superintendent and all other officers, including the regents.[31] Previously, the regents had essentially controlled all appointments, including filling vacancies in their own ranks. The new Republican governor replaced the existing regents and superintendent with Republicans. Six of the new board of regents were black and three white. Reconstruction ultimately failed to protect the rights of the black population, but it did politicize the care of the state's insane to a greater extent than ever before. In the postwar era, the asylum became the source of political patronage and was charged with corruption and extravagance.[32]

The nature and function of the asylum also changed radically. As one can see from Table 7.3, the number of patients exploded after the Civil War, and black patients came to constitute a large proportion of the clientele. Without explicitly abandoning its therapeutic goal, the asylum became a receptacle for large numbers of 'defectives' of both races, whose common denominator was poverty. Emancipation of the slaves transferred responsibility for insane blacks from their owners to the public authorities. Black patients became a significant presence in the institution for the first time. By 1920, they made up nearly half of its clientele, a number close to their proportion of the state's

[31] *Statutes*, vol. 14, 24.
[32] *Charleston Daily News*, 13 January 1869; Journal of the South Carolina House of Representatives, 1869, 27–8; *Columbia Daily Phoenix*, 25 November 1869; J. S. Reynolds, *Reconstruction in South Carolina* (Columbia, SC, 1905), 123–5.

population (51.4 per cent). Over the same period, the number of private patients dropped precipitately. In 1865, the number of paying patients (sixty) was still nearly in balance with the number of paupers (sixty-eight). By 1881, paupers outnumbered paying patients almost twenty to one (464 to twenty-six). The trend continued and became a major political issue. As the number of publicly supported patients increased, politicians justified low appropriations for the asylum by arguing that the state's charity was being abused by families who had enough property to defray the costs of their relatives' care. Governor Benjamin R. Tillman, who had harshly criticized the asylum administration as extravagant during his campaign for office, remarked sarcastically to legislators in 1890 that the proportion of pauper patients was so high 'that we are forced to ask whether only the poor people go crazy'.[33]

Of course, the poor were not the only ones who went mad. The increase in the proportion of pauper patients had a variety of economic and political causes. First, the state's white citizens were poorer than before. Many families who had once been able to pay for the care of their insane relatives at home, in boarding houses, or in public or private asylums could no longer afford to do so. Second, nearly all of the new black patients were unable to pay for their care. Third, a lack of poorhouses and general hospitals meant the state hospital received patients who might have been sent to such institutions. Many counties did not have poorhouses, or their poorhouses had no secure accommodation. Until the end of the nineteenth century, the state hospital was virtually the only hospital in the state outside of Charleston. Finally, a major change in fiscal policy contributed to the sharp increase in pauper patients. In 1871, the state assumed financial responsibility for the care of pauper patients. This freed the institution from the problem of trying to collect fees for pauper patients from the counties. But it also encouraged local officials to commit people, because the state and not the county was paying the cost. Many families, too, were less reluctant to accept assistance from the state than from poor-law authorities. The term 'pauper lunatic' was abandoned in favour of 'beneficiary' during the 1870s, both a sign of changing attitudes towards accepting public care and an additional inducement to such acceptance. By 1900, less than 3 per cent of the patients were paying anything towards the cost of their maintenance at the state hospital.[34]

[33] Quotation from Journal of the South Carolina House of Representatives, 1890, 140; *AR*, 1881, 14, 19; Journal of the South Carolina House of Representatives, 1882, 26–27.

[34] *AR*, 1871, 30, 1883–4, 21, 1891–2, 6, 16, 1894, 13, 1900, 8, 21, 1904, 72. The increase in proportion of black to white patients does not seem to have been affected much by whites' recourse to alternative institutions. The first private asylum in South Carolina, Waverley Sanitarium in Columbia, was not opened until 1915. Some wealthy white families continued to send their insane to institutions in the north after the Civil War as before. But the numbers of South Carolina families that availed themselves of such options seems to have declined. For example, the records

The number of patients grew much faster than the funds appropriated for their maintenance. Between 1875 and 1905, the patient census quadrupled, but the legislative appropriation only doubled, from $70,000 to $140,000. Annual expenditure per patient fell from around $200 to around $100, while prices of food and supplies increased by about 25 per cent. Per capita expenditure for maintenance around the turn of the century was regularly the lowest, or close to the lowest, of American public mental institutions.[35]

The rapid influx of beneficiary patients, combined with low funding, overwhelmed the asylum's facilities. Severe overcrowding, an occasional problem before the Civil War, became chronic. Overcrowding was common to most public asylums in the United States and Europe in the late nineteenth century, and the explanation was much the same everywhere: over time, chronic patients accumulated. The officers constantly complained that the asylum was clogged with chronic and aged patients whose presence excluded the acute insane. More than 18 per cent of the admissions to the state hospital from 1891 to 1911 were diagnosed as imbeciles, idiots, epileptics and inebriates. Between 1875 and 1895 the percentage of resident patients over fifty increased from 11.7 to 27.6.[36]

Faced with a constantly expanding crowd of unpromising patients, the officers repeatedly appealed for additional accommodation. But they were never able to secure enough money from the legislature to keep pace with the demand for space. An example is the completion of the new asylum building begun in the late 1850s. One wing had been erected before construction was suspended by the Civil War. Work resumed in 1870, but was stopped several times by inadequate appropriations or funding delays. The building was not completed until 1885, and then only by financing much of its cost out of appropriations for the patients' maintenance and by using convict and patient labour.[37]

Even after the completion of the new building, many patients, especially the blacks, remained in dangerous and unhealthy structures. During the 1880s, the black women were gradually moved to the original asylum building, which the asylum's officers had for decades condemned as obsolete and unhygienic. The black men remained housed in 'temporary' wooden buildings they had occupied since the late 1860s. The medical superintendents routinely condemned these buildings as firetraps and pleaded with the legislature to replace them with

of the Pennsylvania Hospital in Philadelphia between 1841 and 1865 include those of twenty-seven patients from South Carolina. Between 1866 and 1905, the number of South Carolina patients at that hospital shrank to five. McCandless, *Moonlight, Magnolias, and Madness*, 145, 313 n. 6, 360.

[35] *AR*, 1882–3, 22–5, 1887–8, 15, 57–9, 1900, 15–16; Journal of the South Carolina House of Representatives, 1904, 24; Grob, *Mental Illness in America*, 25–6.

[36] *AR*, 1875–1911; Grob, *Mental Institutions in America*, 307–8; Grob, *Mental Illness and American Society*, 24, 179–88, 195–6.

[37] *AR*, 1870–85.

permanent brick structures, but to little effect. The markedly inferior accommodation allotted to the blacks did not arouse much concern among the white politicians who dominated state government after Reconstruction. Between the late 1870s and 1890s, they stripped blacks of their civil rights and established racial segregation and white supremacy.[38]

The problem of getting permanent facilities for the black insane was complicated by divisions among white politicians over long-term policy towards their care. One group, upset by the proximity of the races as well as the difficulties of managing a multi-racial institution, wanted to establish a separate institution for blacks, a solution adopted by several other southern states shortly after the Civil War. Another group favoured retaining a single institution for both races on grounds of economy and ease of administration. The legislature did not begin to resolve the issue until 1910, when it authorized the purchase of a tract of land several miles from Columbia for the purpose of erecting a second state hospital. Development of this site, known as State Park, was slowed and at times suspended by insufficient appropriations, disagreement over its purposes, and the opposition of Governor Coleman Blease (1910–14). He objected to spending white taxpayers' money on providing a new hospital for blacks. Instead, he proposed to convert the state penitentiary into an asylum for blacks.[39] His opposition was eventually overcome and a transfer of black patients to State Park began in 1914. But development was slow, and the last black patients were not removed from the old state hospital until the 1930s.[40] The failure to develop a second institution for the black insane before the First World War forced the asylum authorities to provide permanent accommodation at the Columbia location. In 1893, the legislature approved construction of a brick building for black men but allocated no money until 1897, when it appropriated a paltry $7,500. Black patients excavated the foundation, convicts made some of the brick, and the demolition of an old brick wall provided the rest. The total cost to the state of the building, designed for 200 patients, was only $21,000.[41]

While the new building for black men was under construction, in 1895, the South Carolina Lunatic Asylum changed its name to the South Carolina State Hospital for the Insane. Superintendent James Woods Babcock proposed the

[38] *AR*, 1877–98; *Report of the Legislative Committee on the State Hospital* (Columbia, SC, 1910), 7; I. A. Newby, *Black Carolinians* (Columbia, SC, 1973), chapter two; Lander, *History of South Carolina*, 106–7.

[39] *AR*, 1880–1, 18, 1887–8, 12–13, 1888–9, 7–8, 1892–3, 6, 11–15; 1902, 12, 1904, 68, 1908, 24, 1910, 5–8; Journal of the South Carolina House of Representatives, 1887, 25; *Report and Proceedings of the Special Legislative Committee to Investigate the State Hospital for the Insane and State Park* (Columbia, SC, 1914), 59–60, 69–70; Hughes, 'Black Mental Illness in Alabama', 441.

[40] *AR*, 1914–35. The state hospitals were integrated in 1964.

[41] *AR*, 1893–8; Notes on Page Ellington, James Woods Babcock Papers, SCL; MBR, 8 February 1912; *Testimony, Taken Before the Legislative Committee to Investigate the State Hospital for the Insane at Columbia* (Columbia, SC, 1909), 393–5.

change to emphasize that the institution was a hospital, not a prison or an almshouse.[42] Although Babcock intended no irony, by the time the 'asylum' became a 'hospital' it was functioning largely as a custodial welfare institution. The percentage of patients discharged as recovered fell from about 30 per cent of those admitted in the 1870s to less than 17 per cent between 1902 and 1911. The hospital's recovery rate was also worse than that in many other institutions. In 1904, a special report of the United States Census Office revealed that the South Carolina State Hospital's recovery rate was about 5 per cent below the national and South Atlantic averages.[43]

The emergence of a custodial regime coincided with the decline of moral treatment. During the late nineteenth century, asylum physicians everywhere were losing faith in the therapeutic effectiveness of moral treatment, as they confronted the accumulation of chronic cases in their institutions. Increasingly, experts on insanity stressed that it was primarily somatic in origin, and that psychological therapies were ineffective.[44] In South Carolina, too, the retreat from moral therapy was an adjustment to the changing nature of the hospital population. But the change in South Carolina was perhaps more sudden than in many other state hospitals because the patient population changed its nature so rapidly and grew so much faster than the facilities or staff. With the violent, suicidal, epileptics, idiots, criminally insane, feeble-minded and senile of all ages mixed through overcrowded wards and segregated by race, a systematic programme of moral therapy became virtually impossible. Traditional amusements such as bowling, billiards, and carriage rides were sacrificed to the demands of retrenchment. In 1909, a legislative investigation claimed that the proportion of the hospital's patients who participated in occupation and amusement was far below the average for American hospitals. Most of the hospital's patients spent their lives sitting in their rooms or wandering about the wards and grounds.[45]

The nature and purpose of employment also changed during the later nineteenth century. It became more economic than therapeutic in nature and also became differentiated by race. After the Civil War, the asylum's officers increasingly tried to cut costs by having patients do much of the institution's cleaning, repair and general maintenance. Patients produced and mended their own clothing, grew their own food, made bricks, dug foundations, and helped erect buildings. As the number of black patients grew, black women patients gradually took over most of the cleaning, laundry and kitchen work, and black

[42] *AR*, 1892–3, 9–10, 1895–6, 8.

[43] *AR*, 1870–7, 1902–13, patient statistics; *Report of Legislative Committee on State Hospital*, 47–8.

[44] Dain, *Concepts of Insanity*, 129–39, 205–6; Grob, *Mental Institutions in America*, 202–4; Rothman, *Discovery of the Asylum*, 265–87.

[45] MBR, 3 January 1878; AR, 1892–93, 17; Report of Legislative Committee on State Hospital, 23, 31–5; Report and Proceedings of the Special Legislative Committee, 363; Testimony, Legislative Committee, 88–9, 160, 303–5.

Table 7.4 *Percentage of deaths of patients under treatment, at the South Carolina State Hospital, at five year intervals, 1890–1915*

Year	All patients	All black patients	All white patients
1890	14	21	9
1895	13	17	9
1900	16	23	10
1905	13	17	9
1910	14	21	8
1915	18	27	11

Note. Averages are rounded to the nearest number.
Source. Annual Report, 1889–90, 1894–5, 1900, 1905, 1910, 1915.

men were assigned most of the menial work outdoors. The investigation of 1909 estimated that 40 per cent of the black patients were employed, but only 24 per cent of the white women and 16 per cent of the white men. When it came to amusement, however, the positions of the races was reversed. The hospital gave white patients cards, a weekly dance, the use of a gramophone, and occasional other entertainments. For black patients, the hospital provided no amusement.[46]

The use of mechanical restraint and seclusion also increased around the turn of the century. The case histories of the period indicate routine use of restraint.[47] Between 1891 and 1914, the hospital probably used mechanical restraint more than at any time since the 1830s, and more than many contemporary American mental institutions. According to the investigating committee of 1909, the average proportion of patients under restraint at any time in American hospitals for the insane was 1 per cent. The committee estimated the proportion at the South Carolina State Hospital at 7 per cent overall, and 10 per cent in the white men's department.[48]

The worst consequence of the deteriorating conditions at the hospital was high mortality rates, especially for black patients. Between 1890 and 1915 mortality averaged about 14 per cent of the patients under treatment. As one can see from Table 7.4, black mortality was more than double that of white patients. These mortality rates were much worse than those of similar institutions in the nation and region. In 1904, a special census report revealed that the mortality rate at

[46] *AR*, 1877–94; MBR, 2 May 1878, 6 February, 3 April 1879, 14 July 1887, 1 August 1908; *Report of Legislative Committee on State Hospital*, 31–5; *Testimony, Legislative Committee*, 61–3, 72–3, 80, 88–90, 132, 158–9, 166–7, 259–61, 279–82, 302–5, 408, 414, 430; *Report and Proceedings of the Special Legislative Committee*, 509; Grob, *Mental Illness and American Society*, 23–4.

[47] SCSH, Case Histories, 14–30; DMH, SCDAH.

[48] Report of Legislative Committee on State Hospital, 40–2, 64–5; Testimony, Legislative Committee, 406–7; Grob, *Mental Illness in America*, 17–19.

the South Carolina State Hospital was more than double the national average, and almost double the average for the South Atlantic states. The black death rate was double the regional average and more than double the national average for blacks in mental hospitals.[49]

The above-average mortality was not entirely the result of substandard hospital conditions. The extreme poverty of the state's population contributed to it. Many patients came to the hospital with severe physical illnesses, some on the verge of death. Nor can the higher black mortality be blamed solely on the hospital environment. Blacks in the general population had much higher mortality and morbidity rates than whites.[50] Yet the hospital's deficiencies undoubtedly increased its mortality rates. A diet heavy in corn meal, molasses and fatback probably contributed to deaths from pellagra, a niacin-deficiency disease whose symptoms include skin lesions, diarrhoea and mental derangement, including psychosis. Pellagra was first diagnosed at the hospital in 1907 and its prevalence may help account for the rapid increase in poor and black patients admitted to the hospital around the turn of the century. The physicians attributed more than 1,100 deaths to pellagra between 1908 and 1914. Overcrowding helped spread contagious diseases such as tuberculosis, the leading cause of death in the hospital around the turn of the century. By 1909, the hospital housed 1,500 patients in buildings designed for 1,000. Many patients slept in corridors or in dark, poorly ventilated basement rooms which the chairman of the State Board of Health compared to dungeons. The original asylum building, congested with black women patients, had a mortality rate of over 34 per cent of the average number of inmates in 1908 and over 28 per cent during the previous five years. In the recently constructed and badly overcrowded building for black men the death rate averaged 27 per cent for the same period.[51]

A high ratio of patients to staff intensified the effects of the hospital's other deficiencies. The number of patients increased much faster than the number of physicians and attendants. The ratio of patients to physicians rose from 105:1 in 1878 to 376:1 in 1909. Over the same period, the ratio of patients to attendants rose from 10:1 to 18:1 in the departments for white women, white men and black women, and to 36:1 in the black men's department. On one of the black men's wards it was 55:1. These ratios were higher than in most other state hospitals in the nation and region. In 1894, the nine state hospitals of New York had an

[49] *AR*, 1890–1915; *Report of Legislative Committee on State Hospital*, 47–50; US Bureau of the Census, *Insane and Feeble-Minded in Hospitals and Institutions, 1904* (Washington, DC, 1906), table 37, 196–8.

[50] E. H. Beardsley, *A History of Neglect: Health Care for Blacks and Mill Workers in the Twentieth-Century South* (Knoxville, Tenn., 1987), 11–41; Newby, *Black Carolinians*, 114–21, 211–17.

[51] *AR*, 1907, 11, 1908–14, statistical tables; *Report of Legislative Committee on State Hospital*, 50–6; *Testimony, Legislative Committee*, 422–4, 432. John Hughes found conditions for blacks in the Alabama State Hospital similar to those in the South Carolina institution, although mortality rates were lower in the former. See Hughes, 'Black Mental Illness in Alabama', 450–5.

average patient-physician ratio of 171:1. In 1923, American mental hospitals averaged 234:1. Around the turn of the century, the ratio of patients to attendants in American mental institutions averaged about 12:1; in the south, it averaged 15:1.[52]

The patients at the South Carolina State Hospital were among the worst cared for in the nation in the early twentieth century. For most of them, especially the blacks, the hospital was unable to fulfil even the minimal expectations of the custodial role. The majority report of the investigating committee of 1909 charged that the hospital's 'custodianship is a menace to the health and life of these afflicted citizens'.[53] Yet it took another damning legislative investigation in 1914 to bring about improvement. The election that year of the Progressive Richard Manning as governor reflected a growing perception that the state had fallen embarrassingly behind the rest of the nation. Capitalizing on this view, Manning secured a series of reforms designed to modernize and improve the administration of public institutions. At the state hospital, mechanical restraint was eliminated, the use of drugs to sedate patients decreased, and opportunities for occupation and recreation increased. By 1920, the mortality rate had fallen to 10 per cent.[54]

Within a few years, practices at the state hospital came closer to those prevailing at similar institutions in other states. Yet the patients' care remained substandard in many respects. Overcrowding continued to be a serious problem. Although mortality rates for patients of both races continued to fall, blacks continued to die at significantly higher rates than whites.[55] Manning's reforms may have brought the hospital closer to the mainstream of American psychiatry, but the founders' goals – a curative asylum for all of South Carolina's insane – remained elusive. Since the Civil War, the institution had experienced a marked deterioration that was difficult to overcome. To a large extent, this development resulted from the state's steep economic decline and destructive political struggles. But it also reflected the fiscal, social, and racial outlook of the white establishment, combined with radical changes in the institution's clientele and function. The result was that an institution established as a house

[52] *AR*, 1876–7, 21, 1877–8, 1891, 1909, 1911, statistical tables; MBR, 5 July 1877, 11 June 1891, 15 June 1893, 18 May 1909; *Report of Legislative Committee on State Hospital*, 24; *Testimony, Legislative Committee*, 423; US Bureau of the Census, *Patients in Hospitals for Mental Diseases in 1923* (Washington, DC, 1926), 240; Grob, *Mental Illness in American Society*, 19.

[53] *Report of Legislative Committee on State Hospital*, 50.

[54] R. M. Burts, *Richard Irvine Manning* (Columbia, SC, 1974), esp. 87–91, 114–15; Correspondence Relating to State Hospital, 1915–18, Governor Richard I. Manning III Papers, SCDAH; AR, 1915, 4–7, 15–27, 1920, 8; Journal of the South Carolina House of Representatives, 1916, 80–4.

[55] A. P. Herring, *Report to the Hon. Richard I. Manning Governor of South Carolina, on the State Hospital for the Insane* (Columbia, SC, 1915), 9; *AR*, 1917, 6, 1921, 11, 19, 1922, 8, 16, 1925, 4, 17, 1926, 6, 18–19, 1930, 8–9; *Annual Report of the State Board of Public Welfare*, 1920, 61, 1921, 53, 1922, 12, 1923, 71–86.

of cure was unable to provide even basic custodial care to patients increasingly marginalized by chronic disease, poverty and race. The history of the South Carolina Lunatic Asylum and State Hospital illustrates how the broader political, economic, social and racial context could profoundly influence the internal workings of mental institutions, and effectively undermine the aims of their founders.

8 The state, family, and the insane in Japan, 1900–1945

Akihito Suzuki

A brief look at the history of psychiatric confinement in Japan from the Meiji Restoration (1868) to the Japanese experience of the Second World War (1941–5) must give a sense of déjà-vu to those familiar with its European counterparts in the nineteenth century. A *cause célèbre* of wrongful confinement led modern Japan to the Mental Patients' Custody Act (1900), its first national legislation for regulating the confinement of lunatics. In 1919, the effort of a few eminent psychiatrists, as well as the initiative of health officials at the central government, led to the Mental Hospitals Act (1919), which promoted hospital-based provision for the insane. Under these two acts, psychiatric provision in pre-war Japan expanded rapidly in the first four decades of the twentieth century, just like its empire in the Far East. Especially when compared with the situation in England in the late eighteenth and early nineteenth century, one is struck by the similarities. When the two countries started to confine the insane on a large scale, with an interval of about one century, they were both in the turmoil of industrialization, which perhaps acted as a kind of predisposing condition to the rise of asylum. Moreover, England and Japan shared three important factors in their creation of asylum-based psychiatric provision: the impetus given by exposé of the abuse of psychiatric confinement, the initiative taken by the central government, and the establishment of a psychiatric profession.[1] The rise of journalism and 'public opinion', the re-definition of the relationship between the central and local governments, and the arrival of a professional society, all left their stamp on the making of asylumdom both in

I would like to thank Dr Kazushige Komine, who has kindly allowed me to consult the archive of Oji Brain Hospital, and Ms Kaoriko Yokozawa, whose efficient assistance has been vital to the conduct of this research. I should also like to thank Professor Ken'ichi Tomobe and Dr Takeshi Nagashima at Keio University, whose encouragement and comments have been invaluable. Parts of this chapter were read at a Seminar for the Research Group of History of Psychiatry and the International Workshop for the History of Psychiatric Hospitals, held at Keio University in December 2000. I should like to extend my thanks to those who attended the seminar and the workshop, particularly to Dr Yasuo Okada. The research for this chapter has been generously funded by the School of Economics of Keio University.

[1] For the situation of psychiatric provision in England in the nineteenth century, A. Scull, *The Most Solitary of Afflictions: Madness and Society in Britain, 1700–1900* (New Haven, Conn., 1993) remains the most comprehensive work.

Japan and England. Such similarities, obvious at a most cursory glance at two randomly selected countries, entice historians to seriously engage in in-depth and comparative socio-cultural studies in the history of psychiatry, one of the principal aims of this volume.

One can as easily spot major disparities between psychiatric provision in pre-war Japan and other European and North American countries. One of the most important differences is the sheer size of institutionalized population. Pre-war Japan confined only a fraction of the mentally disordered within the walls of asylums. In 1919, on the eve of the Mental Hospitals Act, there were only about 3,000 patients confined in mental hospitals, which is a remarkably small number for a nation of the population of about 55 million. England and Wales in the same year had about 35 million population and more than 100,000 asylum inmates.[2] The 1919 Act rapidly increased the number of those confined, but in 1940 the figure reached just around 22,000, still 'lagging far behind' countries in the West. The Second World War and post-war upheavals subsequently paralysed Japanese mental health care. The National Institute of Mental Hygiene reported in 1953 that at that time there were in Japan 18,527 psychiatric beds or 22.6 per 100,000; the corresponding figures for England was 313 beds per 100,000, 278 for USA, and 497 for New Zealand. Even Italy, another defeated power, had 134 psychiatric beds per 100,000.[3] Pre-war Japan thus did not witness the full-bloom 'great confinement', which was a common feature in many western countries discussed in this volume.

The restricted growth of asylum accommodation has been mainly explained by Japan being a 'latecomer'. From the early twentieth century through to the present, there have been abundant discourses that criticize the 'inadequacy' of Japanese psychiatric provision and relate it to the hidden 'backwardness' of a state with impressive military or economic prowess. There is considerable historical truth and moral wisdom in this line of interpretation. This view, however, masks some crucial issues, which have been the subject of intense debate in the recent historiography of psychiatry. Most importantly, criticizing psychiatric provision in pre-war Japan for 'backwardness' tacitly assumes that hospitalization of the insane was an *inevitable* or *natural* step at a certain stage of the evolution of psychiatric social policy. One of the major achievements in the history of psychiatry in the last twenty or thirty years is to have demonstrated that the 'great confinement' was a product of intense conflict, political manoeuvres, and specific historical forces, being far from a pre-destined social phenomenon.[4]

[2] Scull, *The Most Solitary of Afflictions*, 334–74.

[3] National Institute of Mental Hygiene, *Seishin Eisei Shiryo* [Sources for Research in Mental Hygiene], 1 (1953), 21.

[4] Literature on this subject is now too vast to be listed here. For a useful and insightful overview of the latest scholarship, see J. Melling, 'Introduction', in B. Forsythe and J. Melling (eds.), *Insanity, Institutions and Society: A Social History of Madness in Comparative Perspective* (London, 1999).

Without either reducing pre-war Japanese psychiatric provision to a botched attempt at mimicking western practice nor romanticizing it as an 'alternative' to western-style modernity, this chapter will try to analyse its pattern, content and shortcomings within its own context and parameters.

Particular emphasis will be laid on the role of the family. As has been emphasized by Patricia Prestwich in the French context and David Wright in the English one, the major actor in decision-making in psychiatric committals was normally the family.[5] In many cases, the family decided what should be done to control the troublesome and troubled family member; whether he or she should remain at home and under family management, or whether he or she should be sent to an institution to become its inmate, and if so, which type of institution should be utilized. The history of the family and its internal dynamics, its solidarity, and its codes of behaviour for its members, are thus a crucial part of the history of psychiatric institution in many countries in Europe. Moreover, many recent studies in the social history of psychiatry have revealed that the role of the family in psychiatric committal and institutionalization is best understood in the context of the family's interaction with its neighbours, its relationship with the local and the central governments, and its negotiation with the asylum doctors it consulted.[6] This chapter attempts to bring this family-centred contexualist model into the relatively uncharted landscape of Japanese psychiatric provision in the earlier half of the twentieth century.

The major sources this chapter uses are twofold. The first is a series of *Annual Reports of the Department of Hygiene* (*ARH*) which was issued annually by the Central Sanitary Bureau (CSB) in the Ministry of Home Affairs and, after 1938, by the newly created Ministry of Health and Welfare. Although the data published in *ARH* constitute the most basic material for any historian interested in medicine, disease and hygiene in modern Japan, this indispensable set of sources has not been used extensively by historians of Japanese psychiatry.[7] From 1905, *ARH* included detailed statistical tables of mental patients, and from 1928, a list of all mental hospitals and the numbers of patients kept in each hospital, with annual admissions, discharges and deaths. My statistical account below is drawn mainly from compiling the tables published in *ARH*.

[5] D. Wright, 'Getting Out of the Asylum: Understanding the Confinement of the Insane in the Nineteenth Century', *Social History of Medicine* 10 (1997), 137–55; P. Prestwich, 'Family Strategies and Medical Power: "Voluntary" Committal in a Parisian Asylum, 1876–1914', *Journal of Social History* 27 (1994), 799–818. See also chapters by Wright, Moran and Gouglas, and Prestwich, in this volume.

[6] Two recent collections of essays that emphasize this direction are: Forsythe and Melling (eds.), *Insanity, Institutions and Society*; P. Bartlett and D. Wright (eds.), *Outside the Walls of the Asylum: The History of Care in Community 1750–2000* (London, 1999).

[7] *Annual Reports of the Department of Hygiene* for the period 1877–1926 have been reprinted as *Meiji-ki Eisei-kyoku Nenpou* [Annual Report of the Department of Hygiene in Meiji Era] (Tokyo, 1992) and *Taisho-ki Eisei-kyoku Nenpou* [Annual Report of the Department of Hygiene in Taisho Era] (Tokyo, 1993).

These aggregate data are supplemented by statistical analysis of admission registers and individual patients' case records at Oji Brain Hospital (OBH). Established in 1901 in Takinogawa in the northeast suburb of Tokyo, OBH was one of the first private psychiatric hospitals in Tokyo. Its extremely detailed, voluminous and uncatalogued case records from around 1920 to 1945 have survived intact, with other archival materials, which are now held by Komine Research Institute for History of Psychiatry run by Dr Kazushige Komine, a practising psychiatrist and historian of psychiatry based in Tokyo. From this formidable archive, I have been able to examine systematically admission registers between 1925 and 1945 and about ninety case records of the patients discharged in 1935.

As well as these hitherto unexamined materials, this chapter relies heavily on works of two outstanding Japanese scholars. The first is Yasuo Okada, whose voluminous works based on painstaking research into the history of psychiatry in Japan from the late nineteenth century onwards, deserve to be known beyond the small circle of Japanese historians of psychiatry. His history of Matsuzawa Hospital is a 650-page *magnum opus*, that (literally) chronicles the most important psychiatric institution in Japan. It will perhaps remain unsurpassed in its exemplary care about historical details and firm grasp of day-to-day activities within the hospital, where Okada himself worked in his youth as a medical doctor.[8] Okada has also excavated numerous important medical articles, newspaper reports and archival materials and has made them available through reprints with invaluable comments.[9] Less voluminous but historiographically more sophisticated are works by Genshiro Hiruta (who is also a practising psychiatrist) whose *Hayari-yamai To Kitsune-tsuki* [Epidemics and Fox-Possession] is an in-depth study of medical and psychiatric history in small villages in northern Japan in the *longue-durée* from 1703 to 1867.[10] Although Hiruta's work examined a very different area and a much earlier age than my own, his insightful arguments turned out to be a constant source of inspiration for this chapter.

The account below is divided into three parts. The first section will briefly examine the two laws which structured psychiatric provision in Japan before the Second World War, namely the Mental Patients' Custody Act (1900) and Mental Hospitals Act (1919). Emphasis will be laid on both the continuity

[8] Y. Okada, *Shisetsu Matsuzawa Byoin-shi: 1879–1980* [Matsuzawa Hospital: A Private History, 1879–1980] (Tokyo, 1981). Other major works of Okada include two biographies of pioneering psychiatrists in Japan: *Kure Shuzo: Sono Shougai To Gyouseki* [The Life and Works of Kure Shuzo] (Kyoto, 1982); *Saito Mokichi No Shougai* [The Life of Saito Mokichi as a Psychiatrist] (Kyoto, 2000).

[9] The most important collection of materials is Y. Okada and S. Sakai (eds.), *Kindai Shomin Seikatsu-shi*, vol. XX, *Byoki / Eisei* [Social History of the Lives of Populace in the Modern Age, vol. XX, Disease and Hygiene] (Tokyo, 1995). Hereafter this work is referred to as *SHDH*.

[10] G. Hiruta, *Hayari-yamai To Kitsune-tsuki* [Epidemics and Fox-Possession] (Tokyo, 1985).

from earlier times and the innovative nature of this legislation. The second part will investigate the statistical profiles of mental patients and hospitals during 1905 to 1941, and examine the rise and fall of various styles of care of the insane. Particular attention will be paid to care and confinement at home, both because official sanctioning of confinement at the patient's own home is a highly idiosyncratic practice if seen from the viewpoint of modern European practice, and because this home custody provided an important prototype for management of the insane practised in other *loci* of care. The third section will cast a brief look at the early history of OBH and its patients, in order to examine some issues discussed in the previous two sections in the context of one specific institution. The limit of the space only allows me to address a few points from this rich archive, and I will be very selective in my choice of the insights drawn from the intimate life stories of madness at mental hospitals in Japan in the early twentieth century.

Home or hospital?: Mental Patients' Custody Act (1900) and Mental Hospitals Act (1919)

Like early modern European countries, the apparently sole concern of public authorities over lunatics in Japan in earlier times was to prevent the harm done by the insane in the administrative context and to regulate penalties for crimes done under reduced mental capacity in the forensic context.[11] Particularly prominent was the fear of lunatics at large, wandering on the city street and the country field, committing or threatening violence on the person and property of others, setting fire to buildings, and so on. Hiruta's study has revealed that lunatics (who were sometimes, but not very often, regarded as possessed by a fox or other animal) in small northern villages in the Edo period (1603–1867) were put in *sashiko*, or a makeshift cage set up next to their own house. Especially furious ones were further handcuffed or chained in their cages.[12] In Edo, recourse was taken to the prisons and poorhouses for keeping vagrants and the sick poor, as well as putting the patient in the cage at his or her own house.[13] Lunatics were seen and treated in the framework of their threat of violence against themselves and others, and confinement and bodily restraint were the order of the day. In short, they posed a problem to the police, which is a familiar background to understanding insanity in many societies.

[11] For an overview of the concept of medical and legal insanity in earlier Japan, see G. Hiruta, 'Nihon No Seishin-iryo-shi' [History of Japanese Psychiatry], in M. Matsushita and G. Hiruta (eds.), *Seishin-iryo No Rekishi* [History of Psychiatric Practice] (Tokyo, 1999), 35–64.

[12] Hiruta, *Hayari-yamai To Kitsune-tsuki*, 56–126.

[13] H. Kuwahara and K. Itahara, 'Edo-jidai Kouki Ni Okeru Seishin Shougaisha No Shogu, [Management of the Mentally Handicapped in Late Edo Period], parts I–V, *Shakai Mondai Kenkyu* [Studies in Social Problems], 48 (1998–9), 41–59; 49 (1999–2000), 93–111 and 183–200; 50 (2000–1), 79–94 and no. 2, 1–45.

The Meiji Restoration in 1868 put an end to the semi-feudal rule of the Tokugawa Shogunate (1603–1867) and restored the emperor as the head of the centralized state modelled after western countries (particularly Germany). The restoration did not immediately change policies towards the insane. The emphasis on the necessity of controlling the violence of lunatics persisted, with strong associations of insanity with ferocity of animals on the loose. Early police rules of the city of Tokyo soon after the Meiji Restoration put the rules for the regulation of lunatics next to those about unrestrained and dangerous animals on the street, such as oxen, horses and mad dogs.[14] Through successive legislation between 1878 and 1884, the basic pattern for administrative control over dangerous lunatics was completed in Tokyo. When necessary, lunatics were ordered to be kept in custody either in their own house or a hospital, and the local police visited the place of their abode once a month. Although details remain to be investigated, evidence suggests that similar rules were established by local governments in other areas.[15]

Perhaps the most important element in this concern of the policing of the lunatics at large was that of the responsibility of the family for the management of its disorderly member. Although the evidence is patchy, Hiruta's examples from small villages in the Edo period and Okada's evidence from the already highly modernized and then rapidly westernized metropolis concurred on this point of the family's duty to be vigilant over their dangerous insane member. In one of the villages studied by Hiruta, when a lunatic son escaped from his father's house and killed two villagers and himself, the father was punished severely for the neglect of his duty. Half of the father's property was confiscated and he was expelled from the village.[16] The 1882 Old Criminal Code fined between 50 sen and 1 yen 50 sen 'those who failed to perform the duty of the custody of the mad and let them wander on the street'. As Okada has rightly pointed out, the lunacy problem in late nineteenth-century Japan was thus characterized by the convergence of two elements – the police and the family.[17] Perhaps the most striking element is the ease with which a prison-like facility was created in a private house with the sanction, and at the instigation, of public authority.

Predictably enough, the tradition of incarceration at home made the situation open to a type of abuse familiar in the history of psychiatry, namely wrongful confinement, or shutting up a person on the false pretence that he or she was

[14] Y. Okada, 'Seishin Eisei Hou' [Mental Hygiene Act] in *Gendai Seishin Igaku Taikei* [An Outline of Modern Psychiatry], vol. v-c (Tokyo, 1977), 351–97, 353–4.

[15] See the speech of Toshio Saito in the House of Commons in the Imperial Diet, *Teikoku Gikai Shugi-in Giji Sokkiroku* [Parliamentary Debates in the House of Commons] (Tokyo, 1979–85), 19 February 1900.

[16] Hiruta, *Hayari-yamai To Kitsune-tsuki*, 112–13.

[17] Okada, 'Seishin Eisei Hou', 353–5.

insane.[18] In 1885, the nation was shocked by the Soma case, in which a former feudal lord ('*Daimyo*') was confined under dubious pretence. The Soma case revealed the glaring defect of the system and exemplified how easily one could be confined illegally. A call for new legislation quickly gained momentum. In 1898 a governmental committee was appointed to study psychiatric laws in western countries, particularly those of England, and one high government official with a medical background agitated for a major reframing of the regulations of the insane in the light of trends in western medicine.[19] Another motive for the new legislation was to demonstrate legal maturity of Japan towards the western countries, which had still maintained a colonialist tariff policy and allowed only semi-independent status to Japan on the justification that the Japanese legal system was not modern enough for it to be granted an autonomous status in the international community. These concerns culminated in the Mental Patients' Custody Act in 1900.

This first piece of national legislation regulating the confinement of the insane aimed both at the prevention of wrongful confinement and the secure custody of lunatics. For the former purpose, the Act took a straightforward but fresh approach, by newly criminalizing unjust or improper confinement, and set heavy fines and penalties on those who detained a sane person or improperly confined an insane person, and on those doctors who issued an improper certificate of lunacy.[20] In contrast, in order to achieve the second aim, the Act not so much created something new as codified the old practice. In essence, the Act demanded that if one wanted to have a lunatic confined, one should do so by appointing a 'custodian', who was responsible for the provision, care and confinement of the patient. Only the custodian was allowed to confine the lunatic, and he or she could do so only with the permission of the authority of the local government of city, town or village. When a competent custodian could not be found, the administrative head of the city, town or village in which the patient lived, would assume the status of the custodian. The place of confinement should be licensed by the administrative head and should meet special requirements for the safe custody of the patients: normally, the place was either the custodian's house or a mental hospital or mental ward of a general

[18] From the numerous studies of wrongful confinement in England, see particularly R. Porter, *The Social History of Madness* (London, 1987); M. Clark, 'Law, Liberty and Psychiatry in Victorian Britain; an Historical Survey and Commentary, c. 1840 – c. 1890', in L. de Goei and J. Vijselaar (eds.), *Proceedings of the First European Congress on the History of Psychiatry and Mental Health Care* (Rotterdam, 1993), 187–93; P. McCandless, 'Liberty and Lunacy: The Victorians and Wrongful Confinement', in A. Scull (ed.), *Madhouses, Mad-Doctors, and Madmen: the Social History of Psychiatry in the Victorian Era* (London, 1981), 339–62.

[19] Okada, 'Seishin Eisei Hou', 354–5.

[20] *Teikoku Gikai Kizoku-in Giji Sokkiroku* [Parliamentary Debates in the House of Lords] (Tokyo, 1979–1985), no.12, 20 January 1900 and no. 21, 10 February 1900.

hospital. If the custodian wanted to confine the patient at his or her own house, he should do so first by seeking permission from the local administrative head and making a petition which included a detailed plan of the place and cage. Local police and/or doctors were ordered to inspect the place of confinement 'as often as necessity arises' (how often varied from place to place). To ensure that the patient should not escape and do harm to others, he or she kept under custody at the custodian's house was put in a cage set up there. Perhaps both to allow light and air to the place and to facilitate vigilance over the confined, a latticework, with a window to serve food for the patient, seems to have been a norm. This meant extremely high visibility of the patient in confinement, and those now in their sixties or seventies still retain vivid memories of chilling horror and dark fascination when they saw a furious patient through a lattice cage.

Despite the rhetoric of modernization and protection of human rights surrounding this piece of legislation, a core part of the Mental Patients' Custody Act was a national and legal confirmation of a long-standing local semi-customary practice. First, the custody was done in a small social and administrative unit, exclusively the business of the family and/or the local authority (city, town and village). It determined who should pay the cost for confining the patient, and whether in a hospital or at a private house. In contrast, the involvement of the central or intermediate local government ('*fuken*' or prefecture) in the implementation of the Act was nil. Secondly, the pattern of the family asking the local authority for permission to confine the patient in their own house had a long tradition. As Hiruta's work has demonstrated, since the Edo period, putting a lunatic in a *sashiko* was by no means a purely private business, and there existed a fairly strict procedure: those who wanted to set up a *sashiko* had to first ask the permission of the village authority who, at the petition, examined the patient and the place of confinement and referred the case to the legal court of the *han* (a semi-independent feudal state) for the final approval.[21] *Mutatis mutandis*, this procedure was exactly what the Mental Patients' Custody Act decreed. Particularly important here is the question over who should be a custodian of the patient. Although the 'legal guardian' was named at the top of the list of possible custodians, Earl Ogimachi's explanation to the House of Lords in 1900 reveals that this was made merely for legal cosmetic purposes, in order to make this law consistent with other civil codes. The government conceived of the business of custody primarily as the private discretion of the head of the household over matters within the household, rather than that of a guardian, who could hail from outside the household and whose power was

[21] Hiruta, *Hayari-yamai To Kitsune-tsuki*, 107–11. Cases of procedure in a similar spirit are to be found also in Edo. See Yasuo Okada, 'Edo-ki No Seishin-ka-iryo' [Psychiatry in the Edo Period], in Masa-aki Matsushita and Genshoiro Hiruta (eds.) *Rinsho Seishin Igaku Kouza* [*Encyclopedia of Clinical Psychiatry*] Supplement 1 (Tokyo, 1999), 232–3; Kuwahara and Itahara, 'Edo-jidai Kouki Ni Okeru Seishin Shougaisha No Shogu'.

based upon contract.[22] The Custody Act thus officially sanctioned the 'natural' power and responsibility of the head of household over his family members, as well as nationally codifying the long-standing local and customary rules of the domestic confinement of lunatics under public control. An important aspect of the Mental Patients' Custody Act is that it did not aim for the encouragement or enforcement of psychiatric confinement. Actually, the Act made confinement more difficult, both by penalizing improper detention and by setting a standard for home custody, the cost for which (alteration to the building to meet the standard, latticework, lock and so on) was to be met by the custodian. The goal of the Act was the *regulation* of confinement, and to make it both legal and strict. The promotion or numerical increase of confinement was neither the stated aim nor the likely effect of the Act, as I shall discuss in detail below.

The expansion of confinement in hospital was exactly what the Mental Hospitals Act attempted. Again, several factors seem to have converged to effect its creation. The most prominent was the glaring fact that psychiatric provision was badly in short supply. In the House of Commons, Takejirou Tokonami, then Minister of Home Affairs, repeatedly cited the figure that only 4,000 out of 60,000 mental patients were confined in hospitals, and emphasized that this rate of confinement was far below the western standard and a 'national shame'. Another reason frequently raised at the parliament was the wretched situation of some of the patients under home custody, revealed by a massive four-part report written by the team of Shuzo Kure and Goro Kashida, the former being the leading figure in psychiatry in Japan: Professor of Psychiatry at Tokyo Imperial University and the head of the Tokyo Metropolitan Hospital at Matsuzawa (hereafter Matsuzawa Hospital), then the only public asylum in Japan. The report was far from a work of sensationalistic journalism, but was published in a leading medical journal at that time, based on a painstaking, detailed and rigorous survey of about 400 cases of home custody, conducted intermittently from 1910 to 1914.[23] I shall come back to this enormous piece of work below, but suffice it to say that Kure's condemnation of cases of home custody was no doubt a part of the almost universal strategy of psychiatrists to medicalize the realm of the care of the insane and exclude non-medical, lay or unqualified practitioners. The fact that the report included a critical and condemnatory survey of the practice of religious and folk healing of insanity betrays the ultimate motive of the authors of the report. Although their condemnation of the

[22] *Teikoku Gikai Kizoku-in Giji Sokkiroku*, 10 February 1900.

[23] S. Kure and G. Kashida, 'Seishin Byosha Sitaku Kanchi: Jikkyo Oyobi Sono Toukei-Teki Kansatsu' [Home Custody of Mental Patients: Its Situations and Its Statistical Observations], *Tokyo Igaku-kai Zasshi* [Journal of the Medical Society of Tokyo], 32 (1918), cases 521–56, 609–49, 693–720, 762–806. A work in a similar vein is S. Ishikawa, 'Seishin-byousha No Kanchi Ni Tsuite' [On the Custody of Mental Patients], *Kokka Igaku-kai Zasshi* [Journal of State Medicine] 236 (1906), 779–90. Kure and Kashida's work has been reprinted and published by Seishin-igaku Shinkeigaku Koten Kankou-kai in 1973 and re-issued in 2000.

practice of confining the patient in a private house was sincere, and almost certainly motivated by genuine concern over the plight of the patients confined in a cage and exposed to the gaze of neighbours, emphasizing the horror of home custody was, at the same time, a convenient lever towards creating hospitals, which meant stable and prestigious jobs for psychiatrists and the enhancement of their role in the medical machinery of the state.[24]

Another push towards hospitalization came from a renewed and revitalized fear of the danger posed by unconfined patients. This time, however, the image of the dangerous lunatic was not the traditional one of the wild animal, but rather that of criminal monomaniacs. The fear became more intense, because these monomaniacs were apparently normal except for one single issue and thus difficult to spot, unlike the all-too-obvious savageries of classic maniacs. At the House of Commons in 1918, Kiichi Saito, an MP and the founder of Aoyama Brain Hospital, delivered a long speech accusing the government of leaving numerous dangerous lunatics at large.[25] Amid the jeering of 'Shorter! Shorter!' and 'Can't see what you mean!', Saito conjured up the dark threat posed by homicidal monomaniacs, arson-monomaniacs, theft-monomaniacs and rape-monomaniacs, the number of which were all allegedly increasing. Although the government's explanation of the purpose of the new Bill was less hysterical than Saito's panic-mongering, it nevertheless frequently referred to a handful of criminal lunatics (estimated at about 150) and talked about a plan of erecting a national hospital for confining dangerous monomaniacs.[26] The scare raised by the supposedly rapid increase of dangerous monomaniacs who lurked on the street played perhaps an important role in passing the Bill in 1919.

With these concerns as the major driving forces, the Mental Hospitals Act (1919) was conceived in a very different spirit from the Custody Act. While the earlier Custody Act was centred around the prevention of wrongful confinement and the regulation of psychiatric custody, the major aim of the new Act was the *expansion* of hospital-based public provision for mental patients. To achieve this goal, the Act empowered the Minister of Home Affairs to order the prefectures to build public asylums in which poor patients were to be kept, and that half of the cost for building the hospital and one-sixth of the cost for maintaining the patients would be covered by the central government. From the viewpoint of both central and prefectural governments, however, it must have been deemed unrealistic to expect speedy completion of a nationwide system of hospital-based provision based only on purely public resources, for there existed only

[24] The classic studies of the history of psychiatry from this perspective in nineteenth-century Britain and France are, respectively, Scull, *The Most Solitary of Afflictions* and J. Goldstein, *Console and Classify: The French Psychiatric Profession in the Nineteenth Century* (Cambridge, 1987).

[25] *Teikoku Gikai Shugi-in Giji Sokkiroku*, 6 March 1918, a question entitled 'Why Does the Government Leave Numerous Mental Patients at Large When They Disturb the Public Order?'

[26] *Ibid.*, 23 February 1919.

one public asylum, which was in Tokyo and housed about 450 patients in 1918. In contrast, in the same year there already existed fifty-seven private psychiatric hospitals, with total capacity for about 4,000 patients.[27] Most crucially, many of the private mental hospitals admitted patients whose cost for staying at the hospital was paid by their local authority, either through the Mental Patients' Custody Act or otherwise. From the viewpoint of private psychiatric hospitals, keeping public patients brought the benefit of stable income from long-term stay, while from that of public authority, they provided a place to confine dangerously insane patients whose family could not take sufficient care of them.[28] In short, there already existed a large mixed sector in psychiatric provision. The Mental Hospitals Act codified this practice of confining patients in privately run asylums at public cost. Some private asylums were allotted a certain number of 'substitute' ('*daiyo*') beds, and were arranged to accept public patients up to that number. Private mental hospitals thus appointed were called 'substitute hospitals', which were to become the major provider of the care for the insane in the next couple of decades. The manifest goal of all these procedures was to *expand* hospitalization, an aim which was virtually absent in the Mental Patients' Custody Act.

Another important departure from the previous Act was the role of the state, prefectures and the asylum doctor. As noted above, the Custody Act conceptualized the control of lunatics on a small social scale. The business was done by the family, relatives, neighbours and city, town or village. The Mental Hospitals Act put the care in a larger social frame, namely that of the central state and the prefecture. The cost of the provision for lunatics now would be met from the budget of the prefecture, with help from the central government as noted above. The head of prefecture now possessed enhanced powers to admit or discharge a patient to public and substitute mental hospitals.[29]

Yet another beneficiary of the new Act was the asylum doctor. Before, doctors had little power over committal and discharge, which were at the discretion of the custodian of the patient. Under the new Act, medical power increased considerably. The doctor was now able to admit or discharge the patient only with the sanction of the head of the prefecture. Having fended off the suggestion of a system of external inspection by members of the city council, the Act granted the asylum doctor tremendously increased power.[30] Likewise, the Japanese Association of Psychiatrists, a group established by Kure, acted as a

[27] Kure and Kashida, 'Seishin Byosha Sitaku Kanchi', case 524.

[28] The mutual dependence and benefit of the public and private asylums in Tokyo is satirically described in the serialized articles in *Yomiuri Shinbun* [Yomiuri News]. See *SHDH*, 183–223.

[29] For an account of the fierce debate at the Diet over this enormous discretion given to the head of prefecture, see M. Yuasa, 'Lectures on the Mental Hospitals Act', *Shinkei-gaku Zasshi* [Journal of Neurology] 19 (1920), 488–94 and 543–50.

[30] *Ibid.*

kind of professional consultant group for the law makers at the Central Sanitary Bureau and executor of the policy at the Metropolitan Police, and had a certain say both in legislation and administration of the Act.[31] Although the archive of the association reveals more humiliation and bitter compromise of the doctors vis-à-vis the high-ranked civil servants at CSB and the Metropolitan Police, the chance to negotiate with those who virtually ran Japan and Tokyo was obviously greeted by its members aspiring for a secure place in the machinery of the state.[32] The role and the power of psychiatrists in the new Act thus grew to a considerable extent, when compared with those defined in the Custody Act, which referred to doctors mainly in the context of punishing their misconduct. The new Act thus signalled the rise of the state, the prefecture, and the psychiatrist vis-à-vis the family and the local government.

Progress of confinement or persistence of domestic care? Statistical analysis of psychiatric patients

In order both to assess the impact of the two Acts and to detect the trends which prompted the legislation, as well as to describe the general picture of psychiatric patients in early twentieth-century Japan, I should like to turn to the presentation of statistical data and their analysis, compiled from tables in the *ARH*.

Any discussion of psychiatric provision should start with a basic question, namely, the number of patients. Figure 8.1 represents the numbers and rate (per 10,000) of the known mental patients from 1905 to 1941, published in *ARH*. The figures were calculated by health officials by adding up the numbers of three categories of patients, namely: (1) those who were found by a visiting police officer as insane but regarded as 'unnecessary to be confined'; (2) those who were confined (either at home or in institution) under the Mental Patients' Custody Act; and (3) from 1919, those who were put into mental hospitals under the Mental Hospitals Act.[33] During the period under consideration, both the actual number of the patients and the rate per population sharply increased, the former having more than trebled and the latter having doubled. This growth is rather difficult to interpret, like so many data in psychiatric epidemiology. Did the increase in figures or rates represent real increase in disease occurrence, lowered tolerance towards the behaviour of the insane, or more effective detection? Although many early twentieth-century Japanese commentators on

[31] K. Komine (ed.), *Nihon Seishin Byou-i Kyoukai Kiji* [Archives of the Japanese Association of Psychiatrists], (Tokyo, 1974).

[32] *Ibid.*, 1–45.

[33] In addition, a small number of patients (about 100–200) who were 'temporarily confined' were added to the figure. Also note that the figures in the tables published in *ARH* did not include those who were put in a place of confinement through a procedure based neither on the Mental Patients' Custody Act nor the Mental Hospitals Act. For a further discussion on this omission, see note 51 below.

psychiatric issues maintained that the occurrence of mental disease was growing in accordance with the modernization of Japan, the Central Sanitary Bureau was well aware that this type of argument did not hold true, even with a most cursory look at the evidence. In its first report on mental patients in 1905, the CSB pointed out that although major urban areas such as Tokyo or Kyoto showed relatively higher rates, the correlation between mental disease rate and the extent of modernization ended just there. Osaka, with the second largest city in Japan, had the second *lowest* rate, while rural areas such as Iwate, Saga and Okinawa had higher rates.[34] The pattern of regional variations in mental disease rate continued to baffle the simple scenario of the link between civilization and madness until the end of my period.

More promising is the line of argument that sees the growth in rate as a result of increased awareness of those troubled in mind. Without adhering to the simple scenario of linking civilization with intolerance toward madness, there exist several reasons to suppose that some part of the increase of mental disease during this period resulted from more effective detection of lunatics. Among the factors which might have contributed to making the insane more visible, the most powerful one is the public cult of emperor. The colossal ceremonies for the burial of the old emperor and the accession of a new one involved a massive policing of the entire population all over the nation and colonies. [35] In order to prevent any disruption of the sacrosanct sobriety of the ceremonies, an enormous number of local and special policemen were deployed to remove the slightest possibility of disorder. At the accession of Emperor Hirohito, according to a study by Yutaka Fujino, the particular target of scrutiny was the 'Other' in the Japanese society at that time – Koreans, socialists, the urban poor, those suffering from communicable diseases and the insane. To take an example of Hyogo, a round-up survey of *all* households for mental patients and other undesirable members of society was completed in ten days. Wandering madmen were sought during three days from 7.00 a.m. to 4.00 p.m., which led to the discovery of 229 new patients.[36] Also imperial rituals of a more modest size were accompanied by intensive search for mental patients. Indeed, in 1939 a doctor in Hiroshima tried to interpret the high patients' rate of the prefecture (twice as much as the national average) to frequent visits of the royal family to important shrines in the area.[37] The erection of the mausoleum of Emperor

[34] *ARH* (1905), 59–60.

[35] Central Sanitary Bureau, *Showa Tairei Eisei Kiroku* [The Hygienic Report at the Showa Grand Ceremony of Accession], (1929), reprinted in *Showa Tairei Kiroku Shiryo* [The Records of the Showa Grand Ceremony of Accession], introduction by H. Nishi, F. Ogino, and Y. Fujino (Tokyo, 1990).

[36] Y. Fujino, 'Showa Tairei To Minshu No Seikatsu To Kenko' [The Showa Grand Accession Ceremony and People's Health]', in *Showa Tairei Kiroku Shiryo*, 65–86.

[37] 'A Round Table for Mental Patients' Protection Day', *Wako* [Harmonious Light] 6 (1939), 91–108, 97. One police officer at the round table, however, refuted this argument by citing the low rate of Mie, another prefecture visited frequently by the royal family.

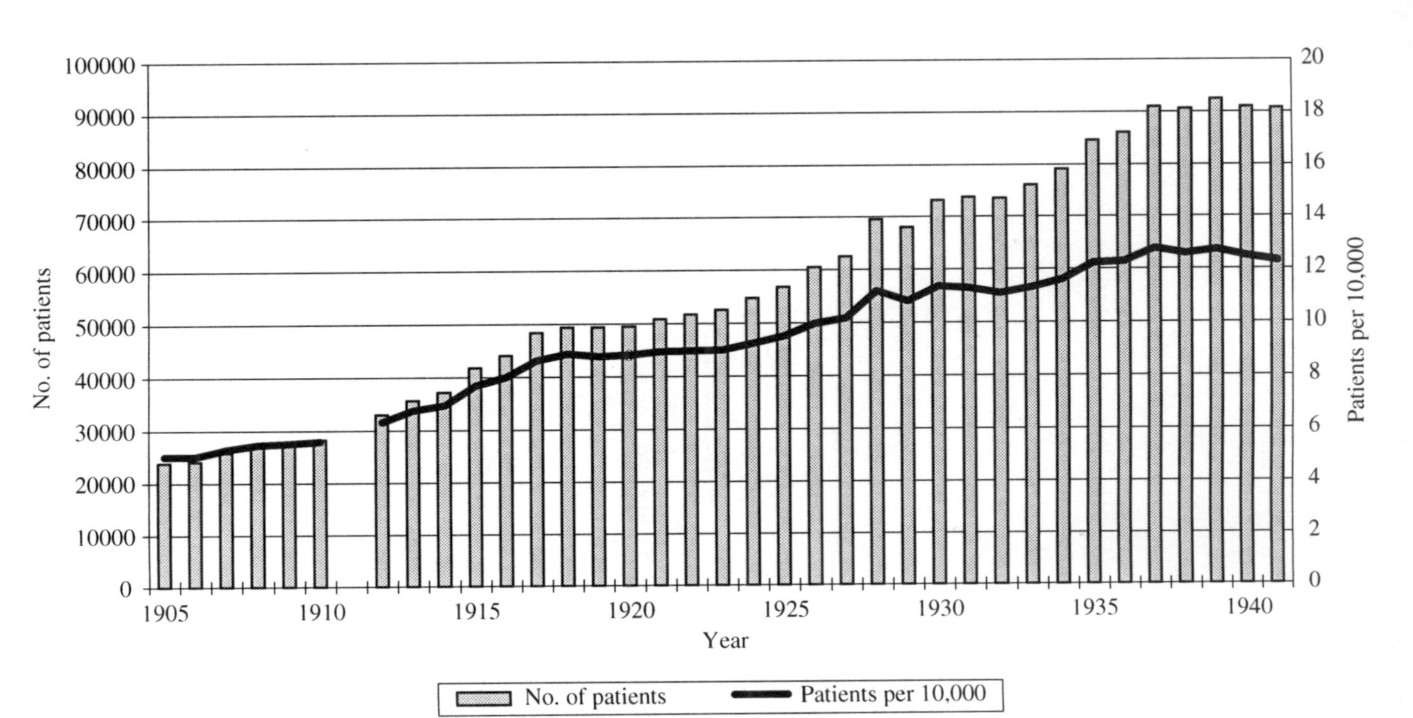

Figure 8.1 – Number of patients

Yoshihito in 1926 in Tama Hill in the west of Tokyo and the increased concern of the police to guard the sacred place, led to tighter regulation of patients staying for hydrotherapy at waterfalls in Takao, a mountainous region close to Tama.[38] An increasingly intense cult of the emperor in the early twentieth century, and the resultant intense scrutiny of the population must have contributed to the steady rise of the number and rate of the patients.

On closer examination, however, it turns out that the extent of the contribution of the search for the undesirable elements of the population was not very great. Figure 8.1 shows two periods of rapid increase in the reported rate of mental disease, namely in 1911–18 and 1924–35. Although the beginning of the first sharp rise coincided with the coronation of Emperor Yoshihito, the start of the second rise was prior to the accession to the throne of Emperor Hirohito. Moreover, it should be noted that those two intense phases of massive surveys did not necessarily result in the increase in institutionalization. Figure 8.2 shows: the number of patients hospitalized; the number of the patients confined in 'other places'; the rate of those hospitalized against the entire patients' population; and the rate of those in 'other places' against the entire patients' population, again from 1905 to 1941.[39] Although the number grew almost steadily, the rate of confinement both in the hospital or at 'other places' remained almost stable during the two phases of rapid growth of the number of the patients. In fact, the rate of those put under home custody sharply declined during the first rapid rise of the number of registered patients. This almost certainly suggests that the intense search of the population in preparation for the sacred ceremonies did not lead to confinement of the insane, either at home or in hospital. At least at the national level, *pace* Fujino, the direct impact of imperial rituals on the *confinement* of the insane does not seem to have been great.

The patients under home-custody will be our next subject of attention. Although the number of the cases of home custody slowly grew, their proportion to the entire patient population was in constant decline during the entire period (Figure 8.2). If examined vis-à-vis hospitalization, the growth in home custody was clearly outstripped by hospital provision. During the period 1905 to 1940, home-custody cases only doubled, while the hospitalized population grew 9.2 times. Certainly the Japanese psychiatric provision in the early twentieth century

[38] W. Omata, *Seishin-byoin No Kigen* [Origins of Psychiatric Hospitals], 2 vols. (Tokyo, 1998–2000), I, 67–8. Omata's work is based on staggeringly ambitious research, collecting numerous pieces of patchy evidence from more than a millennium of Japanese history.

[39] The exact details of the category of 'those confined at other places' remain elusive. Perhaps they included those kept at municipal poorhouses, as well as the home-custody cases. Numerically speaking, no doubt the most important contribution was made by home-custody cases. Fragmentary evidence suggests that municipal poorhouses were far from a major player in the provision for the insane. In 1918 Kure estimated that about 200 patients were confined in such places all over Japan, while in the same year 4,750 patients were confined 'at other places'. Kure and Kashida, 'Seishin Byosha Sitaku Kanchi', case 524.

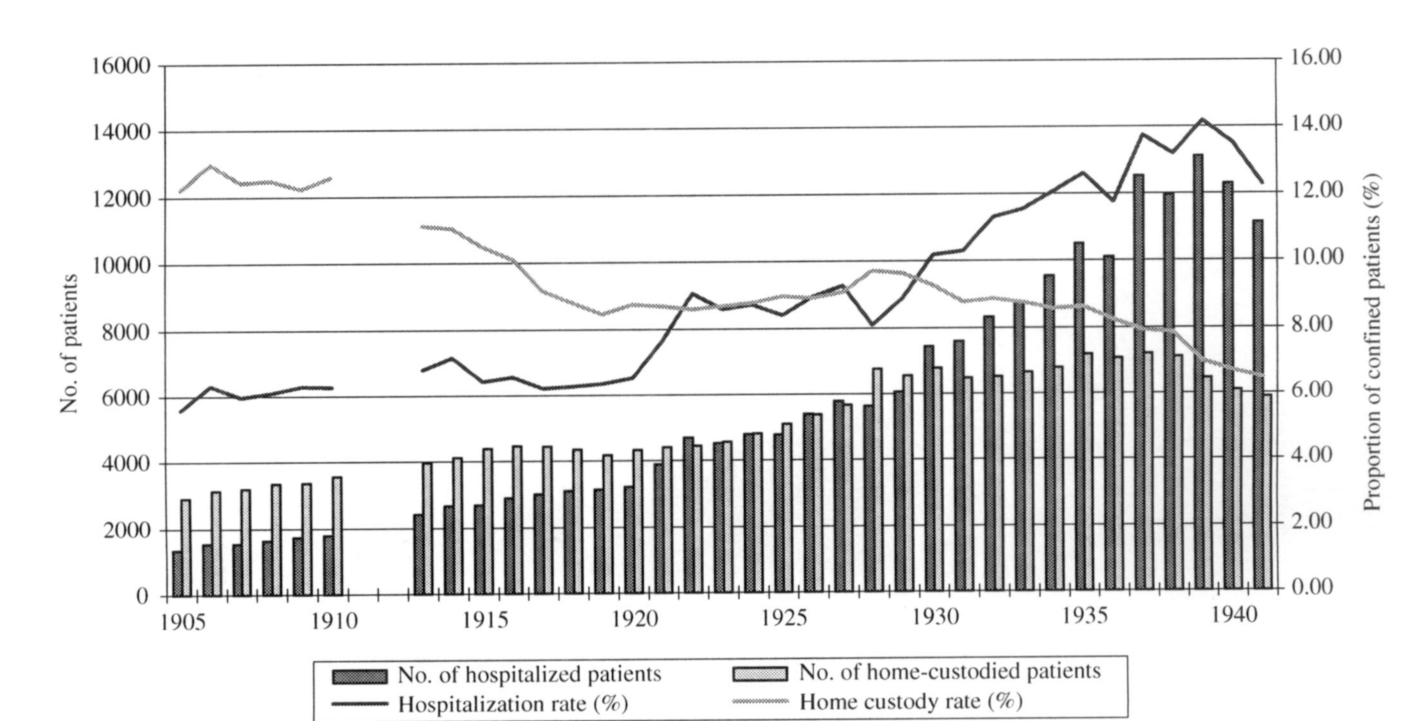

Figure 8.2 – Patients confined in hospitals and at home

witnessed a marked shift in the locus of the confinement of the insane, from their own private home to the hospital governed by a doctor. Especially important was the period of the most sharp decline in the custody rate, which took place from about 1911 to 1920. This suggests that, proportionally speaking, home custody was a means of confinement which was becoming increasingly unpopular, even *before* the passing of the Mental Hospitals Act. Perhaps this decline in home custody had something to do with the urbanization of early twentieth-century Japan. Although there is no conclusive evidence at the national level, home custody seems to have been a rural phenomenon, while hospitalization was an urban solution to the problem posed by insanity.[40] Exactly why urbanization prompted confinement in a hospital instead of at one's own house is unclear. Availability of mental hospitals in cities, sheer lack of space to set up a cage in terraced-houses in major cities, urban sensibility and sense of privacy, and the internalization into urban mentality of the cultural hegemony of medical discourse – all these factors might have contributed to the relative decline of home custody.

Nevertheless, one should not mistake the early signs and dawn of asylumdom with its full arrival: the number of home-custody cases continued to grow until the late 1930s, and it remained a crucial part of the psychiatric provision during the entire period under consideration. Hospitalized patients outnumbered those confined in 'other places' only in the early 1930s, and one-third of confined patients were still placed at home in 1940. Although the private-house-custody is still a historical *terra incognita*, there existed a colossal and invaluable survey conducted by Shuzo Kure as noted above.[41] Kure sent fifteen students and assistants of his to fifteen prefectures to personally visit about 360 patients under home custody, interviewing the patients and their families, sketching the plan of the place of the custody, and photographing the cages as well as the patients. The surveyors spent several days or a few weeks in each area during the summer. Apart from a few cases, the majority of cases were clearly in rural areas. As noted above, the apparent ultimate motive of this survey was to demonstrate the shortcoming of home custody and promote hospital or asylum care, a finding which is more than predictable from the professor of medicine who had studied in Germany and who was at the time involved in the installment of a non-restraint system at Tokyo Metropolitan Asylum at Matsuzawa. The general impression one gets from the survey is, however, not propaganda for asylum care and against home custody, but an

[40] Although no correlation can be statistically established at the national level between urbanization and the ratio of hospitalization or home custody, two prefectures with remarkable high hospitalization rate remained Tokyo and Osaka.

[41] Kure and Kashida, 'Seishin Byosha Sitaku Kanchi'. For a detailed account about the making of this paper, see Y. Okada, K. Komine, S. Yoshioka, 'The Making of "Home Custody of Mental Patients"', *Rinsho Seishin Igaku* [Journal of Clinical Psychiatry] 13 (1984), 1457–69.

Table 8.1 *Types of places of custody*

Detached cottage	8
A room in another house	19
Warehouse	12
Shed	43
In the main house	43
Extension to the main house	18
Extension to the warehouse	4
Unfloored part of the main house	7
Kitchen	3
Others	5
Not in the private abode	13
Total	175

even-minded and rigorous social survey, which requires some closer attention here.

Reflecting that home custody was practised widely across the social classes, the situation and quality of home custody varied greatly. The place of custody varied from purpose-built detached house within the same premises, refurbished warehouse close to the main house ('*dozo*', a common feature for wealthy agricultural households), a part of the main living space ('zashiki'), to a part of a shed ('*mono-oki*', a place to store tools and straws), a shabby extension to the main house, a cage set up in an unfloored part ('*doma*') of the house (see Table 8.1). Although the space of the cage also varied according to the wealth, space availability, and perhaps the extent of compassion towards the patient, mostly each cage was about 3.3 m^2. Personal care of the patient by the family also varied. Some families treated the patient with exemplary kindness and attention, others did not hide their hope to be able to get rid of the patient. One thing in common in those hugely varying places of confinement was the wooden bars or lattice, which seems to have been almost compulsory in order to be approved by the local police. Although seeing a combination of a cage and private home is surreally shocking to our modern sensibility, the lattice allowed the light and the air to the space. Also it secured high visibility of the patient, which must have facilitated vigilance and supervision, as well as exposed the antics of lunatics through the latticework to neighbours and visitors to the house.

An important insight is gained through the analysis of the reasons given for the custody (see Table 8.2). The reasons given by the families reveal an important aspect which has not been fully addressed by historians who have studied the Custody Act. It is true that some of the patients were obviously put under home custody through the concern of the police to confine disorderly and dangerous elements and to secure public order. The reasons attributed to 'public order' in the list above are, however, a decided minority. 'Offence against public

Table 8.2 *Reasons given for home custody*

Domestic violence against person and property	145
Violence to persons and property of others	90
Wandering and vagrancy	72
Arson and threat of arson	31
Offence against public morals	9
Attempted suicide	9
Intrusion into field and woods	7
Intrusion into public places	6
Violence at religious places	2
Public disobedience	1
Others	43
Total	415

morals', 'intrusion into public places', 'violence against religious places' and 'public disobedience' comprise only 4 per cent of the total. By far the largest category is that of 'domestic violence', which suggests that the major motive for putting the patient in custody at home was for the family to protect themselves from the violence and disturbance of an insane person who lived with them, and to facilitate the management of the unruly and dangerous member. The second largest category was that of violence against the person and property of others. In the context of a rural community, this 'others' must have meant 'neighbours'. A culture of domestic responsibility as well as law made the family members responsible for the safe-keeping of their insane family member. If the lunatic at large committed some misdemeanour, the family would be morally blamed by their neighbours, as well as facing the possibility of criminal persecution. The predominance of these two categories clearly demonstrates that home-custody cases were prompted by concrete concerns generated in a small social world of the family and the local community. The initiative for home custody mainly came more from the interaction, negotiation and shared beliefs between the family and the local neighbourhood than from the dictates of public authorities. The role of the public authority, with its concern over public security, was perhaps that of *encouragement* and *sanctioning* of the family's recourse to home custody, not its *enforcement*.

Another important insight gained from this report concerns the attitude of the surveying medical students to the practice of home custody. Despite the sonorous condemnatory tone assumed by their mentor Kure in the end-product of the survey, some students did not universally find signs of glaring cruelty, abuse, or neglect in the places they visited. Actually, their reports suggest their ambivalent attitude towards asylum care and hospitalization. One of the visiting students, Tamao Saito wrote about the area he had visited:

> This prefecture is a place of small industries, with the gap between the rich and the poor still small. Each household has modest property, and the people's behaviour is not very competitive. Accordingly, chronic mental patients are taken care of by their neighbours, and a few wander on the street. This state, however, will not continue for long. The population will grow year by year, highways will be opened, and major industries will arise. Then, if the poor and the weak become insane, their only help will come from the public and the state.[42]

Here, one can sense a kind of nostalgia for a 'traditional society' which modern Japan was quickly losing. Saito appears to have believed that, in this mythical world of the traditional society, people had been kind to each other, and the able helped the unable within their community. He was, however, sure that this idyllic society would before long be washed away by the merciless advent of capitalism. The public psychiatric hospital was, Saito seems to have believed, only necessary in the harsh society whose ominous arrival was impending. He was a half-hearted modernizer, so to speak.

We should now direct our attention to the mental hospital, Saito's antidote against an evil capitalist society. Figure 8.2 clearly shows the rapid growth of psychiatric institutions during our period. The absolute number of institutionalized patients grew more than nine times, and the rate of those hospitalized per population grew nearly three times. It also establishes the impact of the Mental Hospitals Act, which took effect in 1923. The rate of institutionalization had stagnated until 1922, after which the rate increased rapidly until the late 1930s.

After the Mental Hospitals Act, there existed three categories of mental hospitals. First, there were public hospitals, which were maintained at the cost of prefectures with help from the state, and which accepted (mainly) public patients. Second, there were substitute hospitals, which were maintained as private businesses and accepted private patients, 'substitute' patients supported by the prefecture and the state, and public patients whose fee was paid by the city, town or village. Third, there were private hospitals, some of which accepted only private patients, but many of them keeping both private and public patients, but no substitute patients.

The first public psychiatric hospital in Japan opened in Kyoto in 1875. It was situated on the premises of Nanzenji Temple. This venture lasted only for seven years and in 1882 was sold to a doctor who renamed it Kyoto Private Mental Hospital (later Kawagoe Hospital). Before 1920, the only public psychiatric hospital in operation was Tokyo Metropolitan Mental Hospital formerly in Ueno and Sugamo and then at Matsuzawa, in the western outskirts of the suburb. This housed about 350–450 patients between 1905 and 1920, about 700 patients in the 1920s and, after 1930, about 1,000 patients. In 1926 a second public asylum with a capacity for 300 patients opened in Osaka, which soon expanded its capacity

[42] Cited in *Ibid.*, 1464.

to 450. These were followed by one in Kanagawa (1929), Fukuoka (1931), Kagoshima (1931), Aichi (1932) and Hyogo (1937), all created through the Mental Hospitals Act.[43] By 1940, there were seven public hospitals operating. These public asylums, however, housed a distinctively small share of patients. In terms of their size many remained modest: Tokyo excepted, each housed only 120–400 patients, whereas in England the average number of inmates per public asylum exceeded 1,000 in the early twentieth century. The limited provision at public asylums means that the burden of the mass-hospitalization was born by privately run mental hospitals, either substitute or private ones. In 1928 there existed about seventy privately run mental hospitals, and in 1941 the number grew to 160. The ratio of patients at public hospitals to all institutionalized ones remained below 20 per cent between 1921 and 1941. Public mental hospitals in Japan played a decidedly small role in actually carrying the burden of the growth of institutionalization in the early twentieth century. The task was done by privately run hospitals.

The conceptual prototype of these numerous private institutions for the insane was inns built around religious places of healing, such as Shintoist shrines, Buddhist temples, and waterfalls.[44] During the Edo era, a highly developed market economy and the completion of major highways both by the Tokugawa Shogunate and local feudal lords facilitated a travel and tourist industry. Numerous hot-spring baths boasted inns built around it for those who sought the pleasure of bathing as well as cure from various chronic diseases. Profit-making institutions built around places of religious-medical healing were a common business in Edo Japan. In the late eighteenth and early nineteenth century, the Daiunji Temple in Iwakura in northern Kyoto had several inns for those seeking a cure from diseases of the head and the eye. These inns merged with each other and developed into a private mental hospital established there in 1884.[45] Similar patterns abounded elsewhere, in which a guesthouse catering for the patients staying at the place of religious healing of insanity evolved into a centre of care or private asylum.[46] The ways in which early private psychiatric hospitals in Tokyo attempted to attract patients betrayed their inn-like character, although many of them did not have their origins in the hotel business. Tokyo Mental Hospital, the largest private asylum in 1903, sent touts to the gates of the Tokyo Metropolitan Asylum, in order to catch and steer the patients to their own institution, and Tokyo Brain Hospital set up a large billboard next to the public asylum, stating 'IN-PATIENTS INFORMATION OFFICE'.[47]

[43] Except in Kagoshima, these prefectures were all urbanized and industrialized areas.
[44] Omata, *Seishin-byoin No Kigen*, I, 47–192.
[45] N. Kato, 'Iwakura Wo Shu-To-Shita Minkan Ni Okeru Seishin Iryou-Shi' [History of Folk Psychiatric Therapy, with Particular Emphasis on Iwakura], in Matsushita and Hiruta (eds.), *Seishin-iryo No Rekin*, 237–50.
[46] Omata, *Seishin-byoin No Kigen*, I, *passim*.
[47] *SHDH*, 207–8, 214–15.

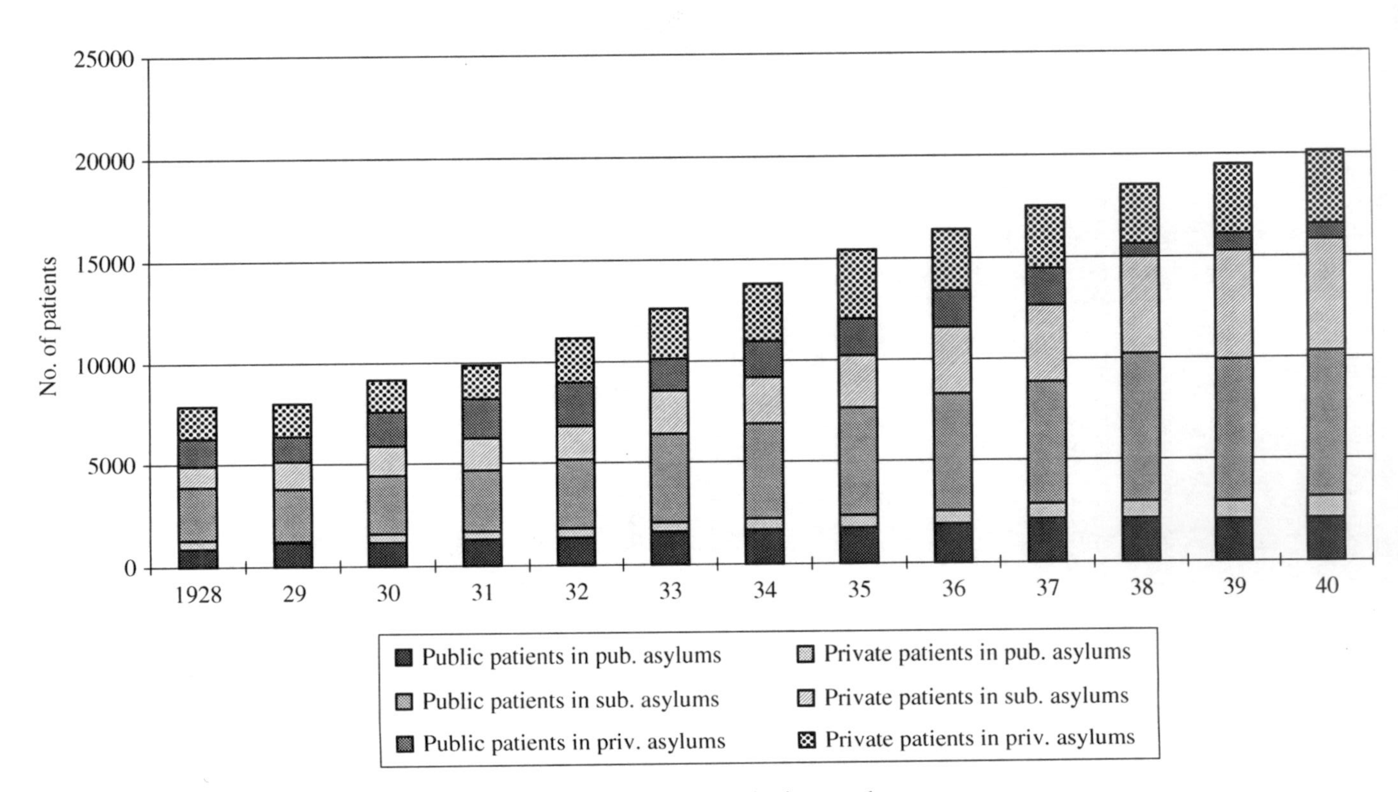

Figure 8.3 – Public and private patients in public, substitute, and private asylums

These touting and advertising activities suggest not only the commercial nature of the early private psychiatric hospitals in Tokyo but also the close liaison between the public sector and the private one in psychiatric provision. The public asylum in Tokyo played the role of the magnet attracting the patients, some of whom were directed to the profit-making sector, in a way very similar to the situation in eighteenth-century London, where two public hospitals for the insane (Bethlem and St Luke's) stimulated the growth of private madhouses.[48] For private institutions in Tokyo, personal connection with public officers was crucial to secure the patients. An officer of the Hygiene Department of the Metropolitan Police contributed money for the foundation of Toyama Hospital for Lunatics, and Tokyo Mental Hospital forged close ties with the hygiene officers at the boroughs of the metropolis.[49] In 1906, about one-third of 601 patients hospitalized by the cities, towns and villages in Tokyo were sent to the private hospitals. In 1918, on the eve of the Mental Hospitals Act, the rate increased to about one-half.[50] The 'substitution' clause of the 1919 Act confirmed this close interdependence between the private and public hospitals, and under this Act both sectors grew hand in hand, creating a large mixed sector in psychiatric provision. The public sector needed the private facilities in order to supplement its severely limited provision, and the private sector wanted the supply of patients and income from the public sector.

The mutual stimulation between the public and private sector seems to have been the main engine behind the increased institutionalization of the insane. Figure 8.3 itemizes the patients in mental hospitals into six categories: patients supported in public asylum at public cost and paying patients in public asylums; patients in substitute asylums at public cost (including substitute patients and other public patients) and paying patients in substitute asylums; patients supported in private asylums at public cost and paying patients in private asylums. Both paying and public patients at public asylums grew steadily but slowly, whereas private asylums kept increasingly larger number of patients until the late 1930s, when they changed status to 'substitute' hospitals and their numbers dwindled. The largest increase in terms of numbers came from respectively public and paying patients kept at substitute asylums. If we lump together the substitute and non-substitute private asylums into 'commercial' and examine the *proportions* of paying and public patients kept there as well as those for public asylums, we get the results shown in Figure 8.4. Purely public patients (patients kept at public asylums at public cost) remained stable or declined very slightly in terms of proportion. The contribution of the mixed sector, i.e., public patients kept at commercial (both substitute and non-substitute) asylums

[48] For the close link between charity and voluntary hospitals for the insane and the profit-making institutions in England in the eighteenth century, see R. Porter, *Mind-Forg'd Manacles: A History of Madness in England from the Restoration to the Regency* (London, 1987).

[49] *SHDH*, 196 and 208. [50] *ARH* (1906) and *ARH* (1918).

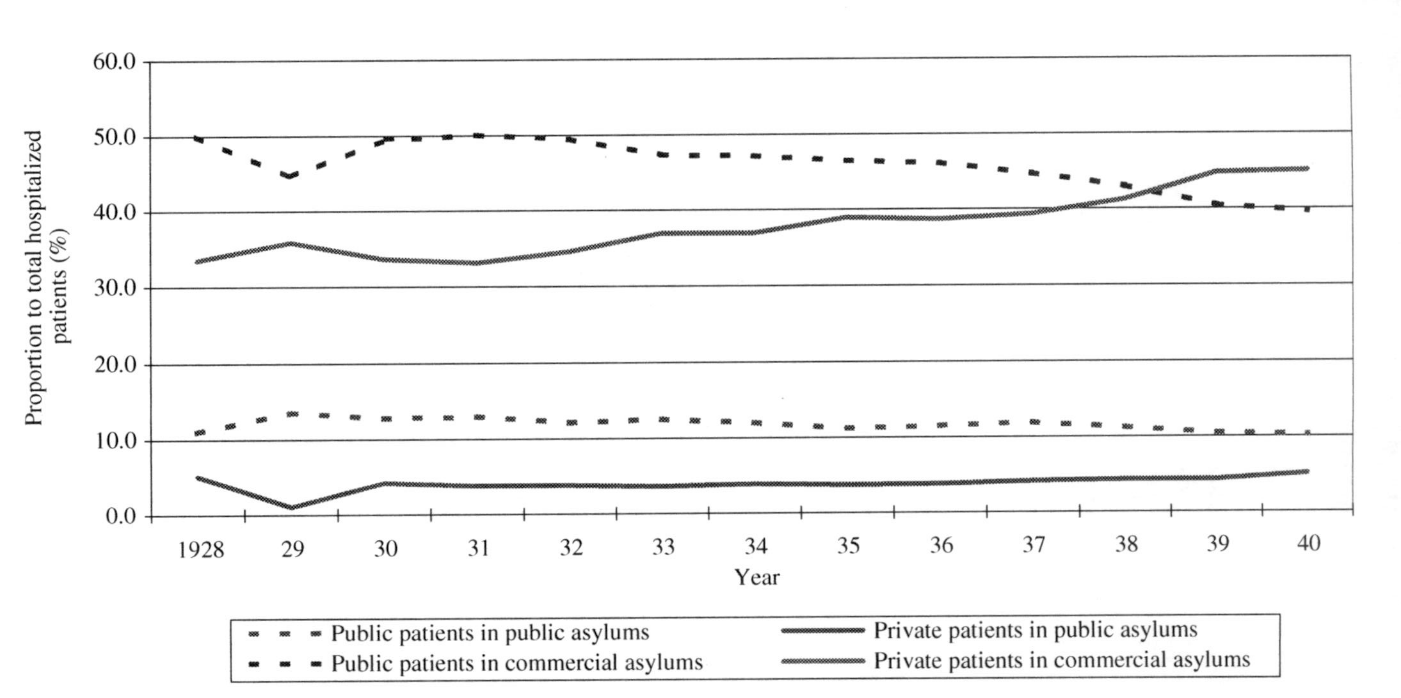

Figure 8.4 – Proportions of public and private patients in public and commercial asylums

declined gradually but more markedly. The sector that had the largest share in 1939 was the private sector, private patients staying at commercial asylums. The numerical and proportional growth of the private sector tells a hitherto little noticed factor in the rise of mental hospitals in pre-war Japan, namely the emergence of a large number of people who were ready to pay significant sums of money to be treated there. In other words, the growth of the clients with *demand* for psychiatric service made the greatest contribution to the making of a society that segregated a large number of the insane.

It should be emphasized, however, that one should not call this a 'great confinement'. Figures 8.5 and 8.6 show the numbers of entire known patients, and those patients regarded as not needing confinement or custody, with the latter's ratios to the former from the years 1905 to 1927 and from the years 1928 to 1941.[51] These two figures conclusively show the persistence of informal domestic care without recourse to either hospital or home custody, long after the Mental Patients' Custody Act and Mental Hospitals Act. Although the ratio declined steadily from 79.5 per cent to 70.8 per cent from 1928 to 1941, at the end of the period, we are still talking about a society that put only about 30 per cent of publicly recognized psychiatric patients in confinement.

The everyday lives of those who were left to the family or 'at large' are hard to know, and here again the survey by Kure throws invaluable light on the lives of the largest category of patients. Kure's students were able to find and interview eight patients publicly recognized as insane but not in home custody.[52] One of them, a lower civil servant, was forced to live on rotten *tatamis* and his son 'does not treat the father with kindness', although the patient was not under

[51] Figure 8.5 and Figure 8.6 represent slightly different categories of patients. Figure 8.5 represents only those confined under the two Acts. Figure 8.6 represents those actually confined in mental hospitals. The general table of mental patients of the *ARH* had included only the number of patients hospitalized, confined, or put in custody either through the Mental Patients' Custody Act or through the Mental Hospitals Act, until 1939. In that year, however, the table started to list patients who were 'hospitalized or confined NOT through the two laws'. This is not a minor group of patients: 9,979 patients coming under this category in 1939. Looked at in detail, this group could be broken into: (1) private patients in private and substitute mental hospitals, (2) public patients supported in private and substitute mental hospitals outside the two laws, (3) patients in the psychiatric wards of hospitals for medical schools, (4) patients in the psychiatric wards of general hospitals, (5) patients maintained in non-medical places of confinement, and (6) patients maintained in nursing homes at temples, shrines and waterfalls. From 1929 on, the exact figures for the sum of categories (1) and (2) were available in table of mental hospitals annually published in *ARH*. The numbers of categories (3) – (6) were, however, unavailable in any of *ARH*. Fortunately, we have exact figures for the numbers of the four types of institutions and their capacity for the years 1929 and 1935, and of the actual number of patients for 1929. Figures for 1929 are taken from the Department of Hygiene's *Survey of the Places for Confining Mental Patients* (1929). Those for 1935 are taken from Osama Kan, 'Hon-Pou Ni Okeru Seishin-byosha Narabini Kore Ni Kinetsu Seru Seishin-Ijousha Ni Kansuru Chousa' [Statistical Survey of Mental Patients and Similar Mentally Abnormals in Japan], *Shinkei-gaku Zasshi* 41 (1937), 793–884. See table 8.5.

[52] Kure and Kashida, 'Seishin Byosha Sitaku Kanchi', cases 106–14.

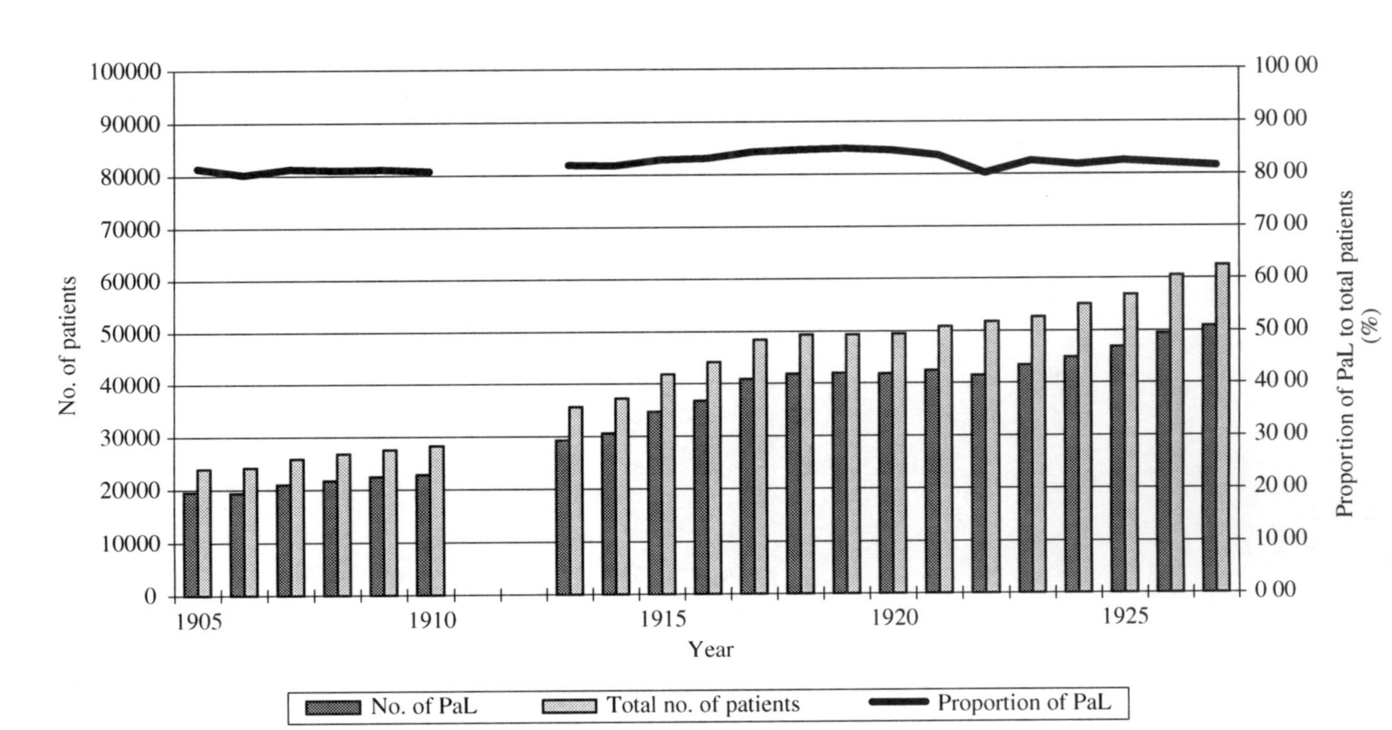

Figure 8.5 – Patients at large (PaL), 1905–1927

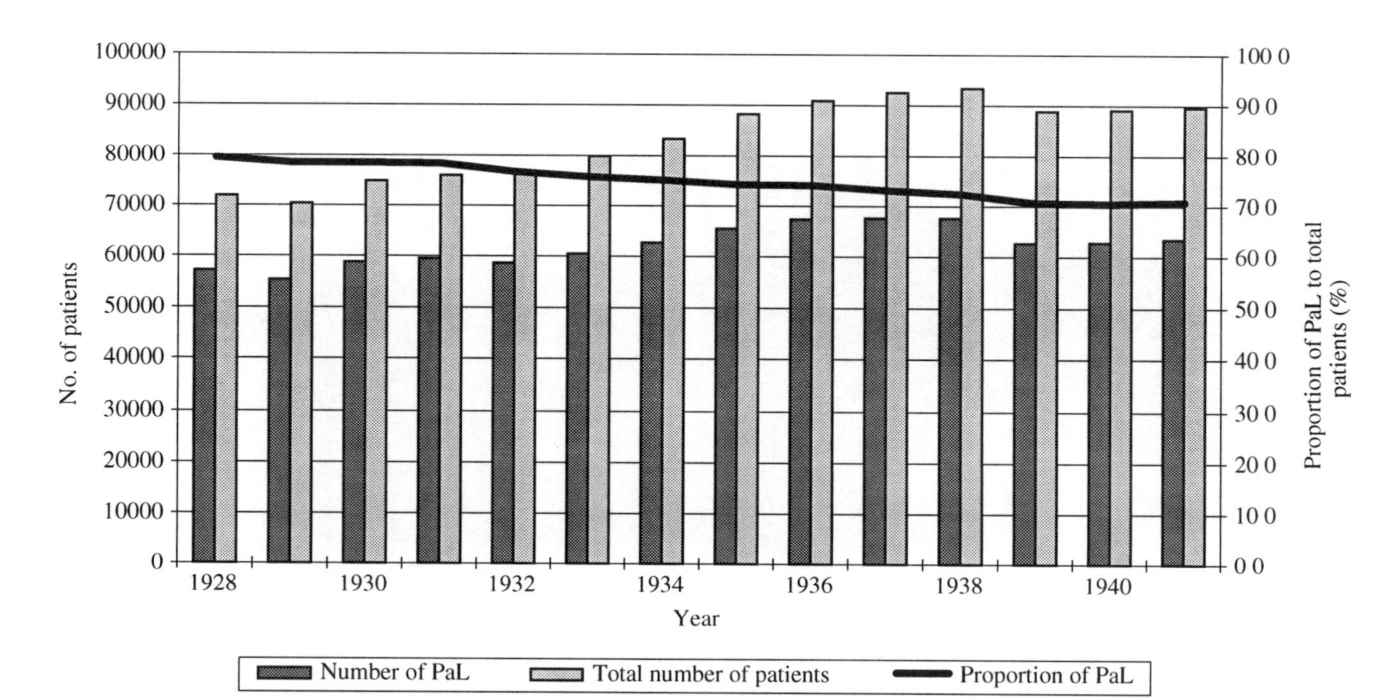

Figure 8.6 – Patients at large 1928–1941

threatening or instrumental coercion. Another patient was reported that 'he does not receive any particularly kind treatment from the family due to his disease', and quarrels with his younger brothers and sisters. Despite these somewhat critical comments, the surveyors did not find widespread glaring abuse or cruelty. Also it should be emphasized that the demonstration of particularly praiseworthy care was rare. In one case, the brother of a female patient made her open and run a haberdasher's shop without any hope of making profit, which must have been conceived both as a kind of work therapy and as a way to keep the outlook of normality. Such generosity was, however, an isolated expression of special concern in the eight samples of Kure, and is highly unlikely to be prevalent in the vast number of the mental patients at large. The common picture emerging from Kure's sample is a life of dependent, sub-normal, troubled, but essentially the same life as the rest of the family. Many patients were 'left to themselves' and the families 'don't meddle with the patients'. They were made to do what work they could. One of them volunteered for outside work without telling her family, for she wanted to be useful.[53] One was explicitly stated to be taking meals with the family.[54] They were not treated either with particular kindness or with particular cruelty. As an disabled person living with able family members, they may have been a source of irritation and no doubt received at least occasional scolding or even punishment. But conflicts or minor punishments were a part of normality in life, without the status of 'sick role', and they do not necessarily mean active cruelty on the side of the family. Rather, they suggest the normal life that the patients were *both allowed and forced* to live.

Oji Brain Hospital and its patients

Against the background of psychiatric provision sketched above, we can now examine the Oji Brain Hospital and its patients. OBH was one of several psychiatric ventures which were started by doctors and entrepreneurs in Tokyo around 1900, counting on the increase in the demand for hospital-based psychiatric service created by the Mental Patients' Custody Act in the metropolis.[55] The only material we now have about its beginning in 1901 is the exposé of abuses, corruption and negligence which occurred at seven psychiatric hospitals in Tokyo, serialized in the *Yomiuri News*, a popular daily paper.[56] Naturally, a historian should treat such an account with caution, but there is a ring of truth in the reports. The beginning of OBH told by *Yomiuri* epitomized one of the origins of Japanese private psychiatric provisions: the intersection of medicine and hotel, a business of providing accommodation associated with a medical facility.

[53] Accordingly, she boasted about being very useful, although in reality her contribution was very small. This led to frequent quarrels with her sister-in-law, who was sometimes hit by the patient. Kure and Kashida, 'Seishin Byosha Sitaku Kanchi', case 110.

[54] *Ibid.*, case 107. [55] *SHDH*, 183–223. [56] *SHDH*, 202–3.

Zenjiro Komine had owned an inn mainly catering for those who travelled to and stayed in Tokyo to be treated at University of Tokyo Hospital. This innkeeper and a few doctors working at another asylum in Tokyo joined forces to start a brand-new and purpose-built psychiatric hospital in Oji, a remote agricultural suburb of Tokyo. *Yomiuri*'s comical account suggests that there were personal disagreements between the governor of the hospital (Komine) and its medical staff, mainly due to the shortage of patients and the very small profit that the hospital was able to make in its early years. The report also satirized Komine as a parvenu rising from an innkeeper to a hospital governor in its depiction of his daily inspection of the hospital kitchen:

> [Komine] threw off his stylish waistcoat to inspect the kitchen of the hospital, and immediately the governor's eyes are fixed on pickles' portions to be served to the patients. He could not tolerate such a generous serving of the pickles. Such was his penny-pinching interests in this kind of matters that he not only scolded the cook but also cut the pickles himself as thin as possible.[57]

Allowing for a certain amount of comic licence, this sketch effectively captures the characteristics of early OBH and many other early psychiatric ventures in Tokyo, that is the mentality of a small family trade, for whose survival the apportioning of pickles made a great difference. Despite such lowly behaviour and personality of its owner (or, perhaps because of his disarmingly unpretentious down-to-earth character), the *Yomiuri* reporter was somewhat sympathetic, hoping that the hospital's perseverance under chronic financial crisis would some day be rewarded.[58]

Yomiuri's ironic and satirical well-wishing materialized. From such an inauspicious beginning, OBH made a meteoric rise in the social ladder. Shigeyuki Komine, Zenjiro's adopted son, studied medicine at Saisei Gakusha, a private medical school in Tokyo, and worked briefly at Tokyo Metropolitan Asylum until 1908, when he assumed the post of OBH's medical superintendent around the age of twenty-five.[59] Shigeyuki was an able superintendent, a well-read medical scientist, and he became a leading member of the profession. Perhaps with Shigeyuki's arrival, OBH had finally solved its chronic problem of the absence of a medical officer and started to flourish. When the Mental Hospitals Act was passed in 1919, OBH was one of the first that was appointed as a substitute hospital, an honour that only eight other private mental hospitals in Japan were able to enjoy at that time. In the mid-1920s, a brand-new three-storied western building, complete with recessed arches, was erected at the old premise, which housed the Komine Research Institute and newly added wards for private patients. Patients from all over Japan flocked to the hospital, whose success allowed the Komine Institute to conduct serious research into

[57] *SHDH*, 205–6. [58] *SHDH*, 207. [59] *Nihon Seishin Byou-i Kyoukai Kiji*, 74.

Table 8.3 *Admissions of private and substitute patients at Oji Brain Hospital (1935)*

	Male	Female	Total
Private	272	119	391
Substitute	43	34	77
Total	315	153	468

Table 8.4 *Lengths of stay of the discharged Oji Brain Hospital patients*

Time Length	Private Male		Private Female		Substitute Male		Substitute Female	
	No.	%	No.	%	No.	%	No.	%
0–10 days	22	16	7	9	0	0	0	0
11–30 days	29	21	14	19	0	0	0	0
1–3 months	46	33	29	39	2	12	1	11
3–9 months	21	15	11	15	1	6	3	33
6–9 months	6	4	3	4	1	6	0	0
9–12 months	4	3	2	3	2	12	0	0
1–2 years	6	4	3	4	3	18	1	11
1–4 years	3	2	3	4	5	29	3	33
over 4 years	2	1	2	3	3	18	1	11
Total	139	100	74	100	17	100	9	100

most up-to-date therapeutics, such as malarial therapy and insulin-coma therapy. Recognition by professional brethrens soon followed. Shigeyuki played important roles in psychiatric associations, and in 1932 was elected a Representative Governor of the Japanese Association of Public and Substitute Mental Hospitals. Komine and his hospital thus represented the flourishing private practice and the upward mobility of Japanese psychiatry in the early twentieth century.

As a Substitute Hospital, OBH accepted both private and public patients. In terms of the number of patients staying at the end of the year 1935, OBH housed fifty-five private patients and 108 public patients, and the ratio between the private and public patients remained about 1:2 during the period of 1927–41. In terms of the number of admissions, however, private patients vastly outnumbered public patients, with their ratio being 5:1 (see Table 8.3 and Table 8.4).[60]

This higher number of private admissions was certainly because the private patients at OBH stayed for much shorter periods than the public patients. The

[60] The male : female ratio of the patients admitted in 1927 was about 32 per cent, not very different from the national average of institutionalized patients.

Table 8.5 *Capacities of places other than mental hospitals, 1929 and 1935*

	Medical schools	General hospitals	Confinement houses	Nursing homes	Total capacity	Total patients
1929	856	192	439	708	2195	1365
1935	1237	127	423	1074	2861	1802[a]

[a]an estimated figure

distribution of the length of stay among the patients discharged (including cured, relieved, uncured, and dead patients) in 1935 is given in Table 8.4. This shows that a long-term stay (more than half a year) was rare among the private patients, while public patients tended to stay much longer. The median length of stay for private patients discharged in 1935 was forty-four days (average being 134) for male patients, and sixty-three days (average = 183) for females, while the corresponding figures for public patients was 717 (average = 869) for male and 404 (average = 575) for female patients. For the majority of private patients, their experience of OBH was a trial for a relatively shorter period than an extended stay. Figures published in the *ARH* mask a large number of private patients who had been admitted but discharged before they were counted into the statistical tables. Experimenting with a spell at a psychiatric hospital was a more widespread practice among relatively well-off sectors of the society than tables in *ARH* suggest.

The general pattern of the patients' experience of OBH briefly discussed above suggests that OBH's major role for the private patients was that of *supplementing* care given at another site, which was often the patient's own home. Numerous cases indicate that hospitalization to OBH was a trial to see whether the patient would recover at the beginning of the disease, an emergency measure to cope with a particularly difficult manifestation of the disease, or a change of scene inserted in long-term domestic care. M.N. first stayed at OBH in about 1921, secondly in 1930 for three months, and thirdly in 1935 for another three months. For the most time during his disease, he stayed at home and often troubled his family and neighbours by his violence.[61] G.T. had been insane for about twenty-five years when he was admitted to OBH, during which time he worked as a farmer at his own house. C.M. had been mentally ill and had stayed at her own house for about eight years before her admission: she quarrelled and occasionally wounded another insane sister of hers during the period. M.G.'s three years of mental disease before admission to OBH had been spent mainly at her own house, interrupted by two brief terms of hospitalization and one longer stay at her relative's in the country for change of air. N.A. had been a

[61] Case Record (hereafter CR) of Oji Brain Hospital for M.N., discharged on 10 September 1935.

highly troublesome patient for about one-and-a-half years, during which she went out naked and used abusive language to her family and neighbours. She had experienced a brief spell at another private asylum, from which she had been discharged 'for financial reasons' and had stayed at her own home until admitted to OBH.[62] Although increasing numbers of private patients were hospitalized, for many of them, staying at an asylum seems to have been shorter episodes inserted in the major framework of care at home.

Conclusion

My account above has done little more than scratch the surface of the vast and rich area of psychiatry in Japan in the early twentieth century. Much of my argument remains tentative, to be further examined through research into psychiatric archives. Having said that, however, in the light of what I have argued above I should like to offer one historiographical point which might be pertinent to the theme of this volume.

The point is about the origin of the major driving forces of psychiatric modernization. The two major laws, the Mental Patients' Custody Act and the Mental Hospitals Act, provided the basic framework for psychiatric provision until the post-war Mental Hygiene Act (1950). Although they were conceptualized in very different spirits and aims, it should be emphasized that these two Acts had one thing in common. Note well that home custody and the mixed sector of psychiatric hospitalization, respectively the core part of each piece of legislation, had already been a well-established practice *before* the Acts codified them. In their crucial aspects, these pieces of legislation followed what people then were practising, not the other way round. The historiographical implication of this pattern of the law heeding the practice is that one should look for factors other than legal or legislative for the *real* driving force of historical change in Japanese psychiatric provision. Social and cultural forces 'from below' created trends, patterns and models of the care of the insane, some of which were selected to be confirmed and encouraged by the law 'from above'. Another implication is that one should not overestimate the influence 'from the West', which has been one of the major frameworks within which the history of Japanese psychiatry during the period under review has been described. Instead of the select few who were enlightened and westernized and who drafted Bills and rules or wrote textbooks of psychiatry, a mass of patients, families and neighbours set the basic trends of psychiatric provision, some of which in turn were selectively codified by the elite with the ambition to 'Westernize' Japanese psychiatry. Needless to

[62] CR for G.T., discharged (dead) on 10 November 1935; CR for C.M., discharged (dead) on 28 February 1935; CR for G.M., discharged on 31 December 1939; CR for N.A., discharged on 7 May 1935.

say, the influence from the West was tremendous in almost every aspect of life in Japan at that time and the care for the insane was no exception. This does not mean, however, the programme of Westernization was the major driving force in the making of basic paradigms of Japanese psychiatric provision.

Instead of seeing the progress of psychiatric confinement in early twentieth-century Japan as an attempt to import western systems of care of the insane, this chapter set out to understand it within the complex interaction and negotiation between various basic social units, such as the family, community, local and central governments, and psychiatrists. This chapter has also tried, whenever possible, to throw light on the larger contexts in which those interactions took place, such as the market economy, urbanization, totalitarian policing of the population, and medicalization. I can only hope that my attempt will encourage future researchers in the comparative social history of psychiatry to tackle big issues at which this chapter is only able to hint.

9 The limits of psychiatric reform in Argentina, 1890–1946

Jonathan D. Ablard

In 1934, *La Nación*, Argentina's oldest daily newspaper, reported that every public institution for the insane and mentally retarded in the republic was severely overcrowded. The National Hospital for the Female Insane (Hospital Nacional de Alienadas, hereafter the HNA), with a capacity for 1,600 patients, cared for over 3,000. The men's Hospice of the Virgin of Mercy (Hospicio de las Mercedes, hereafter the Hospicio), was 890 patients over its 1,100 bed limit. Overcrowding was even more dire in the country's rural facilities, several of which had been designed to relieve urban hospitals.[1]

While the crisis had in fact been long in the making, the 1930s marked a new low point in the public image of the hospitals. In 1910 – Argentina's centennial year – these same hospitals enjoyed reputations as being advanced medical institutions. The 1908 visit of Georges Clemenceau, future president of France, to the Hospicio's rural satellite facility, and his report of the trip in 1910, is emblematic of Argentina's prospects. The future French president reported that the ten-year-old National Colony for the Insane was a 'model for the older peoples' of Europe to emulate. Forcible restraints and isolation cells were absent, and patients lived in modern, spacious and comfortable cottage-style dormitories. The daily schedule revolved around work therapy that kept all able bodied busy, productive and mentally focused.[2] Similar reports, many coming from other European observers, echoed Clemenceau's optimism.[3] Several years earlier, the Argentine government had established a commission to supervise the development of a national network of regional asylums (as well as other medical institutions) to meet the needs of the republic's fast growing, but geographically scattered, population.

[1] 'La hospitalización de alienados en el país constituye un serio problema de solución inmediata', *La Nación*, 26 January 1934.

[2] G. Clemenceau, *South America Today: A Study of Conditions, Social, Political and Commercial in Argentina, Uruguay and Brazil* (New York, 1911), 123–32.

[3] C. Lupati Guelfi, *Vida Argentina (versión española)* (Barcelona, 1910), 146–68; J. Huret, *En Argentine: De Buenos Aires au Gran Chaco* (Paris, 1911), 100–7; A. Meyer Arana, *La Caridad en Buenos Aires* (Buenos Aires, 1911); 'Hommage au professeur Cabred', *L'Hygiene Mentale* 7: 22 (July–August 1927).

Optimism about Argentina's psychiatric facilities mirrored a more general positive outlook about the republic's prospects in the realm of international business. Fuelled by an export economy that rivalled that of Canada or Australia, Argentina received massive immigration from Europe. From the mid-nineteenth century until the eve of the Second World War it transformed a largely *mestizo* and mulatto population into one that was primarily of European stock.[4] In the process, Argentina became Latin America's most likely contender to become a 'developed' nation.[5] Prosperity, and the elite's affinity for French culture, led Buenos Aires to earn in the late nineteenth century the nickname 'Paris of Latin America'.[6] During this same period, Argentine professionals, intellectuals and scholars looked to Europe, and particularly France, Britain and Germany, for models to build a modern, efficient and prosperous society and state.

The neglect suffered by patients, and the inability of doctors to modernize the hospitals, represented the failure of Argentine psychiatrists to live up to their aspirations to reach modern levels of care. This chapter seeks to explore the causes and consequences of the collapse of Argentina's public psychiatric hospitals. Careful attention to the existing archival evidence from the hospitals – as well as close readings of published sources – underscores the extent to which psychiatrists fell far short of their expressed desire to modernize and expand Argentina's public psychiatry. The root of their failure lies as much in the state's unwillingness to properly fund existing hospitals, as in the lack of a national network of institutions.

Origins

Argentina became independent of Spain in 1810 but quickly plunged into a series of civil and international wars that did not end until the 1850s. As a result, national consolidation and the creation of a viable national state were delayed. Not surprisingly, this period – dominated by the rule of local strongmen

4 The impact of immigration on Argentine politics, culture and society was profound, in part because per capita, Argentina received far more newcomers than any other nation. See J. Moya, *Cousins and Strangers: Spanish Immigrants in Buenos Aires, 1850–1930* (Los Angeles, 1998). Most immigrants were from Italy, followed by Spain, France, Northern Europe, Eastern Europe, and the Middle East.

5 The question of Argentina's economic rise, and then steady decline after 1930 is an area of much scholarly debate. See J. Carlos Korol and H. Sábato, 'Incomplete Industrialization: An Argentine Obsession', *Latin American Research Review* 25 (1999), 7–30.

6 On the influence of French architecture and urban planning between 1850 and 1914, see J. R. Scobie, *Buenos Aires: Plaza to Suburb, 1870–1910* (New York: 1974). The Argentine elite was also interested in convincing Europe that Argentina was worthy of joining the ranks of 'civilized' nations. See I. E. Fey, 'Peddling the Pampas: Argentina at the Paris Universal Exposition of 1889', in W. H. Beezley and L. A. Curcio-Nagy (eds.), *Latin American Popular Culture: An Introduction* (Wilmington, Del., 2000), 61–85.

(*caudillos*) – witnessed few lasting developments in the realm of medicine, education or social welfare.

In 1852, the most powerful of the strongmen, Juan Manuel de Rosas, fell to forces who supported the national unification of Argentina and its reorientation towards European and North American models of economic and social development. Although Argentina's constitution was established in 1853, effective national integration was delayed until 1880 when Buenos Aires became a federal district and the national capital. It seems evident that the delay in effective national integration had a profound impact on the later course of the hospitals.

The ascendant liberal elite of Buenos Aires, not yet in full control of the entire country, chose to portray Rosas as the embodiment of Argentina's bestial and barbaric underside.[7] Generations of *porteño* (residents of Buenos Aires) elite viewed the Rosas period as one in which civil disorder, political violence, and provincial rivalry had essentially unhinged the native populace. To his well-placed critics, eager to win European approval, Rosas had upset the Argentine social order by aligning himself with the Indian, mestizo, and Afro-Argentine majority. Similarly, Rosas had demonstrated a disdain for European culture through his public preference for *creole* (that is, native) cultural mores. The reformers were further motivated by the fact that Rosas had repressed embryonic reforms of the country's medical and charitable organizations.[8]

While national political consolidation continued to elude Argentina, urban political elites (primarily in Buenos Aires) nevertheless began to create institutions of social welfare. To that end, the new government established the elite women's Society of Beneficence (SB) and its male counterpart, the Philanthropic Commission, to oversee the development of health and welfare institutions.[9] The government also reopened the country's only school of medicine.[10] These measures, as well as the reform of education policy and the articulation in the national constitution of a policy to encourage European immigration, were part of a larger elite project to reorient Argentina towards Europe.

In this endeavour, health care, and particularly care for the mentally ill, was of critical importance to the national rejuvenation project. Argentine scholar

[7] J. Lynch, *Argentine Dictator: Juan Manuel de Rosas, 1829–1852* (Oxford, 1981); R. Salvatore, 'Death and Democracy: Capital Punishment after the Fall of Rosas', *Work Papers/Documents de Trabajo*, no. 43 (Buenos Aires, 1997), 1–29.

[8] E. A. Balbo, translator, 'Classic Text no. 6 – Dissertation on Acute Mania, Diego Alcorta', *History of Psychiatry* 2 (1991), 207.

[9] The society operated all of Buenos Aires's health and welfare institutions devoted to children and women until 1947. The Philanhropic Commission, by contrast, dropped out of sight.

[10] E. A. Balbo, 'El Manicomio en el Alienismo Argentino', *A sclepio* 40 (1988), 153. All three of these institutions had been created in the late 1820s, during a brief period of liberal reform, but had been quickly thereafter closed by Rosas.

Hugo Vezzetti has argued that the new government viewed the early decades of independence as a period of barbarism, in which racial, class and gender distinctions had eroded along with traditional hierarchies.[11] According to some contemporary observers, the years of civil unrest caused increased mental imbalance, particularly among women and non-Europeans of both sexes. Worse still, it was commonly believed that Rosas had consciously recruited lunatics (usually non-white) to serve as his henchmen.[12]

Argentina's asylums developed coterminously with the establishment of a coherent national state and grew out of a larger effort by the elite of Buenos Aires to restore their political and moral capital after decades of civil and international war, and dictatorship. Reformers thus understood that creating hospitals was part of a larger project that would help Argentina to earn the reputation of a 'civilized' nation. With direction from the newly created charitable organizations, the city of Buenos Aires constructed separate asylums for women and men.

The women's asylum, which started in an old convent on the edge of town, opened in 1854. Over the next four decades the asylum grew with little planning, as the city around it grew in size and population. By 1890, the hospital was a series of buildings connected by covered passageways, and was considered unsanitary and dangerous to patients and staff alike.[13] In 1880, Buenos Aires was federalized, the Society of Beneficence, which ran the asylum, was placed under the Ministry of the Interior and the asylum was renamed Hospital Nacional de Alienadas. Finally, in 1898 the national government placed the society and its institutions under the auspices of the Ministry of Foreign Relations and Religion.[14] There it remained until 1947, when the recently elected government of Juan Perón placed all health facilities under the newly created Ministry of Health and Welfare.[15]

[11] H. Vezzetti, *La Locura en la Argentina* [Madness in Argentina] (Buenos Aires, 1985). Recent scholarship has softened the differences between the two periods, and noted that the post-1853 regime exaggerated the change for its political purposes. See Salvatore, 'Death and Democracy'.

[12] E. A. Balbo, 'Argentinian Alienism from 1852 to 1918', *History of Psychiatry* 2 (1991), 181–92; Vezzetti, *La Locura en la Argentina*; and the earlier work of the same title by J. Ingenieros, *La Locura en la Argentina* (Buenos Aires, 1920), 136; J. Salessi, *Médicos, Maleantes y Maricas: Higiene, Criminología y Homosexualidad en la Construcción de la Nación Argentina 1871–1914* [Doctors, Crooks and Gays: Hygiene, Criminology and Homosexuality in the Construction of the Argentine Nation] (Buenos Aires, 2000).

[13] For a celebratory history of the hospital's early years, see Arana, *La Caridad en Buenos Aires*. Recent scholarship includes C. Jeffress Little, 'The Society of Beneficence in Buenos Aires, 1823–1900', PhD thesis, Temple University, Philadelphia (1980).

[14] Despite nationalization, the hospital continued to be administered by the private Society of Beneficence who named medical directors and kept the books. Furthermore, until the late 1960s, the principle nurses of the hospital came from a female Catholic order. See Archivo Hermanas de Caridad-Hijas de Maria Señora del Huerto-Casa Provincial (Buenos Aires).

[15] K. Mead, 'Oligarchs, Doctors and Nuns: Public Health and Beneficence in Buenos Aires, 1880–1914', PhD thesis, University of California, Santa Barbara (1994), 239. On the history of

The men's asylum followed a similar trajectory, although it was always more directly connected to the national state. In 1863 the city of Buenos Aires, under consultation with the Philanthropic Commission, constructed the Casa de Dementes (House for the Insane) across the street from the women's asylum.[16] By the 1880s, the hospital had deteriorated so severely that the city ordered a total renovation that was completed in 1887. The hospital was renamed Hospicio de las Mercedes in honour of the Virgin of Mercy, patroness of convicts and the mentally ill. Facing the same pressures as the women's hospital, the Hospicio quickly returned to its earlier state of decay and, by 1904, the city relinquished authority of the institution to the federal government's ministry of Foreign Relations where it remained until 1947.[17]

Ideological foundations

Despite the best efforts of many public health doctors, health policy proved self-limiting by the very course of Argentina's political and economic development. Resistance to the creation of a comprehensive health system was largely a reflection of the liberal doctrine adopted by the Argentine elite that called for fiscal restraint in public expenditure. As Argentina's export-based economy expanded, traditional elites maintained an ambivalent view of social-welfare projects. To become modern, Argentina needed to have a public-health network. Yet, as Ernest Crider has pointed out 'By implication . . . such assistance was limited to that necessary to preserve the social structure and the prominent role of the oligarchy; therefore it could not lead to fundamental changes in the living standards and quality of health care extended to the poor.'[18]

One way to maintain this precarious balance between largely contradictory goals was to delegate as much of the public welfare responsibility to private, religious and charitable organizations. A case in point was the placement of the hospitals under the Ministry of Foreign Relations, as well as the large influence of the Society of Beneficence which ran most of the hospitals devoted to the care of women and children in Buenos Aires. The arrangement was a source of constant frustration to many public-health doctors who wished to create a unified, national and modern health-care system. Likewise, wherever possible, the state used either inexpensive labour, such as female, religious or

psychiatric care in Argentina after 1946, see J. Ablard, 'Madness in Buenos Aires: Psychiatry, Society, and the State in Argentina, 1890–1983', PhD thesis, University of New Mexico (2000).

[16] Ingenieros, *La Locura en la Argentina*, 198; M. Sbarbi, 'Reseña historica del Hospicio de las Mercedes', *Acta Neuropsiquiátrica* 6 (1960), 420; L. Meyer, 'Los Comienzos del Hospicio de las Mercedes', *Acta Psiquiátrica Psicológico de America Látina* 33 (1987), 338–9.

[17] *Memoria del Intendecia Municipal* (Año 1903, Administración del Sr Alberto Casares) (Buenos Aires, 1904), 81; E. Allen Crider, 'Modernization and Human Welfare: The Asistencia Pública and Buenos Aires, 1883–1910', PhD thesis, Ohio State University (1976), 61.

[18] *Ibid.*, 227.

even prisoners. These cost cutting practices ultimately prevented the development and maintenance of modern hospitals.[19]

The Argentine elite's parsimony in social projects was reinforced by massive European immigration that exercised a profound influence on both psychiatric theory and practice in Argentina. Since the 1850s, a cornerstone of Argentina's modernization project was to attract European immigrants to fill in the vast 'empty' spaces of the interior. But the plan backfired and Argentina received far fewer northern Europeans prepared to settle in the countryside and many more southern and later eastern Europeans who tended to gravitate to the cities.[20]

As immigration steadily increased between the 1880s and 1920s the *porteño* public-health system was unable to keep up with the pace of urban growth.[21] Yet, Argentina, ravaged by serious but preventable epidemics as late as 1900, needed to provide a basic level of sanitation and health care to make immigration appealing.[22] Although public-health bureaucracies had made crucial strides in the eradication of infectious disease (a key to attracting immigrants) many elites feared that an overly generous welfare system would provide an unwanted 'pull' factor for Europeans in search of an easy life in the New World.[23]

As early as the 1870s medical and lay writers began to identify the problem of mental disturbance among recent arrivals. The directors of the capital's two asylums were the first to call attention to the problem of the *loco inmigrante*, the insane immigrant.[24] In large part, this was because both hospitals, well into the 1930s, housed a greater number of foreign-born than native patients.[25] Most were from Italy and Spain, but there were sizeable numbers from the Middle East, Russia, France and Central Europe.[26] In 1906, for example, future director of the Hospicio, José T. Borda, complained that recent arrivals from

[19] R. González Leandri, 'Médicos, Damas y Funcionarios: Acuerdos y Tensiones en la Creación de la Asistencia Pública de la Ciudad de Buenos Aires', in J. Luís Peset (ed.), *Ciencia, Vida y Espacio en Iberoamerica* (Madrid, 1989), 85; Crider, 'Modernization and Human Welfare', 38–39; Mead, 'Oligarchs, Doctors, and Nuns', 42–52.

[20] C. E. Solberg, *Immigration and Nationalism: Argentina and Chile, 1890–1914* (Austin, Tex., 1970), 93–116; Moya, *Cousins and Strangers*, 55–6. By 1914, a little over half of Buenos Aires's population was foreign born.

[21] Crider, 'Modernization and Human Welfare', 46.

[22] C. Andres Escudé, 'Health in Buenos Aires in the Second Half of the Nineteenth Century', in D. C. M. Platt (ed.), *Social Welfare, 1850–1950: Australia, Argentina and Canada Compared* (London, 1989), 69. Crider, 'Modernization and Human Welfare', 228–30. Crider also argues that public health was intended to help acclimatize new arrivals.

[23] Vezzetti, *La Locura en la Argentina*, 26; Crider, 'Modernization and Human Welfare', 180–95.

[24] Vezzetti, *La Locura en la Argentina*, 185–91.

[25] Sociedad de Beneficencia (SB), *Memoria* (1900, 1905, 1910, 1915, 1920, 1925, 1935, 1940) and *Hospicio* (1893, 1913, 1916, 1917, 1918, 1919, 1925, 1926, 1927, 1931, 1932, 1935, 1936, 1937, 1939).

[26] Ministerio de Relaciones Exteriores y Culto, Comisión Asesora de Asilos y Hospitales Regionales, Hospicio de las Mercedes: *Memoria Médico-Administrativo Correspondiente al Año 1939* (Buenos Aires, 1940), 5–15; Sociedad de Beneficencia de la Capital, *Memorias* (1900–43).

Europe tended to fill mental asylums rather than the countryside.[27] Several years later, writer Lucas Arrayagaray alerted readers that both asylums and gaols were full of immigrants who, shortly after arriving to Argentina, either committed a crime or went insane.[28]

Doctors offered a host of explanations for immigrants' predisposition to insanity. A common theme was their excessive desire for wealth, combined with love of drink, social dislocation, and the disappointment when immigrants' dreams of a new and better life went unfulfilled.[29] For many writers, immigration was tightly woven into the wider problem of modernity which, they argued, destabilized people's mental health. Paradoxically, mental illness was a price that nations paid for modernization, urbanization and immigration.[30] Perhaps the most common connection between social condition and medical diagnosis, however, was found in the image of the alcoholic male immigrant. Of the 7,339 alcoholics admitted to the Hospicio between 1899 and 1913, for example, 76 per cent were foreign born.[31]

Behind medical concern over unchecked immigration lay a more generalized fear for the future of the Argentina 'race'. Degenerationist models had been popularized in Argentina since the 1870s, and remained current well into the 1940s.[32] Fear of the biological and psychiatric danger posed by immigrant groups was aggravated by the fact that, unlike the United States or Canada, Argentina exercised little control over who was allowed to enter the country. Of the over 600,000 immigrants who came to Argentina between 1907 and 1910, for example, only sixty-five were excluded for mental or physical defects.[33] Unregulated immigration, and overcrowded hospitals also led to growing concerns during the 1920s and 1930s over the large number of mentally ill persons who presumably had evaded confinement to psychiatric hospitals.[34]

Ambivalence about immigrants, and fear of creating a welfare state, influenced the material and therapeutic life of the hospitals. Since their creation, both hospitals engaged in a moral therapeutic regimen that was heavily weighted towards productive and lucrative work. Argentine alienists began to implement

[27] J. T. Borda, 'Inmigrantes enfermos', *El Diario*, 12 January 1906.

[28] R. Huertas García-Alejo, 'La Aportación de la Escuela Argentina al Concepto de Criminal Nato', in Peset (ed.), *Ciencia, Vida y Espacio en Iberoamérica*, 111.

[29] Vezzetti, *La Locura en la Argentina*, 185–91; Little, 'The Society of Beneficence in Buenos Aires, 1823–1900', 242–3.

[30] See Archivo General de la Nación-Sociedad de Beneficiencia-Hospital Nacional de Alienadas, Legajo 203, Expediente 5057 (1924–6) (hereafter AGN-SB-HNA). Dr Luís Esteves Balado to Señor Director del Hospital Nacional de Alienadas, Dr Julio G. Nogues.

[31] *Memoria del Hospicio de las Mercedes Correspondiente al Año 1913, Memoria del Ministerio de Relaciones Exteriores y Culto Correspondiente al Años 1913–14* (Buenos Aires, 1914), 564.

[32] M. Ben Plotkin, 'Freud, Politics, and the Porteños: The Reception of Psychoanalysis in Buenos Aires, 1910–1943', *Hispanic American Historical Review* 77 (1997), 49–50.

[33] E. A. Zimmermann, 'Racial Ideas and Social Reform: Argentina, 1890–1916', *Hispanic American Historical Review* 72 (1992), 37.

[34] Ingenieros, *La Locura en la Argentina*, 234–5.

these regimens in their asylums by the 1870s.[35] Like their counterparts elsewhere, doctors understood moral therapy to include the fostering of safe and home-like environments, isolation of patients from harmful influences, and the development of activities that would redirect patients back to mental health.[36] The lack of appropriate indoor space for recreational activities hindered moral therapy's full development at the women's hospital; the men's hospital's programme was curtailed, among other factors, by a lack of appropriate indoor spaces and by the absence of a perimeter wall to prevent outdoor activities from turning into mass escapes.[37]

Work therapy broadly understood to include a variety of tasks, was successfully employed at the hospitals. Male patients essentially built the Casa de Dementes, and the asylum's first director created a number of patient workshops that continued to earn revenue for the hospital at least through the 1940s.[38] Female patients sewed uniforms for the military, and were actively engaged in a range of domestic tasks that allowed the hospital to continue running.[39] In addition to work that fulfilled putative therapeutic ends, the hospitals also utilized patients as unpaid orderlies.[40] Because of these challenges, as well as for reasons of hospital demographics and broader ideological reasons, the most prominent and long-lasting aspect of this approach was work therapy. In addition to justifying work therapy as a crucial component of recovery, directors also credited workshops for providing indispensable financial boosts to paltry hospital budgets.[41] This was particularly true of male patients who engaged in a number of lucrative enterprises and workshops. In addition, their labour was instrumental to the construction of the original Hospicio de las Mercedes in the 1870s and the construction of a men's rural facility that opened in 1899.[42]

[35] On moral therapy, see G. Grob, *The Mad Among Us: A History of the Care of America's Mentally Ill* (New York, 1994), 91.

[36] Little, 'The Society of Beneficence in Buenos Aires, 1823–1900', 259–61. See R. Porter, *The Greatest Benefit of Mankind: A Medical History of Humanity* (New York, 1997), 502. Already by 1838, the French alienist Jean-Etienne Dominique Esquirol was arguing for the therapeutic benefits of isolation. *Memoria del Hospicio de las Mercedes Correspondiente al Año 1893* (Buenos Aires, 1900), 31.

[37] *Hospicio* (1893), 14; Crider, 'Modernization and Human Welfare', 61, 103–4, 151; Little, 'The Society of Beneficence in Buenos Aires', 263.

[38] *Memoria del Hospicio de las Mercedes Correspondiente al Año 1893* (Buenos Aires, 1900), 4.

[39] Ingenieros, *La Locura en la* Argentina, 189.

[40] Crider, 'Modernization and Human Welfare', 151.

[41] The economic incentive of these programmes was so overpowering that the annual report of the men's hospital's rural colony read more like a business report than of a hospital. Colonia Nacional de Alienados: *Memoria Médico-Administrativo Correspondiente a los Años 1908–1910* (Buenos Aires, 1911).

[42] Vezzetti, *La Locura en la Argentina*, 75; Meyer, 'Los Comienzos del Hospicio de las Mercedes', 339; L. Iacoponi, 'El Hospital Interzonal Colonia Dr D. Cabred y el Método Open Door para asistencia y rehabilitación de pacientes psiquiátricos', in *Centenario de la Fundación: Hospital Interzonal Psiquiátrico 'Colonia Dr Domingo Cabred'* (Province of Buenos Aires, 1999), 64. Iacoponi argues that patients were selected for placement in the colony as much for medical

This emphasis on work therapy, which lasted well into the 1940s, dovetails the broader historical relationship between public health (and psychiatry in particular) and immigration. As previously mentioned, Argentine statesmen and elites recognized that a modern public-health system, and particularly the systematic control of epidemics, would make Argentina more appealing to migrants.[43] On the other hand, the government and leading intellectuals feared that an overly generous system would attract malingerers and so-called 'defectives' in search of the easy life. This concern was particularly marked with respect to psychiatric care, where, as psychiatrists looking at the problem of malingering readily admitted, it was not always easy to distinguish the insane from the simulator. Briefly, work therapy was understood as a means to distinguish the sick from the lazy.[44]

Work therapy was not the only manifestation of elite desire to curtail expenses. As was true of many other institutions for women and children, the voluntary work of elite women and of Catholic sisters bolstered faltering budgets at the HNA. First, the hospital was administered free of charge by the elite women's Society of Beneficence. Although many doctors considered this to be an anachronistic arrangement, the society's prominent position in the delivery of public-health care enjoyed government support well into the 1940s. A cornerstone of their administration was the work of an Italian based Catholic order, the Sisters of Charity. The sisters proved tireless supervisors of the largely untrained staff and served also to check pilfering, at a fraction of what it would have cost to hire professional nurses.[45]

Hospital reform

Despite an increasingly strong ideological imperative to re-order Argentine society and contain and control the growing immigrant population, by the 1880s, the men's and women's asylums were plagued by overcrowding, obsolete physical plants, an inability humanely to segregate violent patients, shortages of doctors and trained staff, and unreliable state support. In point of fact, the hospitals had existed in a state of perpetual crisis since their foundation. They had grown in physical size only through the haphazard addition of new buildings and wings and lacked any formal hospital design. The absence of clear legal

reasons as for particular skills that they possessed. Abuse of patient labour in the late nineteenth century is implied by the need in 1893 to regulate its practice. See 'Proyecto de Reglamentación del Trabajo y Peculio de los Alienados del Hospicio de las Mercedes', in *Memoria del Hospicio de las Mercedes Correspondiente al Año 1893*.

43 Escudé, 'Health in Buenos Aires in the Second Half of the Nineteenth Century'.

44 Vezzetti, *La Locura en la Argentina*, 70–9.

45 Archivo Hermanas de Caridad-Hijas de María Señora del Huerto, Hospital Nacional de Alienadas.

guidelines for the confinement, treatment and release of persons thought to be insane further complicated the system's ability to treat patients effectively.

Reforms of public health began to gain momentum by the early 1880s, spurred in large part by a series of fairly serious epidemics in earlier decades. In 1883, the municipal Public Assistance was created and charged to deliver public-health services to the city of Buenos Aires. It ran several of the city's hospitals, including the Hospicio until 1905, and also supervised preventative public-health programmes. Financial and political upheaval in the 1890s further reinforced the impulse to reform all public institutions in Argentina.[46] The imperative to modernize the hospitals was also impelled by the fact that they were located in a neighbourhood that originally was on the city's outskirts but as the city grew up and around the old asylums, it became increasingly difficult to ignore their presence.[47]

At the height of the push to reform and renovate Argentine society, the city's two asylums appointed new directors who each, in their own way, reflected the spirit of the times. Domingo Cabred (director from 1892 to 1916) and Antonio Piñero (director from 1890 to 1905), of the men's and women's asylums respectively, embarked on a series of reforms that proved highly successful in the *short term*. By Argentina's centennial in 1910, for example, the two asylums enjoyed reputations as model institutions, enviable by western European standards of care.[48] While providing important symbolic capital to the Argentine elite and middle classes, the renovated asylum would provide humanitarian treatment for those deemed mentally ill.

Cabred and Piñero received their medical training in Buenos Aires, but before and during their tenures as director, they embarked on voyages to Europe in search of models of psychiatric care that were applicable to Argentina.[49] In addition to visiting a wide assortment of urban and rural hospitals, as well as medical schools, the two doctors appear to have been well versed in European psychiatric literature. Although they seemed rarely if ever to coordinate their

[46] E. A. Zimmermann, *Los Liberales Reformistas: La Cuestión Social en la Argentina, 1890–1916* (Buenos Aires, 1994), 68; 105–8. This period also witnessed increased concern for a whole host of social ills that were often associated with Argentina's transition towards modernity and the market economy. Alcohol abuse, crime, prostitution and sexual deviance, insanity and housing were the principle issues of concern for the new generation of moral reformers. See R. D. Salvatore, 'The Normalization of Economic Life: Representations of the Economy in Golden-Age Buenos Aires, 1890–1913', *Hispanic American Historical Review* 81 (2001), 1–44.

[47] Scobie, *Buenos Aires: Plaza to Suburb, 1870–1910*, 56: 'The presence of hospitals, like that of garrisons and cemeteries, suggested that you were reaching the outskirts and the least desirable residential zones of the city.'

[48] For photographs of the hospitals in 1910, see J. Carlos Stagnaro and J. María Gonzales Chaves, *Hospicio de las Mercedes: 130 Años* (Buenos Aires, 1993); Ministerio de Relaciones Exteriores y Culto, Colonia Nacional de Alienados: *Luján (Provincia de Buenos Aires)* (Buenos Aires, 1910); *Albúm Histórico de la Sociedad de Beneficiencia, 1823–1910* (Buenos Aires, n.d.).

[49] O. Loudet and O. Elías Loudet, *Historia de la Psiquiatría Argentina* (Buenos Aires, 1971).

proposals, they shared a desire to reform the legal, medical and institutional frameworks of Argentine psychiatric care.

One of the most vexing problems facing the hospitals was the absence of a modern legal framework to guide the confinement and release of patients. As matters stood, patients were admitted and released from the hospitals with little or no oversight from the courts. Furthermore, although the national civil code of 1871 included several sections on confinement of the insane, they were unwieldy in their application and in practical terms made emergency confinements possible only by circumvention of the law. As a result of these discrepancies, a large percentage of patients (precise statistics were never produced) did not enjoy the benefit of legal protection against wrongful confinement. Many, perhaps a majority, suffered undue delays in release, and even theft of their possessions while under the care of the hospitals. Unlike in western Europe, the United States or Great Britain, there were no highly publicized cases of wrongful confinement in Argentina that might explain the motivation behind the proposals. It is clear, however, that Piñero, Cabred, and others, believed that a well crafted law would protect doctors from accusations of abuse and give patients and families great confidence in the legal transparency of confinement. Reformers likewise believed that the creation of legal protections for patients would improve Argentina's international reputation, and comparisons with western Europe permeated their discussion.[50]

In addition to concerns about inadequate legal guidelines, Piñero and Cabred had inherited obsolete institutions that had never met contemporary standards of care. The women's asylum, for example, had been forced to close in 1881 when its patient population was nearly double its 200-bed capacity.[51] Cabred's annual report from his first year as director, 1893, noted that the newer wards for tranquil and paying patients, created during an earlier spate of reform by his predecessor Lucio Meléndez, were in good shape. However, he noted that the overall condition of the hospital was poor, and that agitated and violent patients were housed in a primitive cellblock.[52] As such, these reform-minded doctors worried that their institutions' primary function had become custodial and not curative.[53] One of their principal goals, then, was to separate the patients who were deemed curable from those whose condition was believed to be irreversible and chronic. Overcrowding, of course, only made the task of separation all the

[50] See J. D. Ablard, 'Law, Medicine and Confinement in Twentieth-Century Argentina', in M. Ben Plotkin (ed.), *Argentina on the Couch: Psychiatry, State and Society in Modern Argentina* (Albuquerque, N. Mex., 2003).

[51] Ingenieros, *La Locura en la Argentina*, 194; Loudet and Loudet, *Historia de la Psiquiatria Argentina*, 43–6.

[52] *Memoria del Hospicio de las Mercedes Correspondiente al Año 1893*. For conditions under Cabred's predecessor, Lucio Meléndez, see Crider, 'Modernization and Human Welfare', 61.

[53] For the United States, see R. W. Fox, *'So Far Disordered in Mind': Insanity in California, 1870–1930* (Los Angeles, 1978), 40–1.

more difficult to achieve and had inhibited the adequate development of a variety of therapies.[54]

Segregation of patients along diagnostic and behavioural criteria required both the modernization of existing facilities and the creation of new auxiliary hospitals in rural areas. Aware of the development of psychopathic hospitals in western Europe and the United States, the directors sought to transform what they viewed as decaying and inhumane asylums into modern urban institutions for the short-term care of patients. Smaller urban institutions would also serve as teaching hospitals and centres of medical research.[55]

Piñero and Cabred embarked on ambitious reforms of their hospitals. The women's hospital was completely renovated – all of the old buildings were eventually replaced with buildings modelled on the latest hospital designs in Europe. In July 1898, the first buildings were inaugurated with great fanfare and publicity. The plan called for an 800-patient facility that would include a central kitchen, general dormitory wards, as well as wards for the agitated and violent, recreation and work shops, and wards designed exclusively for *pensionistas*, paying-patients.[56] Modifications of the men's hospital were more modest, but also included the development by 1910 of a clinic for acute patients and a faculty of psychiatry.[57]

Obviously, none of these reforms could work without transforming a small and untrained staff into one that could provide professional and reliable supervision of patients. Thus, hand in hand with the push to modernize the physical structures of the institutions, Piñero and Cabred sought to improve the quality of their staffs. Professional nurses were few in Argentina. The women's hospital had traditionally recruited its staff from recently arrived female immigrants – often going down to the docks when ships arrived. Thus, there was a strong imperative to develop training schools on hospital grounds, and improve the professionalism of existing staff.[58]

[54] Little, 'The Society of Beneficence in Buenos Aires', 251–5. For the Hospicio, see *Hospicio* (1893).

[55] Cabred specifically referred to the work of Gresinger and his idea of the small urban clinic. D. Cabred, 'Discurso Inaugural de la Colonia Nacional de Alienados', *Revista de Derecho, Historia y Letras* 1 (Buenos Aires, 1899), 619–20.

[56] Loudet and Loudet, *Historia de la Psiquiatría Argentina*, 211; *La Prensa*, 1 August 1898. The new hospital was designed with wards to accommodate the segregation of patients according to diagnosis *and* class.

[57] D. Cabred, *El Instituto Clínica de Psiquiatría de la Facultad de Medicina de Buenos Aires* (Buenos Aires, 1919). By the 1930s, due to severe overcrowding and staff shortages, the clinic ceased to function as Cabred had hoped.

[58] *Hospicio* (1893), 50; 'En el Hospicio de las Mercedes-Distribución de premios a los enfermeros', *Caras y Caretas* 6: 224 (17 January 1903); Sociedad de Beneficencia de la Capital, *Memoria del Hospital Nacional de Alienadas, 1900* (Buenos Aires, 1901). Training included making sure that all nurses were literate. Hospital regulations imposed fines on staff if patients hurt themselves or ran away.

Piñero and Cabred strongly believed that urban short-term facilities could only thrive if there were rural institutions for the care of chronically mentally ill patients. They hoped that recovery rates for acute patients at the urban institution would improve by being segregated from chronic patients. To that end, Cabred who had visited 'non-restraint' and cottage-system asylums in Scotland and Germany during the 1880s and 1890s, lobbied the national government for financial support. Likewise, Piñero published a treatise in 1893 on the Russian psychiatrist Kovalesky, in which he advocated work therapy and a rural-colony asylum.[59] In 1899 the Colonia Nacional de Alienados (National Colony for Insane Men), was inaugurated.[60] Piñero obtained funding for a similar facility for women in 1904, but it did not open until 1908, several years after the end of the doctor's tenure as director.[61] Well into the 1940s, both institutions operated as auxiliary hospitals to the older urban hospitals. In theory they were intended for patients in need of long-term care, as well as for those for whom some time in the countryside was seen as beneficial prior to their release.

Piñero and Cabred further recognized that one of the principal impediments to the development of their hospitals into modern institutions was that there was no national bureaucracy to address psychiatric or general-health problems. Of paramount concern was the fact that before 1914, there were no psychiatric institutions outside of Gran Buenos Aires, an area that included both the city and province of Buenos Aires.[62] The absence of public asylums in the majority of the nation's provinces was considered a problem of grave consequence for both national development and for the health of the nation's provincial insane. And, because of the geographic concentration of psychiatric resources in Buenos Aires, other provinces sent the mentally ill there for treatment.[63] This practice tended to have deleterious effects on the capacity of urban hospitals to keep patient numbers down, and also tended to encourage the abandonment of provincial patients, who often ended up far from family-support networks. Nationalization of the administration of psychiatric care promised to both expand the reach of psychiatric medicine to areas hitherto ignored, and to help the existing urban hospitals better serve patients within the surrounding areas.

[59] Loudet and Loudet, *Historia de la Psiquiatría Argentina*, 146.

[60] Cabred, 'Discurso Inaugural de la colonia Nacional de Alienados', 615–16. Cabred was particularly impressed by what he referred to as the 'open door' colonies in Scotland: Argyll, Fife, Kinross, Inverness, Haddington and Perth. Cabred also had visited the asylum at Gheel, Belgium.

[61] Loudet and Loudet, *Historia de la Psiquiatría Argentina*, 148.

[62] In addition to the HNA and the Hospicio, the province of Buenos Aires operated a small asylum. It received its first patients in 1881 under the name Melchor Romero. See, E. A. Balbo, 'El Hospital Neuropsiquiátrico 'Melchor Romero' durante los años 1884–1918', in Peset (ed.), *Ciencia, Vida y Espacio en Iberoamerica*, 53–75.

[63] Crider, 'Modernization and Human Welfare', 212–13.

In their proposals, reformers made frequent reference to the French law of 1838, which had nationalized both the criteria for confinement and had also ordered each department to establish an asylum.[64] More generally, they were impressed by the sheer number of asylums in Germany, the United States and England as compared with Argentina. In 1906, at Cabred's urging, the federal government created the Advisory Commission on Regional Asylums and Hospitals.[65] The commission's mission was to develop a plan for a national network of hospitals and asylums and to supervise their operation. During the next two decades, in addition to a number of general hospitals, the commission built a rural colony for the mentally retarded in Torres, Buenos Aires Province and a rural co-ed psychiatric facility in the western province of Córdoba, in the small village of Oliva.[66]

For his part, Piñero's proposal for the reform of Argentina's health system were more ambitious in scope and also more politically controversial. Since his arrival at the HNA, Piñero had joined a long line of well-established medical doctors who challenged Argentina to modernize its medical infrastructure. A particular target was the Society of Beneficence, whose control of *porteño* medical and charitable institutions for women and children was near hegemonic. Although he worked for the society in his capacity as director, Piñero repeatedly challenged its central role in the administration of health-care in Buenos Aires.[67] The conflict with the society came to a head in 1905, when he was relieved of his post. The following year, the doctor issued a stinging attack on the society in the halls of congress in which he urged the centralization of all public-health institutions under the Ministry of the Interior.[68]

Limits of reform

Despite the fact that their efforts coincided with a favourable national economic climate, Cabred and Piñero's reform bore little fruit. Indeed, by the early 1940s, hospital directors were writing memoranda complaining about the same legal, administrative and medical problems that the earlier reformers had tried to correct. Increasingly, psychiatrists and social commentators recognized that without greater financial commitment from the national state, few reforms would

[64] I. Dowbiggin, *Inheriting Madness: Professionalization and Psychiatric Knowledge in Nineteenth-Century France* (Berkeley, 1991). The 1838 law was expanded in 1909.

[65] The commission was Argentina's first effort at national co-ordination of health care until the creation in 1943, of the National Department of Health.

[66] Moses Malamud, *Domingo Cabred* (Buenos Aires, 1972) 39–46; D. Cabred, 'Asilo Colonia Regional de Retardos', *Archivo de Psiquiatría y Criminología* 7 (1908), 735.

[67] Leandri, 'Medicos, Damas y Funcionarios', 85. See also Crider, 'Modernization and Human Welfare', 38–9; Mead, 'Oligarchs, Doctors and Nuns', 42–52.

[68] On his dismissal, see *La Nación*, 10 November 1905; AGN-SB-HNA, Legajo 221, Expediente 'Libro 1905–07', 'Elementos de Juicio para Comprobar las Inexactitudes en que ha Incurrido el Dr A. F. Piñero en su Articulo Publicado por *La Nación* del 8 de Abril 1906'.

prove durable. As a result, during the next three decades patients experienced increased suffering, while doctors saw the status of the profession crumble. The failure to affect a national programme of psychiatric care condemned the existing institutions to overcrowding. Excessive patient population, in turn, prevented the development and implementation of progressive and modern medical techniques. Behind the failure to implement lasting reforms lay ideological and structural causes.

Most devastating to the long-term reform of the hospitals was the continued shortage of provincial and regionally based psychiatric care. This was in fact an old problem, but one that took on increasing importance as Argentina's network of railways developed and as the rural populace increasingly had easier access to Buenos Aires.[69] An early and important success by Cabred's commission had been the opening of the Co-ed Regional Colony Asylum for the Insane in 1914. Located in the village of Oliva in Córdoba province, the Oliva asylum was envisioned within a larger project to provide asylums throughout the interior provinces. Oliva had been designed to fulfil two somewhat contradictory goals; it was to both receive patients from the older urban institutions and also serve as a regional facility for the entire northwest sector of the country. With the plan only partially completed, the Oliva asylum found itself even more overburdened than the older facilities. Reflecting the economic imbalance between the provinces and Buenos Aires, Oliva operated with very meagre government allocations. At the end of 1934, Oliva's patient population was 4,000 but it only had capacity to treat 1,200 men and women.[70] By 1939, Oliva's director reported that 'throughout the hospital's history, [Oliva] has served as an escape valve for the metropolitan hospitals, while also receiving patients from both the capital and all the provinces close to Córdoba'.[71] Ironically, the metropolitan hospitals often complained that Oliva lacked the capacity to handle their patient overflow.[72]

The absence of a comprehensive network of hospitals throughout the republic affected not only Oliva but the metropolitan hospitals as well. Isolated and lacking in services, provinces continued to send their mentally ill on the long journey to Buenos Aires. The women's hospital, for example, found itself constantly reminding northwestern provincial governors to send their female patients to Oliva. The requests were rarely honoured. In 1914, for example, the director of the women's asylum sent out a memorandum to the governors of the northwestern provinces to send their insane to Oliva.[73] Compliance

[69] On the relationship between Buenos Aires and the provinces, see L. Sawyers, *The Other Argentina: The Interior and National Development* (Boulder, Colo., 1996).

[70] *Bulletin of the Oliva Insane Asylum* 3 (March 1935), 1–10.

[71] E. Vidal Abal, 'Twenty-Five Years of the Oliva Asylum', *Bulletin of the Oliva Insane Asylum* 7 (June–August 1939), 243–50.

[72] AGN-SB-HNA, Legajo 90, 'Transfer of Patients, 1914–1933' (19 November 1917).

[73] AGN-SB-HNA, Legajo 233, 'Libro March 15 – December 16, 1915' (11 August 1915).

was sporadic, however, as witnessed by the repetition of the request as late as 1926.[74]

Both hospitals also faced the constant pressure from the province of Buenos Aires. The province had had a provincial facility since the 1880s, but it was quite small, and was located in a rural area.[75] Furthermore, the majority of the province's population resided in urban areas contiguous to the city of Buenos Aires and also linked by transportation networks. In addition to the problem of dumping by provincial police, families frequently sought to have relatives confined in the city hospitals because they were more accessible than their own provincial institution. For these reasons, *bonarenses* (residents of the province of Buenos Aires) were the most numerous of the hospitals' provincial patients.[76]

These migrations (that paralleled the better known migration of provincial peasants to Buenos Aires in search of work) negatively impacted on the medical situation of many patients.[77] Writing on the national context of psychiatric care in 1933, the director of the Hospicio described the impact that long journeys had on patients' mental well being:

> Evidently the insane from all over the Republic come to these hospitals; this is prejudicial to the material and moral condition of the patient. Distance obliges in the majority of the cases, that the patient undergoes the primary evolution of his disease in the heart of the family, and in general is poorly observed and lacks correct treatment; in the best of cases this leads to the individual being brought to the local police station, where the saber replaces the bromide, or the local jail where they are placed in cells while they await transfer to a psychiatric hospital, when the number of insane grows to the point that it is economical. The prejudicial effects of this system are obvious... including the separation of the patient from his family. The proximity of the family is not only helpful for the patient, but also for the doctor for whom this is often the only way to acquire full knowledge of the patient's personal and family antecedents.[78]

Poor distribution of psychiatric care continued well into the late 1950s.[79]

The shortage of hospitals translated into chronic and perennial overcrowding of existing hospitals. The housing of numbers of patients far in excess of hospital capacity produced a vicious cycle – infectious disease spread quickly, curable patients' recovery was delayed by the chaos of the institutions, and also by the inability of doctors to attend to, or even keep track of, their patients. A media exposé of the HNA in 1918, that was based on reports by a former

[74] AGN-SB-HNA, Legajo 199, 'Traslado de Enfermas, 1914–1933', (3 August 1926). Dr Julio Nogués to Señoras Inspectoras.

[75] Balbo, 'El Hospital Neuropsiquiátrico "Melchor Romero" durante los años 1884–1918', 53–75.

[76] SB, *Memoria Correspondiente al Año 1940* (1941), 385.

[77] On the question of the interior provinces' economic development, and its impact on Argentina as a whole, see Sawyers, *The Other Argentina*.

[78] G. Bosch, *El Pavoroso Aspecto de la Locura en la República Argentina* [The Spectre of Madness in the Argentine Republic] (Buenos Aires, 1931).

[79] S. Bermann, 'Análisis de Algunos Datos de Estadística Psiquiátrica', *Acta Neuropsiquiátrica* 5 (1959), 150–60.

doctor, noted that female patients with a range of physical and psychiatric ailments continued to be housed indiscriminately in large dormitories.[80] A 1925 internal memorandum from the women's hospital, written when the institution was almost 800 patients over capacity, eloquently expresses the scope of the problem:

> What would we say of a surgeon who operated in the old chambers of the Men's Hospital – or of a service for sick children where patients with pneumonia, whooping cough, and measles were all mixed together? The same should be said of a psychiatric hospital in which 700 patients are treated in a ward built for 250, where the excited, depressed, persecuted and anxious are all mixed together.[81]

Similarly, a 1923 congressional investigation of the Hospicio, reported that the hospital was unable to separate patients with communicable diseases from the other patients. Likewise, overcrowding and a deteriorating physical plant precluded separate wards for patients with different psychiatric diagnoses.[82]

In this respect, the case of patients who were considered violent was of particular concern. The cell-blocks for violent patients were only razed after both Piñero and Cabred had left their posts; the women's Media Luna (Half-Moon), was only torn down in 1913 and that of the *Hospicio* only in the 1920s.[83] Violence likewise remained a topic of great concern both in the medical literature and in the popular press. In addition to a spate of murders of doctors by patients and of patients by staff, there was also a small uprising in the men's hospital which further brought home the point of the level of chaos inside national psychiatric institutions.[84]

Ironically, efforts to segregate patients were frustrated by delays in the completion of Piñero and Cabred's reforms. As early as 1911, the HNA's director Dr Esteves complained that the as yet unfinished project initiated by his predecessor in 1894 was already obsolete. From his perspective, the plan called for too many dormitories, which would end up housing a population in excess of the hospital's therapeutic capacity. In sum, the old plan reinforced the hospital's tendency to admit large numbers of patients.[85] Even the more modest physical plant reforms at the Hospicio were delayed through the end of Cabred's tenure as director.[86]

80 'Hospicio Nacional de Alienadas – No Hay Legislación Sobre esta Material', *La Unión*, 24 July 1918.

81 AGN-SB-HNA, 1923–1947. 'Expediente 6323', 'Señoras Inspectoras del HNA from Presidenta del SdB, Feb. 7, 1925'.

82 Argentina. Congreso Nacional. Daily Sessions of the Chamber of Deputies. *Diputados* (14 March 1923), 457–60.

83 Balbo, 'Argentinian Alienism from 1852 to 1918', 185.

84 Violent incidents in the hospitals rarely made it to the professional journals. For newspaper reports about violence in the hospitals, see Ablard, 'Madness in Buenos Aires', 184.

85 AGN-SB-HNA, Legajo 221, Expediente 'Libro 1910–1911', Dr José Esteves to Señoras Inspectoras, 16 February 1911.

86 *Memoria Correspondiente al Año 1913–14* (1915), 569–70.

Miserable conditions also hampered the financial solvency of the hospitals. Since their foundation, both hospitals had relied on a small number of paying patients to bolster meagre hospital budgets. As early as the 1880s, the Hospicio's director had advised the city government that improving the hospital's physical conditions would help attract paying patients, whose financial contribution would in turn help keep the hospital solvent.[87] Despite often precarious conditions, both hospitals at the beginning of the twentieth century, were attracting a healthy number of paying patients. At the HNA, for example, 20 per cent of all patients were from first- to fourth-class pensioners. By 1920 – the last year that statistics were kept in this category – however, the number was down to a paltry 7 per cent.[88] At the Hospicio, the administration was unable to acquire the funds necessary to construct a ward for paying patients, which in turn diminished the institution's appeal to paying patients.[89]

Similarly, and in large part because of problems of overcrowding and budget constraints, neither institution was able to offer attractive work conditions or professional opportunities to hired staff or doctors. A 1923 report by the director of the HNA complained that the pay scale for his nurses and aides was lower by a third than at comparable municipal hospitals: 'All the best ones leave, and those who replace them are increasingly worse. Often they are women just off the boat, who have never cared for an ill person. Worse still, our desire not to hire illiterates is made impossible by the slim choices available.'[90] Doctors also suffered, as witnessed by a 1935 women's hospital report on the careers of psychiatrists in public facilities. The report complained that most doctors in public facilities worked without the benefit of a salary.[91] Rural hospitals faced the additional challenge that they were often located in isolated areas that offered few if any amenities that the average middle-class Argentine might expect. Once again, low levels of economic development in the interior worked against the possibility of creating viable hospitals.

The crisis of public psychiatric care, from both the local and national perspective, was aggravated by the failure of psychiatrists to convince congress to pass laws governing the confinement of the insane. No national law governing the confinement of the insane passed congress until 1983 despite repeated efforts by some psychiatrists. The lack of clear legal guidelines, as psychiatrists complained, gave the public all the more reason to suspect hospitals as being

[87] Crider, 'Modernization and Human Welfare', 151.

[88] Sociedad de Beneficencia, *Memorias del Año 1903* (Buenos Aires, 1904); Sociedad de Beneficencia, *Memoria del Año 1920* (Buenos Aires, 1921).

[89] *Hospicio de las Mercedes: Memoria Medico-Administrativa Correspondiente al Año 1925* (Buenos Aires, 1926), 1–13.

[90] AGN-SB-HNA, Legajo 201, Expediente 8051 (30 December 1923). Dr Esteves to the Señoras Inspectoras.

[91] AGN-SB-HNA, Legajo 208, Expediente 8172, 'HNA-Reglamentación de la Carrera Hospitalaria en el Establecimiento' (12 April 1935).

dangerous places. It was, after all, common knowledge, that most patients in public and private institutions were confined and released without the benefit of judicial oversight.[92]

Conclusions

The failure of the national state to expand and improve psychiatric care in Argentina had profound repercussions on both patients and doctors. For patients, continued overcrowding and a shortage of qualified staff and doctors resulted in excessively prolonged hospital stays which were often characterized by physical discomfort, neglect, and even danger. While the ideal of institutionalizing all mentally handicapped persons had great currency in Argentina by the 1930s, the hospitals found themselves unable to accommodate existing patients.

The promise and failure of the *porteño* mental-health system fits into a larger political, administrative, and social history of twentieth-century Argentina. The cause of the hospitals' decline can be attributed to a number of factors of which overcrowding was but a symptom. First, the hospitals operated within an administrative arrangement that probably fostered indifference at top levels of government. Until the creation of the National Department of Health within the Ministry of the Interior in 1943, oversight for the two hospitals was under the Ministry of Foreign Relations and Religion. Although the particular arrangements differed for the two institutions, it is clear that public health had been appended to a ministry with little time or expertise to devote to such matters. This administrative anomaly fitted into a broader pattern of health-care provision in Argentina, wherein authority for medical establishments was widely dispersed among federal and municipal bureaucracies.[93]

Furthermore, the fact that the national government failed to provide direct or financial assistance to the country's most needy had a negative impact on the hospitals' finances and may explain the high ratio of indigent to paying patients in the hospitals. As historian John Fogarty has observed for this period, 'A characteristic of the Argentine approach to welfare is that it was concerned overwhelmingly with the welfare of working people rather than the destitute, and it was generally implemented on an industry-by-industry basis rather than universally.'[94] Even under Juan Perón, social security remained closely linked to workplace and union affiliation, and therefore was limited in its reach.[95] The

[92] Ablard, 'Law, Medicine and Commitment in Twentieth-Century Argentina'.

[93] H. Recalde, *La Salud de los Trabajadores en Buenos Aires (1870–1910) a Través de las Fuentes Medicas* (Buenos Aires, n.d.), 322.

[94] J. Fogarty, 'Social Experiments in Regions of Recent Settlement: Australia, Argentina and Canada', in Platt (ed.) *Social Welfare, 1850–1950: Australia, Argentina and Canada Compared*, 190.

[95] M. Ben Plotkin, 'Politics of Consensus in Peronist Argentina (1943–1955)', PhD thesis, University of California at Berkeley (1992), 290–318.

pattern of neglecting the most needy was even worse in those provinces that did not share the prosperity of Buenos Aires. For economic, political and social reasons many regions of the republic lacked adequate psychiatric facilities and thus contributed every year a large proportion of the hospitals' patients. This trend was probably accentuated by the increasing flood of provincial migrants coming to Buenos Aires in the 1930s in search of work in the burgeoning industrial sector.[96]

Matters were made worse by the fact that much of the hospitals' funding came from national lotteries rather than a fixed and stable budget. Usually, the expansion of a hospital or the construction of a new one was at least partially funded through lotteries. The Santa Fé provincial asylum in Oliveros that opened in the late 1930s, for example, was funded largely from the earnings of the provincial lottery.[97] Likewise, the hospitals remained dependent on fees paid by pensioners and the value of products made by indigent patients.

The hands-off attitude of successive governments to the health needs of its citizens is further illustrated by a series of curious exchanges between the national government and the Society of Beneficence. Despite the precarious existence of the HNA, in 1931, and again in 1942, the Ministry of Finance offered the society administrative responsibilities for all facilities that were currently under the Advisory Commission on Regional Asylums and Hospitals. In both cases, the president of the Society conferred with the Minister of Foreign Relations and Religion and decided that the vagaries of national budgets made additional responsibilities imprudent. The attempt to dump additional health-care responsibilities on the society is shocking since generations of medical doctors had criticized the society's control of so many hospitals and other welfare institutions.[98]

The desire to reorganize and modernize Argentine health care became a political topic throughout the 1930s and 1940s. In September 1933 Socialist deputy Angel M. Giménez proposed the creation of a National Department of Social Welfare to be run out of the Ministry of the Interior. Like other critics, Giménez noted that health care in Argentina was disorganized and lacked co-ordination. For Giménez, the impact of the world depression on Argentina had brought into clearer focus the deficits of health care; increased poverty and desperation had had a residual effect on the functioning of hospitals and asylums.[99] Furthermore, Argentina relied excessively on antiquated notions of

[96] Sawyers, *The Other Argentina*, 183–4.

[97] Argentina. Congreso Nacional. *Diario de Sesiones de la Cámara de Diputados* (12 September 1941), 365. For lottery funding at the HNA, see Sociedad de Beneficencia, *Memoria Correspondiente al Año 1942*, 23.

[98] Sociedad de Beneficencia, *Memoria Correspondiente al Año 1942*, 15–17.

[99] A. Giménez, *Por la Salud Física y Mental del Pueblo* 2 (Buenos Aires, 1938), 50. 'La situación se ha agravado en los últimos tiempos con la profunda crisis, la masa enorme de desocupados y la casi pauperización del proletariado argentino con salarios insuficientes y condiciones de vida precaria.'

charity that, while often well intentioned, did not allow for the development of an efficient, modern and preventative health system. On the last point, Giménez was adamant, and he noted that 'we make [natural] selection in reverse, with the best intentions we bastardize the race. Medical help always arrives too late, when the condition is irreversible.'[100] Giménez, clearly influenced by eugenics, continued that the responsibility of the modern state was to help keep its fit citizens healthy; charity tended only to take notice at the point when an individual was often too far gone to benefit from care.[101]

To correct the situation, Giménez proposed the creation of a national health office that would oversee the diverse health establishments; while respecting the autonomy of such institutions as the Society of Beneficence and private hospitals, the new office would assure better co-ordination of services and hopefully eliminate extraneous or duplicated services. Interestingly, Giménez afforded guarded respect for the work of the society; citing his own earlier attack on the society from 1915, Giménez acknowledged that the society ran good hospitals and that their officers rarely interfered with decisions made by medical professionals.[102] By contrast, Giménez was highly critical of Domingo Cabred's creation, the Advisory Commission of Regional Asylums and Hospitals. The commission had deviated far from its original purpose of proposing and funding the construction of hospitals, and in the last decade had become a 'sui generis office of national public assistance'.[103] Giménez further noted that the commission tended to construct identical institutions, a 'one size fits all' policy that ignored the specialized needs of distinct institutions and their clientele.

The groundswell for a reorganization of national health care was also found in more conservative writers. Eduardo Crespo of the daily *La Nación* complained in 1936 that there was administrative disorganization in all Argentine health care. The hospitals operated under the ministry of Foreign Relations and Religion, an office ill-suited and uninterested in the task. The creation of a new ministry would have required an amendment to the constitution – a task too difficult to risk. Crespo pointed out that half of all patients in *porteño* hospitals were immigrants, and a full quarter came from outside the city limits. Clearly then provincial governments and immigrant and ethnic mutual-aid societies and hospitals were not pulling their weight. In addition, Crespo noted that only 10 per cent of all hospitalizations in Argentina were private, whereas in the United States the figure was closer to 30 per cent.[104]

[100] *Ibid.*, 49. 'Hacemos una selección al revés, con los mejores propósitos bastardeamos la raza, llegando tarde, cuando el mal es irreparable.'

[101] *Ibid.*, 49. [102] *Ibid.*, 56.

[103] *Ibid.*, 51. 'Destinada a asesorar y construir obras, por decretos sucesivos, se le ha dado atribuciones directivas, técnicoadministrativas de los establecimientos que creaba, viniendo a constituir una sui generis asistencia pública nacional.'

[104] E. Crespo, *Nuevos Ensayos Politicos y Administrativos* (Buenos Aires, 1938), 113–24.

Argentine scholar Hugo Vezzetti has asserted that for doctors 'the hospitalized insane are an emblem of the advances of western civilization, and the necessary price for the construction of a modern nation and a vigorous race'.[105] Mid-nineteenth-century social reformers had claimed that barbarity, understood as the antithesis of modern European 'civilization', produced mental imbalance. Yet throughout the first half of the twentieth century, by many psychiatrists' reckoning, Argentine psychiatrists increasingly viewed modernity as the cause of mental imbalance. And as modern as Argentina was, it lacked the resources to alleviate the consequences of its own social transformation.

[105] Vezzetti, *La Locura en la Argentina*, 225.

10 Becoming mad in revolutionary Mexico: mentally ill patients at the General Insane Asylum, Mexico, 1910–1930

Cristina Rivera-Garza

On 18 July 1921, Modesta B., a thirty-five-year-old single woman, arrived at the General Insane Asylum – the largest state institution devoted to the care of mentally ill men, women and children in early twentieth-century Mexico.[1] Days earlier, a police agent had apprehended her because she was allegedly involved in a street fight. As she did not calm down in gaol, authorities called on the expertise of a licensed doctor who diagnosed her as mentally unstable. Modesta B. passionately rejected his diagnosis but, poor and lacking a supportive network, her escalating agitation was only used as further evidence of mental derangement. Following a standard procedure, the government of the Federal District then issued an order requesting her committal to the asylum. Located in Mixcoac – a village to the south of the capital city where members of the elite relaxed over the weekends – the once imposing medical facilities lingered in utter decay.[2] Indeed, little if anything was left of the glamour that surrounded the inaugural ceremony of the massive architectural complex on 1 September 1910 – an act that opened the nationalistic festivities of the centenary of the Independence of Mexico under the vigilant gaze of aging president Porfirio Díaz.[3] During his thirty years in office, General Díaz led the nation through an

[1] 'Modesta B., 1921' from research conducted at the Archive of Public Health, General Insane Asylum, section of clinical files, box: 105, file: 6639, 1921 (hereafter cited as AHSSA), F:MG; S:EC; C:105; Exp:16 (6639). Last names of inmates have been omitted. For a fictional recreation of this patient's life in early twentieth-century Mexico, see C. Rivera-Garza, *Nadie me Verá Llorar* (Mexico, 2000); *No One Will See Me Cry* (Curbstone, 2003).

[2] For an oral history of the Mixcoac neighbourhood in Mexico City, see P. Pensado and L. Correa, *Mixcoac. Un Barrio en la Memoria* (Mexico, 1996). An analysis of urban development in Porfirian Mexico City is Ariel Rodríguez Kuri, *La Experiencia Olvidada. El Ayuntamiento de la Ciudad de México: Política y Gobierno, 1876–1912* (Mexico, 1996). Also, M. Johns, *The City of Mexico in the Age of Díaz* (Austin, Tex., 1997); P. Picatto, 'Urbanistas, Ambulantes, and Mendigos: The Dispute for Urban Space in Mexico City, 1890–1930', in C. A. Aguirre and R. Buffington (eds.), *Reconstructing Criminality in Latin America* (Wilmington, Del., 2000).

[3] The event received a lot of attention. See G. García, *Crónica Oficial de las Fiestas del primer Centerario de la Independencia de México: Apéndice* (Mexico, 1911), 58–60; M. Tenorio-Trillo, '1910 Mexico City: Space and Nation in the City of the Centenario', *Journal of Latin American Studies* 28 (1996), 75–104; I. Ruiz López and D. Morales Heinen, 'Los Primeros Años del Manicomio General de la Castañeda (1910–1940)', *Archivo de Neurociencia* 1(1996), 124–29; S. Ramírez Moreno, *La Asistencia Psiquiátrica en México* (Mexico, 1950); G. Calderón Narváez,

era of rapid social transformation based on growing involvement in the global economy and centralization of political power, which came to an end with the outbreak of the Mexican revolution in November 1910.[4] Drastically affected by the social upheaval, the state asylum faced severe financial deficits that resulted in chronic overcrowding, physical ruin, and poor staffing. This was the place in which Modesta B. became officially mad in that summer of 1921.

As with all human experiences, Modesta B.'s internment was unique. Not all inmates, for example, were so adamant against a doctor's diagnosis and few left the lengthy narrative she produced while in confinement.[5] Yet, admission registers show that her age, civil status, and class standing were common among the population served by the institution. Furthermore, official documents also demonstrate that her pathway into the insane asylum was fairly representative of the bureaucratic and medical procedures used to justify committal during the early revolutionary period in Mexico. While informative, these trends only raise new questions. What were the continuities and discontinuities in strategies of confinement in an institution that bridged the Porfirian regime and the revolution, eras usually construed as oppositional in Mexican history? Did Porfirian ideology, which permeated the planning and construction of the asylum, influence admission policy after 1910? What was revolutionary about the procedures through which men and women became asylum inmates from 1910 to 1930? More generally, how did men and women become mad in early revolutionary Mexico? This chapter attempts to answer these questions by using admission registers and medical files of the asylum from 1910 to 1929, the year in which, mirroring larger social processes, the institution underwent medical and administrative reform.

The case of the Mexican asylum is particularly important to the historiography of confinement because it shows the peculiar malleability of institutions of mental health especially when they emerge, as did the General Insane Asylum, in a time of profound social transformation.[6] An expensive project that

'Hospitales psiquiátricos de México. Desde la Colonia hasta la actualidad', *Revista Mexicana de Neurología y Psiquiatría* 7 (1966), 111–43.

[4] Scholarship on the transition from Porfiriato to revolution is large in modern Mexican history, see J. Hart, *Revolutionary Mexico. The Coming and Process of the Mexican Revolution* (Berkeley, 1989). While controversies abound regarding chronology, most scholars agree that the early period of the revolutionary war extended from 1910 to 1917, the year in which a new Mexican constitution was issued. While the armed conflict tended to decrease in the following years, most scholars have argued that it was not until the presidency of Lázaro Cárdenas, 1934–40, that the revolution came to an end with the emergence of agrarian reform programmes. Still others have argued that, while interrupted, the revolution continues to the present day. See A. Gilly, *La Revolución Interrumpida: Mexico 1910–1920. Una Guerra campesina por la tierra y el poder* (Mexico, 1971).

[5] Modesta B. called her writings 'diplomatic dispatches'.

[6] Historical studies of psychiatric practice in Latin America include P. Farmer, 'The Birth of the *Klinik*: A Cultural History of Haitian Professional Psychiatry', in A. D. Gaines (ed.), *Ethnopsychiatry. The Cultural Construction of Professional and Folk Psychiatrists* (Albany, NY, 1992).

developed over a period of twenty-four years, the asylum embodied almost to perfection the modernizing agenda of the Porfirian regime.[7] Asylum designers who perceived themselves as heralds of a new era in the late nineteenth century emphasized, for example, their skillful use of foreign sources in the conception of the institution. Likewise, as positivist thinkers, they stressed the thorough empirical research that led to the asylum's planning and construction.[8] In the same pragmatic vein, they accentuated the medical over the merely charitable functions of the modern welfare establishment.[9] Furthermore, both its final location in the periphery of Mexico City and its monumental architectural design reflected concerns with order and progress, a dyad intimately associated with the Porfirian era.[10]

Soon, however, the Mexican revolution reshaped the destiny of both asylum and nation. A complex phenomenon that mobilized peasants, workers and members of the middle classes throughout the country, the Mexican revolution left little untouched in the nation.[11] The asylum was no exception. In fact, by 1914, the expanding meadows of the institution became battlefields in the conflict.[12] The conditions of the facilities deteriorated throughout the first decade of the twentieth century and did not improve when the armed phase of the revolution came to an end in 1917 with the issuing of the Mexican constitution. Facing immediate budgetary problems, the General Insane Asylum became a forsaken entity where a reduced number of physicians could hardly treat, much less cure,

Also, A. Ruiz Zevallos, *Psiquiatras y locos. Entre la Modernización Contra los Andes y el Nuevo Proyecto de Modernidad. Perú: 1850–1930* (Lima, 1994). For an intellectual history of psychoanalysis in Argentina, see M. Plotkin, 'Freud, Politics, and the Porteños: The Reception of Psychoanalysis in Buenos Aires, 1910–1943', *Hispanic American Historical Review* 77 (1997), 45–74. A social history of insane asylums in Argentina is J. Ablard, 'Madness in Buenos Aires: Psychiatry, Society, and the State in Argentina, 1890–1983', PhD thesis, University of New Mexico (2000).

[7] For cultural analyses of the process of Porfirian modernization in Mexico, see M. Tenorio-Trillo, *Mexico at the World Fairs: Crafting a Modern Nation* (Berkeley, 1996); W. Beezley, *Judas at the Jockey Club and Other Episodes of Porfirian Mexico* (Lincoln, Nebr., 1987).

[8] For an analysis of positivism in Latin America, see B. Burns, *The Poverty of Progress. Latin America in the Nineteenth Century* (Berkeley, 1980). For the case of Mexico, see C. Hale, *Mexican Liberalism in the Age of Mora 1821–1853* (New Haven, Conn., 1968).

[9] See J. Felix Gutiérrez del Olmo, 'De la caridad a la asistencia. Un enfoque de la pobreza y la marginación en México', in J. Felix Guitérrez del Olmo (ed.), *La Atención Materno-Infantil. Apuntes Para su Historia* (Mexico, 1993), 9–51.

[10] C. Rivera-Garza, 'An Architecture of Mental Health: the Planning and Construction of the General Insane Asylum, 1884–1910', manuscript.

[11] G. Joseph and D. Nugent (eds.), *Everyday Forms of State Formation: Revolution and the Negotiation of Rule in Modern Mexico* (Durham, NC, 1994); A. Knight, *The Mexican Revolution* (Cambridge, 1986); R. Ruiz, *The Great Rebellion: Mexico, 1905–1924* (New York, 1980); Hart, *Revolutionary Mexico*.

[12] 'Diversos. La ocupación del establecimiento por fuerzas zapatistas', AHSSA. F:BP; S:EH; Se:MG; Lg:4; Exp:28. 'Diversos. Tiroteo zapatista', AHSSA. F:BP; S:EH; Se:MG; Lg:4; Exp:19, 1. 'Diversos. Ocupatción Zapatista y Constitucionalista', AHSSA. F:BP; S:EH; Se:MG; Lg: 4; Exp: 37.

increasing numbers of patients suffering mostly from chronic conditions. Even then, the revolutionary regimes led by generals Alvaro Obregón (1920–4) and Plutarco Elías Calles (1928–32) did not opt for closing its doors. Not only had the new constitution included an article granting the right to health care for all Mexicans, but the revolutionary regimes also infused welfare ideology with notions of social justice and reform.[13] Indeed, according to the board responsible for the management of welfare institutions the future of the nation rested on state enterprises, such as the asylum, in which individuals could be treated and rehabilitated in order to become useful citizens. Thus, while the institutional framework of the asylum remained unaltered between 1910 and 1930, a growing focus on the welfare responsibilities of the state brought new social and medical demands to the establishment. In carving a place in revolutionary Mexico, asylum authorities slowly but surely underscored the strategic position of the institution as a centre for mental rehabilitation and social reform, an effort especially evident in the attempts to reorganize the asylum according to the principles of work therapy that began in 1929. This process, however, not only came as a response to external pressures. Asylum actors – patients and doctors, authorities and staff members – actively participated by engaging in an animated debate about the defining characteristics of mental illness and the social meanings of the institution designed to treat it. This debate was particularly visible in the set of official and unofficial rules that determined admission verdicts. Based on an exploration of inmates' profiles, this chapter first argues that, from 1910 to 1930, asylum authorities, police agents, and members of the community transformed the institution to better fit the welfare ideology and practice of early revolutionary Mexico.

While the singularities of the Mexican asylum were many, this establishment also shared basic characteristics with state mental-health institutions worldwide. As in these other asylums, Mexican state agents played an active role in the detection, apprehension and confinement of men and women suspected to be mentally ill.[14] In 1910, for example, some 85 per cent of admissions involved a government order – a document usually issued at the request of the police, welfare authorities, or city gaol officials.[15] This figure remained roughly

[13] For a detailed description, see A. Mazzaferi, 'Public Health and Social Revolution in Mexico: 1870–1930', PhD thesis, Kent State University, Ohio (1968). Also, 'El derecho del hombre a la salud', in J. Alvarez Amezquita *et al.* (eds.), *Historia de la Salubridad y la Asistencia en México* (Mexico, 1960), 72.

[14] Noting different levels of state agents' participation, similar procedures have been detected in California, Ireland, Nigeria, England, and Argentina. See R. W. Fox, *So Far Disordered in Mind. Insanity in California, 1870–1930* (Berkeley, 1978); J. Sadowsky, *Imperial Bedlam. Institutions of Madness in Colonial Southwest Nigeria* (Berkeley, 2000); M. Finnane, *Insanity and the Insane in Post-Famine Ireland* (London, 1981); J. K. Walton, 'Lunacy in the Industrial Revolution: A Study of Asylum Admissions in Lancashire, 1848–1850', *Journal of Social History* 13 (1979), 1–22; J. Ablard, 'Madness in Argentina'.

[15] Based on a random sample of 100 files from 1910. Means of arrival.

unchanged over the next two decades. Nevertheless, in at least 15 per cent of the internments, family members – mothers and fathers, husbands and sisters – played significant roles too, adding complexity to views that represent asylums as mere institutions of state control.[16] Just as in Europe and the United States, these Mexican figures have raised new questions about the social nature of insane asylums as well as the various and contesting ways in which governments and patients have appropriated these medical institutions over time and across cultures.[17] This chapter adds to this debate arguing that, in the Mexican case, both families and the state placed social and medical demands on the asylum, which ultimately defined the roles of the institution as both a site of control and a place of refuge and assistance in early revolutionary Mexico.

Because information about the planning and construction of the Mexican asylum is both scarce in Mexican historiography and relevant to this inquiry, this chapter first traces the medical and social debates that shaped the design and construction of the institution from roughly 1884 to 1910. It then explores the ways in which revolutionary politics affected the medical and welfare services of the establishment. Such analysis leads into further exploration of the actors that gave life to the institution, especially the men and women who became inmates and whose social and demographic characteristics greatly shaped the various roles played by the asylum in society at large. Further, an analysis of asylum diagnoses illustrates the changing ways in which doctors, patients and families defined mental illness in early revolutionary Mexico. Special attention is given to cases of epilepsy, mental retardation and alcoholism, the largest diagnostic groupings during this era. While cases of moral insanity were not as numerous, examination of these files illuminates medical and social discussions about the place of women in revolutionary society. The chapter ultimately argues that asylum inmates' profiles shed light on the negotiation through which state agents and family members defined mental illness, an aspect of growing importance for revolutionary regimes concerned with the reconstruction of the nation. In fact, paraphrasing the discourse of revolutionary welfare authorities that perceived physical health as a national asset, asylum authorities soon claimed that

[16] A representative selection of views of insane asylums as institutions of social control includes M. Foucault, *Madness and Civilization: A History of Insanity in the Age of Reason* (New York, 1965); D. Rothman, *The Discovery of the Asylum. Social Order and Disorder in the New Republic* (London, 1971); E. Goffman, *Asylum: Essays on the Social Situation of Mental Patients and Other Inmates* (New York, 1962).

[17] As research on the history of asylums grows, views of mental-health institutions have become more complex. A representative sample of revisionist literature includes E. Dwyer, *Homes for the Mad. Life Inside Two Nineteenth-Century Asylums* (New Brunswick, NJ, 1987); G. Grob, *Mental Institutions in America: Social Policy to 1875* (Basingstoke, Hants., 1973); A. Digby, *Madness, Morality and Medicine: A Study of the York Retreat 1796–1914* (Cambridge, 1985); L. Bachrach, 'Asylum and Chronically Ill Psychiatric Patients', *American Journal of Psychiatry* 141(1984), 975–8; A. Goldberg, *Sex, Religion and the Making of Modern Madness. The Eberbach Asylum and German Society, 1815–1849* (Oxford, 1999); Sadowsky, *Imperial Bedlam*.

caring for the mental health of the community was central to the future of the country.[18]

The modern asylum: the rise of the General Insane Asylum, 1884–1910

The history of mental-health care in Mexico dates back to the early colonial times when private parties with the support of the Catholic church established the San Hipólito and the Divino Salvador, hospitals devoted to the care of mentally ill men and women respectively.[19] Almost four centuries later, the emergence of the General Insane Asylum in 1910 represented the transition from custody and charity to therapy and correction.[20] While the hagiography of the asylum usually presents this change as a result of a compelling 'moment of realization', the history behind the walls of the modern medical facilities was long and less spontaneous.[21] In fact, the planning of the institution involved a complex process which included translating foreign sources – mostly documents from insane asylums located in the United States and France – and a careful consideration of local conditions in which both the improvement of psychiatric treatments and concerns with the ordering of society played important roles. These social and medical variables produced paradoxical views of institutions for the insane as both sites of control *and* places of refuge. From beginning to end, then, there was an ambivalence as to the exact nature of the asylum project.

The early inception of the asylum took roots in a stable society that enjoyed high rates of economic growth. Known as the golden era of the Porfirian regime, the last two decades of the nineteenth century witnessed the emergence of a myriad of urban, medical and social projects with which Porfirio Díaz and his cabinet – formed by self-named *científicos* (scientists) – expected to ratify the modern character of the regime.[22] The costly drainage works that saved Mexico City from recurrent floods, the erection of public buildings bearing the influence of French architecture, the implementation of the telephone system, and even

[18] A. Torres, 'El Manicomio General', *Revista de la Beneficencia Pública*, 34–8 (1917), 30–2.

[19] For a brief history of the San Hipólito hospital, see J. S. Leiby, 'San Hipólito's Treatment of the Mentally Ill in Mexico City, 1589–1650', *The Historian* 54 (1992), 491–8. For a history of the Divino Salvador hospital, see C. Berkstein Kanarek, 'El hospital Divino Salvador', thesis, Universidad Nacional Autónoma de México, 1981.

[20] See Gutiérrez del Olmo, 'De la caridad a la asistencia', 9–51.

[21] S. Ramírez Moreno, a well-known psychiatrist in early revolutionary Mexico, referred to the inception of the asylum as a 'moment in which Porfirio Díaz realized that something had to be done to treat the insane in proper ways', in *La Asistencia Psiquiátrica en México*, 9. (All Spanish translations into English are by the author.)

[22] The científicos were also known as the wizards of progress. See Tenorio-Trillo, *Mexico at the World Fairs*.

the renaming of the streets turned the capital city into the showcase of the era.[23] However, while supporters were generally optimistic, the rapid pace of social change also produced anxiety and trepidation. Massive land dispossessions in the countryside and industrial growth in urban areas prompted a migration of peasants into Mexico City. Dark-skinned and poor, immigrants soon became a source of concern among city designers and social commentators for whom their ethnicity, class origins and lifestyles not only embodied the antithesis of modernization but also represented a social threat. Porfirian analysts thus unleashed an unprecedented effort to identify and control potentially dangerous members of society, especially targeting criminals, prostitutes, alcoholics and the insane.[24] Committed to the protection of society, experts unabashedly supported the creation of institutions able to contain the pernicious influence of what they perceived as wayward men and women. Authorities of the Public Welfare Administration, which had been secularized since 1861, quickly responded to the challenge. Unlike religious welfare institutions that worked on the principles of charity, Porfirian welfare ideology developed a firm conviction on the benefits of seclusion and the possibilities of correction.

It was in this context that the federal government financed and published El Manicomio (The Insane Asylum), a report written by physician Román Ramírez in 1884, which included an extensive and comparative collection of documents concerning the construction and management of insane asylums in the United States and Europe.[25] Interested in pragmatic information that could be put to use in Mexico, Ramírez's selection of documents was partial to the United States and to therapies that involved seclusion. Thus he included the translation of standards of construction and rules of governance for insane asylums created by the Association of Medical Superintendents of American Institutes for the

[23] See B. A. Tenenbaum, 'Streetwise History: The Paseo de la Reforma and the Porfirian State, 1876–1910', in W. Beezley *et al.* (eds.), *Rituals of Rule. Rituals of Resistance. Public Celebrations and Popular Culture in Mexico* (Wilmington, Del., 1994), 127–50; T. Morgan, 'Proletarians, Politicos, and Patriarchs: The Use and Abuse of Cultural Customs in the Early Industrialization of Mexico City, 1880–1910', in Beezley *et al.* (eds.), *Rituals of Rule, Rituals of Resistance*, 151–72; V. Cuchi Espada, 'La ciudad de México y la Compañía Telefónica Mexicana. La construcción de la red teléfonica, 1881–1902', *Anuario de Espacios Urbanos. Historia. Cultura. Diseño* (Mexico, 1999), 117–60.

[24] C. Roumagnac, *Por los Mundos del Delito. Los Criminales de México. Ensayo de Psicología Criminal* (Mexico, 1904); J. Guerrero, *La Génesis del Crimen en México. Ensayo de Psiquiatría Social* (Mexico, 1901); M. Macedo, *La Criminalidad en México. Medios de Combatirla* (Mexico, 1897). For contemporary studies on criminology see R. Buffington, *Criminal and Citizen in Modern Mexico* (Lincoln, NE, 2000); Aguirre and Buffington (eds.), *Reconstructing Criminality in Latin America*.

[25] See R. Ramírez, *El Manicomio: Informe Escrito por Comisión del Ministro de Fomento* (Mexico, 1884). Ramírez's interest on social aspects of mental health was also developed in his *Resumen de Medicina Legal y Ciencias Conexas para uso de los Estudiantes de las Escuelas de Derecho* (Mexico, 1901).

Insane, a professional organization founded in 1844.[26] Equally relevant were asylum records and superintendent reports from various American institutions, notably the New York Lunatic Asylum, the Illinois and Iowa Hospitals for the Insane, and the Alabama Insane Hospital. Systematic in approach and rich in detail, Ramírez's report was the first introduction to the inner workings of modern mental-health facilities ever to appear in Mexico. Providing information about both mental-health treatments and the management of asylums, Ramírez placed his work in that ambiguous realm in which science and social concerns converged. In translating documents from both areas, Ramírez fulfilled the role of a cultural policy translator – a task of increasing relevance in a regime committed to modernity at all costs.

Ramírez's evident support for mental-health treatments that emphasized social segregation clearly responded to ongoing anxieties about the urban poor. However, increasing concerns among the psychiatric community over the efficiency of existing institutions for the insane also played a role. Indeed, while small in numbers, physicians with working experience at local mental-health hospitals continuously requested improvement of medical treatments and living conditions of the mentally ill. Supported by foreign medical theories, mostly the flexible language of degeneration theory, Mexican physicians published articles in Mexico City medical journals where they explored in detail the deplorable status of the mentally ill.[27] They did not forget, however, to praise the methods used for their treatment abroad. As academic interest in mental pathologies grew, the School of Medicine offered an upper-level elective class on psychiatry for the first time in 1887. Taught by doctor Miguel Alvarado, the director of the Divino Salvador hospital, this class marked the beginning of the psychiatric profession in Mexico.[28] Seven years later, doctor Juan Peón Contreras, a physician from the province of Yucatán, occupied the first

[26] For a history of the AMSAII, today's American Psychiatric Association, see W. E. Barton, *The History and Influence of the American Psychiatric Association* (Washington, DC, 1987). Also, American Psychiatric Association, *One Hundred Years of American Psychiatry* (New York, 1944).

[27] For a study on the emergence and uses of degeneration theory, see I. R. Dowbiggin, *Inheriting Madness. Professionalization and Psychiatric Knowledge in Nineteenth-Century France* (Berkeley, 1991). This author describes degeneration theory as 'a steady though not necessarily irreversible hereditary deterioration over the course of four generations . . . [Including] symptoms such as moral depravity, mania, mental retardation, and sterility. Physicians ascribed a variety of causes to degeneracy, including alcoholism, immorality, poor diet, and unhealthy domestic and occupational conditions. However, the principal cause of degeneracy that physicians cited was heredity.' For an analysis of degeneration theory in the Latin American context, see D. Borges, '"Puffy, ugly, slothful and inert": Degeneration in Brazilian Social Thought, 1880–1940', *Journal of Latin American Studies* 23 (1993), 235–56. Also see C. Rivera-Garza, 'Dangerous Minds: Changing Psychiatric Views of the Mentally Ill in Porfirian Mexico, 1876–1911', *Journal of the History of Medicine and Allied Sciences* 6 (2001), 36–67.

[28] G. Somolinos D'Ardois, *Historia de la Psiquiatría en México* (Mexico, 1976).

official professorship in the field of psychiatry.[29] The asylum project developed conjointly.

By 1896, a group of lawyers, engineers, welfare agents and two physicians who had worked at mental-health hospitals constituted the official board charged with building the asylum. After analysing both international and local conditions, they made recommendations to the authorities that disclosed medical strategies to treat insanity, spatial tactics to prevent contagion, and social policies to preserve the order and progress of society at large – fundamental values of the Porfirian regime.[30] First, they recommended that the asylum be located away from populated areas to create a division between the world of reason and the world of madness, thus avoiding confusion and the possibility of contagion. Second, they recommended that authorities implement a strategy to classify inmates, both medically and spatially, within asylum walls. They advocated the creation of a department of admission and classification in which doctors could observe and examine inmates carefully because, as they were acutely aware, 'insanity lacked a characteristic mark' and could go easily undetected or misdiagnosed.[31] They too backed the division of the asylum into separate wards, each housing inmates suffering from the same ailment. In addition, to protect the finances of the institution, they suggested that inmates be classified according to first and second categories, giving priority to paying inmates. The committee attached blueprints of administrative offices, wards, workshops, libraries and other facilities to illustrate how the architectural design of the asylum reflected medical concerns with classification and order. A confident committee concluded the report by stating that the humanitarian nature of the enterprise would only ratify the levels of modernization already achieved by the Porfirian regime. A year later, the Public Welfare Administration bought the 485,700 square metres of land that eventually housed the asylum in the periphery of the sprawling metropolis.[32]

In the following years, as the Porfiriato became a stable dictatorship, support for public works grew, especially for those projects which, like the asylum,

[29] For a biography of José Peón y Contreras, see F. Fernández del Castillo, *Antología de Escritos Histórico-Médicos*, 2 vols. (Mexico, 1952), 1057–8. Also, E. Aragón, 'Biografía del Dr Juan Peón del Valle (Sr)', *Mis 33 Años de Académico*, I (Mexico, 1943), 99–105.

[30] Named by the Minister of the Interior, General Manuel González Cosío, the 1896 board included a president (Dr Vicente Moreales, inspector from the Welfare System), a secretary (Dr Manuel Alfaro, the creator of the highly controversial 1867 Regulation of Prostitution), and four additional members (Dr Antonio Romero, who had been director of the San Hipólito hospital; Ignacio Vado, a doctor with working experience at the Divino Salvador hospital; Dr Samuel Morales Pereyra; and Luis L. de La Barra, an engineer from the welfare system). See S. Morales Pereyra and A. Romero, 'Exposición y proyecto para construir un manicomio en el Distrito Federal', *Memorias del Segundo Congreso Médico Pan-Americaono verificado en la Ciudad de México, República Mexicana, noviembre 16–19 de 1896* (Mexico, 1898), 888–96.

[31] *Ibid.*, 896.

[32] 'Departamento de construcción y conservación de edificios del Manicomio General,' AHSSA. F:BP; S:EH; Se:MG; Lg:13; Exp:11, 1.

clearly embodied the ideology of the regime. Thus, in 1905 engineers De La Barra, who worked for the welfare system, and Salvador Echegaray drafted a long and persuasively argued document which came to constitute the 'definitive study' leading to the construction of the General Insane Asylum.[33] The narrative strategies used in the document reflected a careful process of cultural negotiation. Divided into four sections – general plan of the asylum, general services, services for inmates and general organization of the asylum – the project included guidelines following a consistent formula. First, they presented a brief yet insightful overview of foreign sources, a section called 'theoretical conditions'. Unlike the 1884 report which emphasized documents from the United States, this section drew heavily from French sources, especially reports from commissions in charge of asylum construction under the rule of famous administrator Baron Georges-Eugène Haussman, the prefect of the department of Seine during the second empire and responsible for the transformation of Paris.[34] Second, in a section titled 'programme', the authors introduced the specific needs of the Mexican setting, resorting when possible to data from the San Hipólito and Divino Salvador hospitals. Lastly, they concluded each proposal with a 'suggested solution', usually a compromise between the former two. Thus, as true modernizing agents, Echegaray and De La Barra drew information from foreign asylums, but they did it in a critical fashion, adapting lessons and experiences to the local conditions of Mexico.

However persuasive the document was, experts in the fields of psychiatry, engineering, criminology as well as state bureaucrats recommended further modifications. Some were concerned with finances, others wanted to dispense with architectural ornamentation, still others strove to acquire the best equipment for the hospital. While numerous, the recommendations made by the Council of Public Buildings and members of a new asylum committee formed by lawyer and criminologist Miguel Macedo, engineer Alberto Robles, and psychiatrist Juan Peón del Valle, did not alter the 1905 document in fundamental ways. Twenty months later, on December 1906, engineer Salvador Echegaray was ready to submit yet another document, with additional blueprints attached, to the Ministry of the Interior.[35] Then, three years later, on June 1908, the Ministry of the Interior and engineer Porfirio Díaz Jr, the son of the president of the republic, signed a contract to begin construction.[36] The asylum would

[33] 'Memoria sobre el proyecto de Manicomio General para la ciudad de México,' AHSSA. F:BP; S:EH; Se:MG; Lg:49; Exp:1.

[34] For an analysis of European urban planning and its impact on the design of Latin American cities, see J. E. Hardoy, 'Theory and Practice of Urban Planning in Europe, 1850–1930: Its Transfer to Latin America', in R. M. Morse and J. E. Hardoy (eds.), *Rethinking the Latin American City* (Baltimore, Md., 1988), 20–49.

[35] 'Modificaciones al proyecto presentado por el ingeniero Don Salvador Echegaray', AHSSA. F:BP; S:EH; Se:MG; Lg:1; Exp:10.

[36] 'Contrato', AHSSA. F:BP; S:EH; Se:MG; Lg:49; Exp:2.

include twenty-five buildings, counting inmates' wards, doctor residences, infirmaries, and the general services building whose imposing façade of classical lines became the hallmark of the institution.

The General Insane Asylum replicated values and hierarchies of the city in which it was built. Echoing fears of disorder and contagion characteristic of the Porfirian political imagination, the physical layout of the institution secured separate areas for men and women, dividing them with fences disguised with bushes and plants to avoid 'the appearance of a gaol'.[37] Mirroring social hierarchies, the asylum also allocated the front areas, those closest to the gardens and the entrance, to paying inmates who lived in single rooms. Behind them, the common wards for indigent inmates began. The 848 female and male inmates arriving from the existing mental-health hospitals in 1910 were distributed in seven wards, including those for tranquil, dangerous, alcoholic, epileptic, idiotic, elderly and paying inmates.[38] Social order was further embedded in the governing rules of the institution, which placed a physician-director at the top, followed by an administrator, the medical staff including both doctors and nurses, and attendants. Written by an inspector from the Public Welfare System and the directors of five hospitals, the asylum regulations of 1913 included rules to provide inmates with the best psychiatric assistance available while securing the administrative order and scientific status of the institution.[39] Enforcing these rules, however, proved a monumental task. The institutional order was limited by the rapidly changing social context in which the asylum emerged.

Revolution and the asylum: 1910–1930

The revolution – a social upheaval that took over one million lives in the country – impacted on the General Insane Asylum shortly after official inauguration. Indeed, without the economic and political investment that gave it birth, the asylum soon faced mounting financial dilemmas, which affected both its administrative and medical branches, forcing a gradual redefinition of the institution as a whole. Rather than the medical and research institute envisioned by modernizing Porfirians, the establishment quickly reverted to its custodial functions.

As in state-run asylums abroad, the Mexican institution soon faced the problem of inmate overpopulation – a phenomena that reflected the growing demand for asylum accommodation during the early revolutionary era. Indeed, careful demographic calculations had resulted in the provision of 1,330 beds in 1910 – 730 reserved for women and 600 for men – but the demand outstripped the

[37] 'Memoria', AHSSA. F:BP; S:EH; Se:MG; Lg:49; Exp:1, 27.

[38] Data appear in J. Luis Patiño Rojas and I. Sierra Mercado, *Cincuenta años de Psiquiatría en el Manicomio General* (Mexico, 1960).

[39] 'Reglamento Interior del Establecimiento, 1913', AHSSA. F:BP; S:EH; Se:MG; Lg:3; Exp:25.

supply of beds by 1911.[40] This gap decreased during the 1920s, but grew back over time.[41] Authorities faced a dilemma. While they acknowledged that the number of inmates had to be reduced, they were also aware that this situation stemmed form the very welfare principles that ruled the institution, including the stipulation to provide care to all individuals regardless of sex, age, religion, and social status.[42] Also, as the most important national institution in the field, the asylum not only admitted boarders from the capital city, but also from the provinces of Mexico and, at times, even from foreign countries. In addition, most admitted patients suffered from chronic conditions that required long periods of hospitalization. These three variables became aggravated by the needs of the revolutionary era. In a time of change and dislocation, where violence and starvation were not rare, the asylum accommodated great numbers of destitute patients who, for the most part, had nowhere else to go.

Social indifference and governmental neglect also affected the physical structure and the quality of the asylum's general services, both of which deteriorated throughout the armed phase of the revolution. For example, by 1916, inspectors from the public welfare system noted that inmates wore inadequate garments and ate small pieces of bread that 'did not even weigh 40 grams'.[43] By 1920, asylum problems went far beyond clothing and food supplies, including the lack of mattresses, electricity and basic medications, as well as leaking roofs and the deterioration of hardwood floors, doors and windows of most buildings.[44] Sensing fertile ground for sensationalist news, journalists visited the asylum and described it as a ravaged landscape, an institution 'in complete desolation, lacking hygiene in the kitchen, providing inmates with poor and scant meals, supplying indigent inmates with miserable clothing. [In sum] wards, isolation rooms, gardens, streets and patios were completely forsaken.'[45]

The ominous state of the institution was not limited to its welfare services. The lack of financial support also compromised its status as a medical establishment, for the scientific personnel soon became insufficient. Despite internal regulations, by 1912 only one intern bore full responsibility for the care and treatment of ninety-eight inmates in the ward of tranquil inmates 'A', a situation that was the norm, rather than the exception throughout the hospital.[46] The limited number of nurses and poorly trained attendants seriously aggravated the problem. Only two years after opening, each asylum nurse took care of an

[40] Data appear in Patiño Rojas and Mercado, *Cincuenta Años de Psiquiatría.*

[41] According to Patiño Rojas and Mercado, the number of enrollments in 1910 was 1, 004. In 1920 confinements went down to 697. By 1930 the asylum housed 1, 001 inmates. Despite this data, complaints against overcrowding remained high throughout this period.

[42] 'Reglamento interior del establecimiento, 1913', 2.

[43] 'Informe de Inspectores, 1916', AHSSA. F:BP; S:D; Se:DG; Lg:18; Exp:21, 132.

[44] 'Informe del Manicomio, 1920', AHSSA. F:BP; S:EH; Se:MG; Lg:10; Exp:24.

[45] 'Diversos. Nota del Universal, 1918', AHSSA. F:BP; S:EH; Se:MG; Lg:8; Exp:27.

[46] 'Presupuesto, 1912', AHSSA. F:BP; S;EH: S:MG; Lg:3; Exp:7, 7.

average of 150 inmates in various wards. Similarly, eighty-six attendants supervised 1,024 inmates, roughly half as many as the director determined were needed to provide adequate attention.[47] Under these circumstances, emphasis on the custodial functions of the institution increased.

Although neglected, the asylum remained open throughout the early revolutionary era fulfilling important welfare functions. As new regimes intently rearticulated welfare ideology and practice to suit the needs of revolutionary society, asylum authorities began the reorganization of the institution based on the principles of work therapy. Led by Samuel Ramírez Moreno and Manuel Guevara Oropeza, physician-directors of the institution between 1928 and 1932, this administrative and medical reform significantly coincided with the emergence of political organizations linked to the state such as the National Revolutionary Party – a national party that, under different names, remained in power to the end of the century.

A view from within: inmate profiles

While designers and authorities infused the asylum project with specific ideologies and aspirations, the General Insane Asylum's varied role in Mexican society was ultimately shaped by the population it served. For some, like Luz M. de S., whose husband confined her against her will trying to divorce her, the asylum was a gaol that reinforced uneven gender relations.[48] For others, like Esperanza T., who became an inmate at three different times in her life, the asylum came to rescue her from the harsh life of a beggar on Mexico City streets.[49] Still for others, like Merino G., who was confined after hitting a municipal president in his hometown, the asylum eventually turned into a liminal arena, namely a place he left when feeling well and returned to in times of need.[50] Some inmates, like Modesta B., found a job in there and, while opportunities to escape abounded, remained within its walls for thirty-five years.[51] For some families, such as the Q.'s, the asylum was the last resort to treat the unruly behaviour of a spirited daughter.[52] Others, like Altagracia G., returned to her family in the province of Aguascalientes recovered from a nervous breakdown.[53] Many more found nothing but death within its walls, a condition resulting from poisoned food and neglect rather than mental illness *per se*. While highly diverse, the population that made the asylum also shared a set of social and medical characteristics which, put together, reflect the various social meanings ascribed to the asylum.

[47] *Ibid.*, 8. [48] 'Luz D. de S., 1911', AHSSA. F:MG; S:EC; C:22; Exp:63.
[49] 'Esperanza G., 1911', AHSSA. F:MG; S:EC; C:101; Exp:15.
[50] 'Merino G., 1921', AHSSA. F:MG; S:EC; C:97; Exp:67.
[51] 'Modesta B., 1921', AHHSA. F:MG; S:EC; C:105; Exp:16 (6339).
[52] 'Guadalupe Q., 1911', AHSSA. F:MG; S:EC; C:6; Exp:35(404).
[53] 'Altagracia F. de E., 1920', AHSSA. F:MG; S:EC; C:106, Exp:12.

During its early years, the asylum's purpose in society was especially open to definition – a situation that both the state and the community used to place a variety of demands on the institution. While Porfirian designers had envisioned the asylum as a medical establishment where both the wealthy and the destitute could secure care, paying boarders constituted a rare minority from the start. Most surely, they found medical assistance at the Lavista Clinic, a private hospital located to the south of Mexico City, or at the small sanatorium owned by psychiatrist Samuel Ramírez Moreno in Coyoacán.[54] Indeed, the state asylum admitted all women and a high percentage of men as free and indigent inmates during the 1910s, a trend that remained roughly unchanged in the following decades.[55] Further, as in asylums in Nigeria, Ireland and Argentina, most inmates were committed involuntarily.[56] A government order preceded the committal of 86 per cent of women and 68 per cent of men during the 1910s. Public-welfare authorities played an active role in the commitment of 2 per cent of female and 6 per cent of male inmates. In addition, prisoners constituted 10 per cent of the male population of the asylum.[57] In these cases, the intervention of the police and welfare officials was crucial in detecting and apprehending people suspected to be mentally ill – a process that, as attested by the case of Modesta B., usually began on the streets and in other institutions of the welfare system. Thus, in relieving the streets from men, women and children deemed insane, the asylum contributed to the social order of city and community.

However, state agents were not always involved in asylum confinements. First, while mostly resting on the judgement of police members or other representatives of state order, governmental requests also involved, at least in some cases, the participation of the family. Cresencia G., for example, came to the asylum after the municipal president of her hometown in the state of Mexico requested her committal.[58] Yet, her family's concern for her mental health – she had become increasingly violent after the death of one of her sons – prompted the official request in the first place. Similar processes were not rare, particularly when impoverished families proved unable to care for relatives or when violent behaviour threatened family dynamics. Thus, families actively participated in confinement procedures even when state authorities officially initiated committal processes. Second, families initiated the internment of 12 per cent of female and 16 per cent of male inmates.[59] In these cases, relatives and neighbours were instrumental in the identification of mental illness and the initial evaluation of

[54] Interview with Mexican psychiatrist Luis Murillo, San Diego, Calif., May 2000.

[55] Based on a random sample of 100 files from 1910. Status of inmates.

[56] For the case of Nigeria, see Sadowsky, *Imperial Bedlam*. For Ireland, see Finnane, *Insanity and the Insane in Post-Famine Ireland*; also N. Scheper-Hughes, *Saints, Scholars, and Schizophrenics. Mental Illness in Rural Ireland* (Berkeley, 1979). In Argentina, the participation of the police was higher in the confinement of men than women. See Ablard, 'Madness in Buenos Aires'.

[57] Based on a random sample of 100 files from 1910. Means of arrival.

[58] 'Cresencia G., 1920,' AHSSA. F:MG; S:EC; C:105; Exp:46.

[59] Based on a random sample of 100 files from 1910. Means of arrival.

methods for treatment. Some came to the asylum as a last resort, looking forward to being relieved from the burden of care. Others brought their relatives to the asylum hoping to find a cure, their faith in the capabilities of modern medicine flickering through letters and telegrams asking for signs of improvement or a discharge date. The pathways into the asylum thus illustrate the varied ways in which state and families appropriated the institution for different purposes, which at times were not necessarily compatible or complementary.

The diversity of functions fulfilled by the asylum reflected the variety of inmates it assisted for, while generally poor, the asylum population was hardly homogenous. First in the observation room and later in the wards, psychiatrists came in contact with the cargo carrier who worked for a couple of cents in Mexico city markets and with the singer hit by misfortune. They talked with the eloquent, if somewhat misguided, pharmacist, and with the tailor and shoemaker whose skills came in handy in the establishment. Likewise, they evaluated the mental health of students and teachers, washerwomen and prostitutes. While indeed the contingent of unskilled workers formed by day labourers, street pedlars and clerks was more numerous, the asylum also admitted artisans as well as professionals from the middle classes, such as lawyers and teachers.[60] While the occupations of female inmates were not as varied – some 60 per cent of women were responsible for unpaid domestic chores – they also included domestic servants, seamstresses and washerwomen. Those listed as unemployed – 16 per cent – were generally prostitutes, an occupation mindful administrators did not dare to acknowledge.[61] Perhaps, as psychiatrist John Conolly once claimed, insanity was indeed a 'great leveller', but in the case of Mexico, the social dislocation brought about by the revolutionary war clearly contributed to this process.

As in state asylums serving destitute patients in Ireland and England, New York and California, the Mexican asylum most frequently admitted people in the early and middle stages of their adult life.[62] Only 6 per cent of the asylum population was under the age of twenty, and only 10 per cent above the age of seventy. Most inmates ages' ranged from twenty to forty years. While some variation occurred during the first three decades of the twentieth century – female inmates were mostly in their twenties and thus relatively younger during the 1920s while, during the 1930s, they were mostly in their forties – age-specific admission rates remained nearly unaltered.[63] The relative youth

60 Based on a random sample of 100 files from 1910. Occupations of men.

61 Based on a random sample of 100 files from 1910. Occupations of women.

62 Similar trends are noted by Finnane, *Insanity and the Insane*, 130; Walton, 'Lunacy in the Industrial Revolution'. Also, J. K. Walton, 'Casting Out and Bringing Back in Victorian England: Pauper Lunatics, 1840–70', in W. F. Bynum, R. Porter, and M. Shepherd (eds.), *Anatomy of Madness. Essays in the History of Psychiatry*, 2 vols. (London, 1985), 132–48; E. Dwyer, *Homes for the Mad*, 244.

63 Based on a random sample of 100 files from 1910. Ages of inmates.

of asylum population was matched by the lack of family support. Indeed, 66 per cent of female and 78 per cent of male inmates were either single or widowed during the 1910s.[64] Further, while most lived in Mexico City, 64 per cent of men and women were born in the provinces of Mexico.[65] Singleness and migration might have not necessarily meant social isolation, but in the context of the Mexican revolution they evidently increased the vulnerability to committal.

During the first three decades of the twentieth century, thus, the insane from Mexico were relatively young and, according to institutional statistics, more likely to be women than men.[66] In the midst of the revolutionary war that brought popular armies from the north and indigenous armies from the south to central Mexico, lack of family support and migration made men and women more susceptible to confinement. Because the social upheaval not only affected the very poor, but also people from other classes, the asylum opened its doors to a range of social groups affected by economic hardship and depravation. The General Insane Asylum fulfilled ambivalent roles in Mexican society. On the one hand, uplifting its status as an institution of social control, the asylum contributed to the urban order of Mexico City by confining people deemed as insane by state authorities. On the other, the General Insane Asylum also proved helpful for families unable to care for relatives suffering from mental illness. The permanence of the asylum as a welfare institution through the early revolutionary period doubtlessly responded to its dual role as both a site of control and a place for social assistance.

A view from within: diagnoses

While mostly fulfilling welfare functions, especially custodial care, the General Insane Asylum also strove to provide inmates with medical attention. Asylum authorities used a basic spectrum of symptoms to place inmates into specific wards with separate buildings for men and women. For example, violent or agitated patients were placed in the ward for dangerous inmates, as were prisoners who required special surveillance. Patients suffering from chronic mental diseases went to the wards for tranquil inmates, indigents to section 'A', located in a back ward, and paying boarders to section 'B', located in front rooms

[64] Based on a random sample of 100 files from 1910. Civil status.

[65] Based on a random sample of 100 files from 1910. Place of birth and place of residence.

[66] Throughout the first three decades of the twentieth century, women were more numerous than men. In 1910, men constituted 38.54 per cent of asylum population, women 57.76 per cent, children 3.68. In 1920, men constituted 33.14 per cent of asylum population, women 57.67 per cent, children 6.59 per cent. In 1930 men constituted 33.46 per cent of asylum population, women 62.63 per cent, children 3.89 per cent. Patiño Rojas and Mercado, *Cincuenta Años de Psiquiatría*.

surrounded by gardens.[67] Unlike asylums that differentiated between the mentally ill and the feeble minded, the Mexican asylum admitted inmates who suffered from mental conditions that 'affected their intelligence', placing them in the ward for imbecile inmates. Epileptics, among whom women predominated, went to the ward of the same name. Alcoholics, among whom men predominated, went to the ward of the same name. Most elderly patients were allocated to a ward for senile inmates. Lastly, while the asylum admitted children, the institution lacked a ward especially devoted to them. The social, medical and spatial classification of inmates attempted to secure the internal order of the institution, validating, in passing, hierarchies prevailing in society at large.

However, while the number of wards was limited, the amount of diagnoses grew over these three decades of the twentieth century. In fact, betraying the lack of standardized psychiatric language, asylum doctors recorded about eighty different diagnoses during the early revolutionary era.[68] Because doctors reached a diagnosis after observing and interviewing inmates and, when possible, his or her families, these medical interpretations of mental health usually involved a social dialogue. Particularly involving the identification of the range of behaviours that led men and women into asylum grounds, these dialogues were hardly transparent or conflict-free. In fact, close scrutiny of the content of these dialogues sheds light on the changing social and medical mores that contributed to the revolutionary definition of mental illness in early twentieth-century Mexico.

Like the San Hipólito and Divino Salvador hospitals, the General Insane Asylum admitted great numbers of male and female inmates suffering from epilepsy during the 1910s – evidence of the lingering legitimacy of Porfirian definitions of mental illness in early twentieth-century Mexico. Indeed, asylum designers, who had had access to the records of pre-revolutionary hospitals, noted that while epileptic patients constituted about 15 per cent of the hospital's population, the figure increased remarkably among female patients. Thus, because 28.41 per cent of women suffered from epilepsy at the Divino Salvador hospital, they planned a larger area for them at the General Insane Asylum.[69] Data from the institution proved them right, at least during the first decade of the twentieth century when 28 per cent of female inmates and 22 per cent of male inmates were

[67] For an analysis of the architecture of the mental-health institution, see Rivera-Garza, 'An Architecture of Mental Health', manuscript.

[68] See Patiño Rojas and Mercado, *Cincuenta Años de Psiquiatría*, 5. For an analysis of psychiatric classificatory issues, see G. E. Barrios, *The History of Mental Symptoms: Descriptive Psychopathology since the Nineteenth Century* (Cambridge, 1996). Also G. E. Barrios, 'Obsessional Disorders during the Nineteenth Century: Terminological and Classificatory Issues', *The Anatomy of Madness* (London, 1985–7), 166–87.

[69] Data on diagnoses from the Divino Salvador and San Hipólito hospitals appear in M. Rivadeneyra, 'Apuntes para la estadística de la locura en México', BA thesis, Escuela Nacional de Medicina de México (1887).

listed as suffering from this condition, becoming the most numerous medical grouping in asylum grounds.[70] In a setting defined by violence and depravation, chronic illnesses such as epilepsy placed especially heavy economic burdens on the family of the patient, which the asylum relieved.[71] The large number of confinements associated with epilepsy also disclosed that the stigma of this illness outlived the Porfirian era. In fact, asylum doctors wrote scant comments in the files of these patients, generally accepting diagnoses made by family members or the police. When time or interest allowed it, doctors detected similar sufferings in the family of the inmate, adding ubiquitous comments about the hereditary legacy of this ailment.[72] Likewise, doctors provided these sufferers with little in terms of treatment, allowing peaceful inmates to work in the workshops of the institution or, when necessary, prescribing sedatives for agitated patients. Nevertheless, doctors became increasingly unwilling to admit or to diagnose inmates with this condition over time. During the 1920s, for example, only 18 per cent of women and 17 per cent of men remained at the institution as epileptics.[73] By the 1930s, female epileptics constituted only 7.52 per cent of asylum population, while male epileptics amounted to 10.86 per cent.[74] While the armed phase of the revolution had dwindled by then, the economic stagnation and agitated negotiation of rule that characterized early revolutionary regimes did not justify such a dramatic decrease in epilepsy diagnoses. Greater awareness about this condition among the psychiatric community contributed to the declining figures, but only to a certain degree.[75] Much more crucial was, however, asylum doctors' waning emphasis on chronic mental illnesses for which they could offer little in terms of treatment and cure.

A similar process occurred regarding the diagnosis of mental retardation and *dementia praecox*. During the first decade of the twentieth century, General Insane Asylum doctors diagnosed a great number of patients with mental retardation – a condition variously referred to as idiocy, mental debility and imbecility. Constituting 16 per cent of female patients and 18 per cent of male

[70] Based on complete entries in the registry books from 1910. Diagnoses for men and women.

[71] See E. Dwyer, 'Stories of Epilepsy', *Hospital Practice*, 30 (1992), 65–92. For a comparative analysis of epilepsy, see A. Kleinman, 'The Social Course of Epilepsy. Chronic Illness as Social Experience in Interior China', in *Writing at the Margin: Discourse Between Anthropology and Medicine* (Berkeley, 1995), 147–72.

[72] Degeneration theory was particularly influential during the Porfirian era. Its influence, however, continued throughout the early revolutionary period. See Rivera-Garza, 'Dangerous Minds'. Also see N. L. Stepan, *The Hour of Eugenics. Race, Gender, and Nation in Latin America* (Ithaca, NY, 1991).

[73] Based on complete entries in registry books from 1920. Diagnoses for men and women.

[74] Based on complete entries in registry books from 1930. Diagnoses for men and women.

[75] Mexican psychiatrist Ignacio Ruiz argues that asylum doctors from the early twentieth century frequently misdiagnosed epilepsy. In his opinion, numbers decreased during the 1920s and 1930s because doctors' interpretative apparatus allowed them to correct past mistakes in these years. Interviews with Ignacio Ruiz, May 1995, Mexico City.

inmates, this set of mental conditions was the second in importance in the Mexican asylum.[76] Because Porfirian doctors working at the San Hipólito and Divino Salvador hospitals failed to include this category in their medical groupings, admissions based on mental retardation reflected the use of new psychiatric categories to classify mental imbalance in revolutionary Mexico. As with epilepsy, however, figures too declined over the next two decades. Likewise, *dementia praecox*, a term coined by Emil Kraeplin, a German psychiatrist who exerted great influence in institutional Mexican psychiatry during the early revolutionary period, affected some 9 per cent of female inmates and 11 per cent of male inmates in 1910.[77] Diagnoses of *dementia praecox* too decreased in the next decade.

By contrast, during the 1920s, and increasingly in the next decade, asylum doctors paid exacting attention to mental illnesses associated with alcohol and drug consumption, two socially originated conditions that they perceived as curable. Because the Divino Salvador and, especially, the San Hipólito hospital predominantly housed alcoholics, this tendency did not constitute a radical break from Porfirian understandings of mental illness.[78] Indeed, medical experts had readily connected alcohol consumption with criminality and mental illness during the Porfirian era.[79] However, the social meanings of alcoholism, and drug addiction for that matter, underwent social scrutiny in revolutionary Mexico. In the context of state reconstruction, revolutionary regimes insistently called for the creation of a social medicine, namely 'a preventive medicine which was a juridical, technical, and administrative branch of the federal government; an adequate tool to protect the physical and mental health of all the citizens of the country and to safeguard their lives when they are threatened by diverse unhealthful causes'.[80] Simultaneously, revolutionary regimes showed increasing interest and growing commitment to eugenic views of the population.[81] Inspired by these projects, doctors pushed, for example, to change the legal status of alcoholism, from an extenuating circumstance in criminal cases to

[76] Based on complete entries in the registry books from 1910. Diagnoses for men and women.

[77] Based on complete entries in the registry books from 1910. Diagnoses for men and women. For an analysis of the evolution of this diagnosis, see P. H. Wender, '*Dementia Praecox*: The Development of the Concept', *American Journal of Psychiatry* 119 (1963), 1143–51.

[78] According to data collected by Rivadeneyra, alcoholics constituted some 55 per cent of the patient population at the *San Hipólito*, a hospital devoted to the care of mentally ill men. While diagnoses of alcoholism only amounted to 6 per cent at the *Divino Salvador*, a hospital devoted to the care of mentally ill women, alcohol consumption was listed as the cause of mental illness in 40 per cent of the cases. See Rivadeneyra, 'Apuntes'.

[79] See P. Piccato, 'El Paso de Venus por el Disco del Sol: Criminality and Alcoholism in the Late Porfiriato', *Mexican Studies/Estudios Mexicanos* 11 (1995), 203–41. Also P. Piccato, 'No es Posible Cerrar los Ojos: El discurso Sobre la Criminalidad y el Alcoholismo Hacia el Fin del Porfiriato', in R. Pérez Monfort (ed), *Hábitos, Normas y Escándalo. Prensa, Criminalidad y Drogas durante el Porfiriato Tardío* (Mexico, 1997), 75–134.

[80] Amezquita *et al.*, (eds.), *Historia de la Salubridad y la Asistencia en México*, 72.

[81] See Stepan, *The Hour of Eugenics*.

an aggravating one. Likewise, in 1919, doctors not only supported a ban on the cultivation of marihuana but also encouraged the incorporation of drug addiction as a health crime in the Mexican penal code.[82] Further, the 1924 Sanitary Code, which helped expand the scope of activities of public-health officials at both local and national levels, importantly included strategies to combat the spread of social diseases, among which they counted alcoholism.[83] Asylum doctors' tendency to diagnose poor inmates as alcoholics took place in a politically charged context in which the physical and mental health of the citizens came to constitute national assets. Indeed, doctors diagnosed 15 per cent of asylum inmates as alcoholics in 1910.[84] By contrast, a decade later, some 41 per cent of inmates received this diagnosis.[85] Thus, the number of inmates listed as alcoholics in 1920 more closely resembled Porfirian diagnosis patterns from the San Hipólito and Divino Salvador hospitals. Further, as their Porfirian counterparts, revolutionary asylum doctors diagnosed 29.37 per cent of male inmates and 11.66 per cent of female with alcoholism – a figure that contributed to the construction of this condition as a typically masculine mental illness.[86]

The principles of social medicine also encouraged revolutionary regimes from the early twentieth century to extirpate practices and behaviours contrary to the health of the community, especially in the terrain of sexuality.[87] Health officials thus conducted an animated debate about the dangers associated with unrestricted sexuality, particularly represented by prostitution, which they saw as conducive to syphilis. In fact, public-health physicians committed to the creation of national programmes against the spread of syphilis and to the criminalization of this illness, collected and published statistics showing that the number of syphilis-related deaths in the country had grown from less than 1 per cent in 1916 to almost 2 per cent in 1925.[88] During the same period, asylum doctors increasingly detected cases of progressive paralysis, the tertiary stage of syphilis, among asylum inmates. In 1910, cases of progressive paralysis constituted a mere 0.47 and 3.79 per cent of diagnoses for female and male inmates respectively.[89] A decade later, doctors diagnosed 4.44 per cent of

[82] Amezquita *et al.*, (eds.), *Historia de la Salubridad y la Asistencia en México*, 145.
[83] *Ibid.*, 250.
[84] Based on complete entries in registry books from 1910. Diagnoses for men and women.
[85] Based on complete entries in registry books from 1920. Diagnoses for men and women.
[86] Based on complete entries in registry books from 1920. Diagnoses for men and women.
[87] See C. Rivera-Garza, 'The Criminalization of the Syphilitic Body. Prostitutes, Health Crimes, and Society in Mexico, 1876–1930', in C. Aguirre and G. Joseph (eds.), *Crime and Punishment in Latin America* (Durham, NC, 2001). Also, K. Bliss, 'The Science of Redemption: Syphilis, Sexual Promiscuity, and Reformism in Revolutionary Mexico City', *Hispanic American Historical Review* 79 (1999), 1–40.
[88] B. Gastélum, 'La persecusión de la sífilis desde el punto de vista de la garantía social,' AHSSA. F:BP; S:AS; Se:DAES; L:1; Exp:11, 7.
[89] Based on complete entries in registry books from 1910. Diagnoses for men and women.

female inmates and 13.94 per cent of male inmates with this condition.[90] By 1930, women suffering from progressive paralysis amounted to 13.24 per cent and men to 16.59 per cent.[91] As with the case of alcoholism, asylum doctors increasingly defined syphilis-related mental illnesses as masculine conditions. Sharing dominant medical views, they perceived women, especially prostitutes, as agents of this illness, and men, all men, as victims of unrestricted female sexuality.

Doctor–patient relationships within asylum walls involved a certain degree of tension and distance. This was especially true between male doctors, who composed the totality of the medical staff at the General Insane Asylum, and female patients.[92] In exploring the medical stories of female inmates, asylum doctors paid special attention to their sexual history, asking questions about menarche, intercourse, abortions and menopause. As Porfirian experts in the past, they believed that there was a connection between the female genitalia and mental illness.[93] This linkage had led to diagnoses of moral insanity among patients of the Divino Salvador hospital and helped produce similar diagnoses during the first decade of the twentieth century in the grounds of the General Insane Asylum.[94]

A term originally coined by English physician James Prichard during the early nineteenth century, moral insanity described a condition in which patients recognized good and evil impulses, but were unable to resist the latter.[95] While no longer in use in most asylums from the early twentieth century, Mexican psychiatrists employed it to explain female behaviours that violated implicit rules of decency and domesticity. Diagnoses of moral insanity only amounted to some 2 per cent among female inmates in 1910, but doctors mentioned it as an important component in cases of alcoholism, violent jealousy, and mental

[90] Based on complete entries in registry books from 1920. Diagnoses for men and women.

[91] Based on complete entries in registry books from 1930. Diagnoses for men and women.

[92] Labour records from the General Insane Asylum show that only one female doctor worked at the institution between 1914 and 1915. She was Rosario M. Ortiz, first an external doctor and, months later, an intern. See, 'Relación de personal de 1914 a 1915,' AHSSA. F:BP; S:EH; Se:MG; Lg:4; Exp:23, 2–3.

[93] On the emergence of Mexican sexual science, see Rivera-Garza, 'The Criminalization of the Syphilitic Body'. A good example of Porfirian views of the relation between sex and mental illness is M. E. Guillén, 'Algunas Reflexiones Sobre la Higiene de la Mujer Durante su Pubertad', BA thesis, Facultad de Medicina de México (1903).

[94] On diagnoses of moral insanity at the General Insane Asylum, see Rivera-Garza 'She Neither Obeyed Nor Respected Anyone: Inmates and Psychiatrists Debate Gender and Class at the General Insane Asylum, Mexico 1910–1930', *Hispanic American Historical Review*, forthcoming.

[95] Prichard defined moral insanity as a form of 'madness consisting in a morbid perversion of the natural feelings, affections, inclinations, temper, habits, moral dispositions, and natural impulses, without any remarkable disorder or defect of the intellect or knowing and reasoning faculties, and particularly without any insane illusion or hallucination'. See J. Prichard, *A Treatise on Insanity and Other Disorders Affecting the Mind* (New York, 1973), 16. Originally published in 1837.

illnesses that involved the category of sex.[96] By 1930, however, asylum doctors no longer diagnosed women with this ailment, a trend that replicated the declining use of this category in foreign psychiatric circles. Morally insane women ceased to exist in early twentieth-century Mexico when feminist discourses that advocated gender equality and more complex understandings of women's place in society grew at a steady pace. For example, in the 1916 Feminist Congress that took place in Yucatán, male and female feminists called for and used a definition of femininity that clearly transcended easy associations between women and sex – the basis on which diagnoses of moral insanity were made in Mexico.[97] Simultaneously, however, the asylum admitted women at higher rates during this decade, amounting to 63 per cent of the inmate population.[98] Nevertheless, asylum doctors who so readily listed men as alcoholics were less self-assured when observing female inmates. Indeed, in 1930, roughly equal numbers of female schizophrenics, epileptics and syphilitics were admitted in asylum wards. Further, the number of female inmates listed as mentally healthy or without diagnoses was as high as each one of the diagnostic groupings mentioned above.[99] Ongoing debates about the female question, which affected revolutionary society at large, seemingly impaired asylum doctors' ability to produce a typically female mental illness. Female inmates played a role in this process. Unlike journalists, writers and political activists who used different arenas to campaign for gender equality, female inmates articulated the stories of their lives to confront or, more accurately, to evade psychiatric labelling. The unfolding of these stories, which usually unnerved doctors, revealed the conflictive domestic context – especially spousal abuse – in which familial diagnoses of mental illness first emerged.[100]

As with Mexican society at large, diagnoses at the General Insane Asylum underwent dramatic, although not linear, change between 1910 and 1930. While Porfirian understanding of mental illness permeated diagnoses during the first decade of the twentieth century, especially in cases of epilepsy, revolutionary practice and ideology more clearly informed medical identification and diagnoses of what they referred to as social mental illnesses during the 1920s and beyond. Yet, at least in the case of alcoholism, such a revolutionary remaking

[96] Based on complete entries in registry books from 1910. Diagnoses for men and women.
[97] A. Macías, *Against All Odds: The Feminist Movement in Mexico to 1940* (Westport, Conn., 1982).
[98] See Patiño Rojas and Mercado, *Cincuenta Años de Psiquiatría en México.*
[99] Based on complete entries in registry books from 1930. Diagnoses for women. According to asylum data 7.52 per cent of female inmates were schizophrenics; 12.18 per cent were epileptics; 11.46 per cent suffered from progressive paralysis; 8.96 per cent were alcoholics; and 8.24 per cent were listed as healthy or without diagnoses. However, different types of schizophrenia were noted, which together amounted to 23.54 per cent of the female population of the asylum.
[100] E. Dwyer noted that, occasionally, asylums were used to shield individuals from highly conflicting families, especially in cases of domestic abuse. See E. Dwyer, *Homes for the Mad*, 94.

of mental illness did not depart from but rather worked within Porfirian psychiatric frameworks. In accordance with the revolutionary commitment to the principles of social medicine, asylum doctors readily identified social mental illnesses, especially alcoholism and, at times, drug addiction. Likewise, reflecting the increasingly debatable status of women in revolutionary society, doctors found it difficult to diagnose growing numbers of female inmates with moral insanity, a dubious category by psychiatric standards of the early twentieth century that, in the Mexican setting, described women who did not ascribe to traditional definitions of female domesticity and submissiveness. Because both state agents and families initiated confinement processes, identifying mental illness before the asylum doctor classified it and treated it, variation in asylum diagnoses reflected changing definitions of accepted and deviant behaviours in early revolutionary Mexico.

Revolution, mental illness, and the nation

While elite Porfirians generally greeted the inauguration of the General Insane Asylum as an achievement of modern science, the black legend of the institution grew at a steady pace in revolutionary Mexico. Nevertheless, both supporters and detractors of the asylum have tended to present it as a homogenous institution. A careful analysis of its history as well as its internal dynamics suggests that the General Insane Asylum was, on the contrary, a highly heterogeneous establishment, fulfilling important roles as an impromptu gaol for drunkards and vagabonds, a welfare centre where destitute patients found custodial care, and a medical establishment where doctors paid close attention to promising cases. In fact, rather than a single institution, the asylum was many institutions at once – a reality that became readily evident in 1968 when the Mexican government replaced the asylum with seven different hospitals. They included a hospital for acute mental patients with 600 beds, a hospital for 200 children, three rural hospitals with 500 beds each, and two home-hospitals for incurable patients with 250 beds each.[101]

While not at the centre of revolutionary debates about the nation's future, the asylum both reflected and illuminated societal change. Porfirian designers, for example, envisioned the asylum as a medical institution where psychiatrists would treat patients from different classes and backgrounds, fulfilling scientific, corrective, and humanitarian functions. Soon, however, as the revolutionary war developed in city and countryside, the asylum proved unable to attract professionals in the field. Further, the asylum mostly served destitute patients habitually suffering from chronic conditions. Lack of funding, scant medication, overcrowding and poor staffing forced the asylum to offer custodial rather than

[101] Calderón Nárvaez, 'Hospitales Psiquiátricos de México', 121.

medical care to its patients. But the 1910 revolution not only affected the asylum; it also reshaped it. Once the armed phase of the revolutionary struggle declined, new regimes ruled the country aiming for the reconstruction of the nation. By the early 1920s, the governments of Alvaro Obregón and Plutarco Elías Calles placed increasing emphasis on welfare establishments, among which the asylum was the largest and most neglected. Asylum authorities and medical staff, who strove to produce a niche for psychiatry in revolutionary society, responded by adapting diagnoses and treatments to the needs of the new era. On the one hand, growing emphasis on the welfare responsibilities of the state and revolutionary commitment to social medicine prompted asylum doctors to diagnose socially originated mental illnesses among asylum inmates. The increasing number of alcoholics and drug addicts among asylum inmates after 1920 serves as a case in point. On the other hand, revolutionary ideology also played a role in the decreasing number of women diagnosed as morally insane inside the asylum – a result of the careful, spirited and increasingly complex scrutiny of women's roles in society. While the latter clearly represented a discontinuity between Porfirian and revolutionary psychiatry, that was not the case of the former. In fact, the growing number of diagnoses of alcoholism during the 1920s more closely resembled Porfirian diagnoses from the San Hipólito and the Divino Salvador hospitals. Asylum doctors' training, which clearly took place in Porfirian Mexico, as well as the modernizing agenda of the revolutionary regimes contributed to this continuity in psychiatric practice in early twentieth-century Mexico. Furthermore, emphasis on the social dimension of illnesses, specifically mental illness in this case, allowed revolutionary doctors and the welfare institutions in which they worked to offer treatments that were both medical and social in nature. Because doctors and bureaucrats perceived social illnesses as fundamentally curable, they were able to present society with a clear and scientifically based avenue for reform.

The asylum, nevertheless, not only reflected social change; it also contributed to shape such change, albeit in subtle, peripheral ways. During the late 1920s, when the era known in Mexican history as the Maximato began, the asylum played an important role in supporting and giving scientific legitimacy to state discourses about work and the place of the poor in early revolutionary Mexico – an issue of political importance in a regime intently trying to co-opt a mobilized working class through the creation of labour unions and peasant leagues. Indeed, under the leadership of strong man and Maximum Chief, General Plutarco Elías Calles, the regime brought new attention to the role of the state not only as protector but also as a moral guide for the destitute. For this reason, welfare authorities from the early revolutionary era used state facilities not only to treat but also to improve, that is to reform, the habits and behaviours of the poor – hereby, the political relevance of the implementation of work therapy throughout the asylum in 1929.

For all the linkages that went from the asylum to society and vice versa, the institution was hardly an apparatus of social control. Lack of funding as well as overcrowding and poor staffing clearly limited the role of the asylum as a model for social order. Also, revolutionary regimes from the early twentieth century used the asylum to produce and propagate revolutionary understandings of health and work but only in indirect ways. The typical inmate from the revolutionary era was not a political activist confined against his or her will, but the destitute patient suffering from chronic illness who placed heavy economic burdens on family or city agencies. In fact, asylum doctors were especially deaf to medical histories that involved political activism and military participation in the revolution, which were not rare in asylum files. Doctors, for example, failed to diagnose, or even to mention, war shock in their diagnoses.[102] Lastly, while the asylum changed to better fit the needs and expectations of society, this accommodation significantly involved the participation of both state *and* families, the major initiators of confinements. Further analysis of these variables at both quantitative and qualitative levels will help clarify the ambivalent roles of the asylum in early revolutionary Mexico.[103]

[102] Kleinman, 'Violence, Culture, and the Politics of Trauma', *Writing at the Margin*, 173–89.
[103] See C. Rivera-Garza, *Mad Narratives: Inmates and Psychiatrists Debate Class, Gender, and Nation at the General Insane Asylum, Mexico, 1910–1930* (Lincoln, Nebr., forthcoming).

11 Psychiatry and confinement in India

Sanjeev Jain

> The establishment of lunatic asylums is indeed a noble work of charity, and will confer greater honor on the names of our Indian rulers than the achievement of their proudest victories.[1]

The history of asylums in India provides an opportunity to study the spread of ideas about mental illness, and notions of care and responsibility for the mentally ill across cultures and time. Although there are suggestions that hospitals have been known in the south Asian region[2] from antiquity, there is little documentary proof of their existence. References for institutions for the sick and needy can be found during the reign of Ashoka (268–231 BC).[3] Travellers' accounts of AD 400 mention similar services established by rich merchants and nobility.[4] Mental hospitals had a long history in the Arab world, and the growing Muslim influence in India lead to the establishing of similar hospitals.[5] However, the prevailing social situations have led some authors to suggest that these were seldom used except by 'soldiers and foreigners'.[6] Medical care in medieval India was based on Ayurvedic (derived from the Charak Samhita and other classical Indian and largely pre-Islamic texts, thus predominantly Hindu)[7] and Unani (the Muslim school of medicine), and derivative systems, delivered by

I would like to thank the Wellcome Trust and the Wellcome Institute of History of Medicine, the Commonwealth Trust, and the Department of State Archives, Government of Karnataka for help in preparing this manuscript. I would also like to thank Dr Vivek Benegal, Dr Satish Chandra, Dr Melvin Silva, Mr D. M. Joseph, Mr G. Vidyadhar, and Mr C. C. Silva for help with material and suggestions.

1 W. Forbes, 'Review of Practical Remarks on Insanity in India', *Psychological Medicine and Mental Pathology* 6 (1853), 356–67.

2 L. P. Verma, 'History of Psychiatry in India and Pakistan', *Indian Journal of Neurology and Psychiatry* 4 (1953), 138–64.

3 S. Dhammika, *The Edicts of Ashoka* (Kandy, Sri Lanka, 1993).

4 Fa Hein, *A Record of Buddhist Kingdoms*, trans. J. Legge (Oxford, 1886).

5 J. G. Howells (ed.), *World History of Psychiatry* (New York, 1968).

6 L. P. Verma, 'Psychiatry in Unani Medicine', *Indian Journal of Social Psychiatry* 11 (1995), 10–15.

7 D. Wujastyk, *The Roots of Ayurveda* (New Delhi, 1998).

professionals trained by study and apprenticeship. These professionals were most often attached to the court, or provided services for a fee. Religious and caste divisions perhaps did not allow a public space for uniform treatment for the ill to exist.

The growing European influence in the second half of the millennium had a profound impact. During the first half of this period, between 1500 and 1750, there was a growing awareness of 'European' medicine. European practitioners were often attached to the courts of kings all over India, including the Mughal Emperor. The Portuguese established a hospital in Goa, which served the needs of their sailors and soldiers. Garcia d'Orta, perhaps one of the earliest European physicians in India, established a herbarium, and was renowned for his medical skills, and published his colloquies in 1563.[8] He interacted with the local Indian physicians and learnt about the Indian pharmacopoeias. However, after his death it was discovered that he was a Jew, and had transgressed existing laws regarding the travel of Jews on Portuguese ships. His body was exhumed and burnt at the stake.[9] It is also suggested that his consorting with the 'heathens' and acknowledging their knowledge systems could have added to his heretical status. It is not known whether the Portuguese hospitals made a specific provision for the mentally ill. Several other European doctors did make their mark in India. Manucci, a Venetian, became a self-proclaimed doctor, towards the end of the seventeenth century, and describes treating a few mentally ill patients with leeches, cupping and various native medicines, often with success.[10]

The British gained ascendancy over all the other European powers in India by the end of the eighteenth century. The East India Company is alleged to have obtained permission to set up a trading post in Calcutta, which proved to be the most important for its long term interests, as a favour for medical help provided by Boughton, an English physician, to ladies of the Court of Shah Shuja, the brother of the Mughal Emperor, in 1638.[11] Almost a century later, another physician, William Hamilton, was to provide medical help to the then Emperor, and was rewarded with further concessions in Madras and Surat.[12] Throughout the colonial period, medicine and politics would continue to be linked.

Perhaps the first establishment for treating the mentally ill was the one established by Surgeon George M. Kenderline in Calcutta in 1787. However, it

[8] J. Barros, 'Garcia Da Orta – his Life and Researches in India', in B. V. Subbarayappa and S. R. N. Murthy (eds.), *Scientific Heritage of India, Mythic Society* (Bangalore, 1986).

[9] C. R. Boxer, *Two Pioneers of Tropical Medicine*, Wellcome Historical Medical Library, Lecture Series 1 (London, 1963).

[10] N. Manucci, *A Pepys of Moghul India* (Srishti, 1999).

[11] Lt. Gen. Sir Bennett Hancie, 'The Development and Goal of Western Medicine in the Indian Sub-Continent (Sir George Birdwood Memorial Lecture)', *Journal of the Royal Society of Arts* 25 (1949).

[12] D. G. Crawford, *History of the Indian Medical Service 1600–1913* (London, 1914).

could not be granted 'official recognition' as the surgeon had been previously dismissed from service for neglect of duty. Soon after, William Dick in Calcutta established a private asylum for 'insane officers and men, and civilians of various stations', in 1788. Others in Bombay and Madras followed. The asylum at Madras was ordered to be built in 1793, for sixteen patients, and given a generous endowment and land, on the provision that no rent was to be paid as long as the building was devoted to public purposes.[13] Assistant Surgeon Valentine Connolly, wrote to the medical board saying that: 'want of an asylum on the coast has been long a matter of regret, and in some instances it has been attended with dreadful consequences'. Suggestions for the asylum included detailed plans for buildings and staff, with a payment from the company for each patient admitted to the asylum. Connolly later 'privatized' this arrangement and began paying a rent of pagodas 825 to the company, and finally sold it to Surgeon James Dalton in 1807 for pagodas 26,000. It was long known as Dalton's madhouse, and is now part of the medical college.[14] By this time the asylum accommodated fifty-four Europeans, and a staff of fifteen keepers. Asylums within Bengal (Murshidabad, Dacca), Madras (Chittoor, Tiruchirapalli and Masulipatanam) and the Bombay (Colaba) presidencies were set up.

Prior to this, patients, especially those whose symptoms lasted for more than a year, were to be transported to England. John Reading, a doctor in Chingleput near Madras, writing to George McCartney,[15] recommends that one Mr Porter, who has been suffering from a maniacal complaint, be sent home. He also mentions that several such patients now live in Madras. The hot tropical climate was often to blame, and a voyage home held out the promise of a cure. Allegations of exorbitant charges and corruption in contracting private hospitals were already being made.[16] Financial irregularities, and overcharging the company (expenses are proportionate to the number of surgeons, rather than the number of sick) for the care of the ill was a frequent concern, as were poor maintenance and misuse. Prompted by this, the East India Company in 1802 ordered asylums to be built for the wandering insane in all its territories. Indian kingdoms were not very encouraging. However, Hoenigberger, a German doctor, who travelled overland and lived in Punjab, did establish a small asylum in Lahore early in the nineteenth century.[17] This was paid for by the court at Lahore, but had been ordered by the British Commissioner. It existed for a few years, and was staffed by European doctors, but it fell into disuse once Hoenigberger returned to Europe.

[13] H. D. Love, *Vestiges of Old Madras 1640–1800* (Madras, 1996).

[14] D. V. S. Reddy, *The Beginnings of Modern Medicine in Madras* (Calcutta, 1947).

[15] George McCartney 1737–1806, First Earl McCartney, Governor and President of Fort St George, Madras. Correspondence and papers concerning medical services at Madras 1782–7. MS 5746, Wellcome Library, London.

[16] *Ibid.* James Hodges to McCartney, 17 April 1783.

[17] J. M. Hoenigberger, *Thirty Five Years in the East* (London, 1852).

In this period, before the Indian Mutiny (which occurred in May 1857), approximately thirteen asylums had been established in various parts of the company's dominions. By this time, the British directly controlled several large portions (the Calcutta, Madras and Bombay presidencies) and had administrative control over other areas through bilateral agreements between them and the rulers of independent states. The sub-continent was thus broadly divided between princely or native kingdoms, and the British possessions. The contrasting outcomes in the various asylums, and differences in the cost of maintaining them, led to one of the first official enquiries in 1818, and has been summarized earlier.[18] The select committee had raised similar issues in the United Kingdom in 1815/16. The concept of the asylum was defined as a 'retreat, providing for the tender care and recovery of a class of innocent persons suffering from the severest of afflictions to which humanity is exposed'.[19] Gross deviations from this noble aim were observed. Most asylums were seen to be a cluster of ill-constructed and poorly maintained buildings, resembling gaols rather than asylums. Conditions within were deplorable, with indifferent staff, unwholesome food, inadequate clinical classification and care.

The native clientele of the asylums

While most asylums were established as distinct establishments, in Bangalore in southern India, a somewhat different approach was taken. The city of Bangalore, lying almost in the centre of peninsular India, was of considerable strategic importance. It was also noted for its 'salubrious climate' and was used as a convalescent base for the Crimean, Afghan, the First and Second World Wars. It was felt that 'the climate is particularly congenial to the European constitution; sores quickly take on a healthy action, and convalescence from acute diseases is rapid, often in a remarkable degree; and the protracted convalescence and low chronic state of disease, seen in other parts of India, are seldom met with at this station'.[20] The British had moved the Mysore division of the Madras army here from Seringapatanam, after the 'White mutiny' of 1809, and also because of the high rates of fevers at that place. A large cantonment was built, which had elaborate hospitals. Dragoons (cavalry), European infantry, Indian soldiers and civilians were provided with separate hospitals. The Garrison hospital, which had both European and Indian soldiers, was considered the 'best'.

[18] W. Ernst, 'The Establishment of "Native Lunatic Asylums" in Early 19th Century British India', in G. J. Meulenbeld, D. Wujastyk and E. Forsten (eds.), *Studies on Indian Medical History* (Groningen, 1987).

[19] *Ibid.*

[20] 'McPherson Report on the Medical Topography and Statistics of the Provinces of Malabar and Canara' (Madras, 1844).

The city was administered between 1831 and 1881 by appointed Commissioners, one of whom was Sir Mark Cubbon. A man of considerable foresight, he initiated a number of; public-health services. At the time of rendition[21] in 1881, when the administration reverted to the Maharaja of Mysore, a total of three general hospitals, seventeen dispensaries, two maternity hospitals, eight gaol dispensaries, ten railway hospitals and two special asylums (leper and lunatic) had been established in the kingdom.[22] Rates of various diseases were quite high. Dysentery, hepatitis and delirium tremens were frequent causes of illness in European soldiers. Between 1829 and 1838, in the 15,590 European soldiers, the commonest diseases were syphilis (25 per cent), wounds and injuries, dysentery, fever, hepatitis and chest diseases. The Indian troops (70,000) had much lower rates of illness. Fever, diarrhoea, wounds, chest diseases, rheumatism and syphilis (2.1 per cent) were recorded, but not at the high rates as were noted for the European soldiers. Excessive drinking and 'wanton' behaviour were often blamed for the high rates of hepatitis. Ebrietas (drunkenness) was recorded as a diagnosis for more than a hundred European soldiers every year between 1834 and 1838, but not even once for an Indian soldier.[23]

Dr Smith, who appears to have been a physician to Sir Mark Cubbon, in addition to being a public doctor, began his diary in 1833. He provides one of the first detailed case notes of psychiatric diseases, and suspects that a large proportion of them are caused by organic factors.[24] He describes patients who show depressive symptoms, progress to dementia and after death are discovered to have inflammatory changes in the brain, or spicules surrounded by inflammation.[25] These provide the first descriptions of neurocysticercosis, which was formally described only several years later.

Dr Smith's casebook has several case histories. One patient became suspicious of European and native officials and shot dead a native in order to force attention upon himself. Another maintained an exemplary life in the office for fourteen years, but was otherwise 'eccentric to the point of madness' and suddenly became acutely disturbed and Dr Smith was 'obliged to put him in a straitjacket'. Of the 138 patients with mania treated by Dr Smith, thirty-eight

[21] *Rendition*: the kingdom of Mysore was under direct British rule through a commissioner between 1831 and 1881. At the time of taking control in 1831, Britain had contracted that the kingdom would revert to native rule in fifty years. As a result, the administration reverted to the Maharaja of Mysore in that year (1881), but the British retained control of large tracts of land, and part of the city of Bangalore (the cantonment).

[22] B. L. Rice, 'Mysore: A Gazetteer Compiled for Government' (London, 1887).

[23] 'McPherson Report on the Medical Topography and Statistics of the Provinces of Malabar and Canara', 20.

[24] S. Jain, P. Murthy and S. K. Shankar, 'Neuropsychiatric Perspectives from 19th Century India: The Diaries of Charles Smith', *History of Psychiatry* (forthcoming).

[25] Charles Irving Smith, commonplace book, containing medical notes, MS 7367, Wellcome Library, London.

died, putting these symptoms at par with ascites and paralysis in terms of prognosis. Given the number of mentally ill patients that he treated at the Hospital for Peons, Paupers and Soldiers, he was able to convince Sir Mark Cubbon about the need to establish a ward for the mentally ill at this hospital in 1847, and eventually an asylum.[26] In 1850, the asylum was moved out of the hospital into the gaol, and subsequently a new building was constructed on an elevation near a large lake. This facility, and its successors, would have an important role in the growth of psychiatry in India. He was not averse to using native medicines. He prescribed limejuice and pepper for an attack of rheumatism to Mark Cubbon, and strongly recommended coconut water as a blood purifier.

Asylum reports form the bulk of historical sources of psychiatry in India. The asylum in Delhi, as the report pointed out in 1870,[27] was situated just outside the ramparts, close to the gaol and Feroze Shah's tomb. This asylum lay in the path of the mutineers marching towards Delhi from Meerut, and on 11 May 1857, all 110 inmates escaped.[28] After the mutiny, the asylum was reorganized and lasted until 1861, when it was moved to Lahore. Bad conditions, and the 'barbarous practice of using jails' as asylums was often a cause for complaint. Chemical and bacteriological examination of the water supply in 1867 revealed that the water was unfit, and new sources were identified. In the 1850s, G. Paton introduced a very strict discipline in the Delhi Asylum. Servants could be dismissed if the wards were dirty. Tobacco was to be given only when patients performed active work. Food was diversified, so that those who worked got better food than those who did not. The patients were employed in laying out and maintaining extensive gardens. Economic incentives could also be offered. Medical treatment consisted of blistering the head and neck, cold and warm baths, and tonic and aperient medicines (both native and European). In 1873, the superintendent of the Delhi Asylum opines that the 'insane lose all caste prejudices', and thus could be housed in common wards. This is important, as western hospitals, and doctors in general, were viewed as unclean under orthodox beliefs. The superintendent, however, lamented that there was one *baniya* (a member of the merchant caste), who was very rigid and refused to accept food from his hand (being British). Some of the other Hindu patients refused meat, but accepted chapattis (bread) from everyone. In 1873, Mr and Mrs Gilson, a British couple, lived with the patients, and shared the food. They got glowing tributes by successive superintendents, and much regret was expressed upon their transfer to Agra.

It was felt that 'amusement helps to cure lunacy as [much as] anything else, besides having a humanising effect on the violent patients'. An orchestra by the patients was organized, with a sitar, tabla, etc. and there was 'much singing in

[26] *Ibid*.
[27] Annual Report of the asylum at Delhi,1867, V/24/1718, India Office Library, London.
[28] Annual Report of the asylum at Delhi, 1872, V/24/1719, India Office Library, London.

the wards'.[29] Pets were a particular passion, and the patients maintained cats, pigeons and monkeys in the wards. On prominent festival days like Dussehra, the patients were dressed up in fine clothes, several bullock carts were hired, and they were all taken to the fair in front of the Red Fort. It was felt that contact with the wider community would be effective in reducing the prejudice against the insane. By 1877, it was reported that there was a gradual improvement in the quality of the deputy superintendents, and in time, it would be possible to bring the asylum as near to the English standard as the circumstances of the country permit. There seems to have been some attention to administrative probity, as some of the British staff was suspended for laxity in discipline, or stealing money from patients.

Clinical descriptions are also quite illuminating. In 1877, an Irish soldier claimed that he was a general and alleged that the government had stolen his pay and spent it on oranges.[30] He converted to Islam and announced at the Jama Masjid (the main mosque of Delhi where the Mughal emperors offered prayers) that the Russians were on their way, and that all Muslims should get ready to help them. He was admitted to the asylum, but there was a public outcry, as it was felt that he was being considered insane for converting to Islam, while conversions to Christianity were not similarly viewed. Faced with an uncomfortable situation, and with the population of Delhi 'excitable', Mr H. was quickly transferred to Colaba in Bombay, where there was a holding asylum for Europeans while they were on their way to the Ealing Asylum in England. Another instance is of a Sikh soldier, who was admitted in 1883 after being caught eating the dead body of a child. The soldier explained that he belonged to a particular sect, that forbade him to work or beg for food, and he was supposed to eat whatever providence brought his way. Walking along the riverside, he saw some jackals eating the body, and after chasing them away, he did the same. It was decided that he was not insane, and he was set free. Other clinical vignettes describe behaviour in some detail, suggesting a close interaction between the doctors and the patients.

Elsewhere in northern India, for instance, in the Punjab, asylums had a chequered history.[31] After the annexation of Punjab in 1848–9, the twelve patients in the asylum set up by Hoenigberger were handed over to the British. After much debate, disused barracks in Anarkali in Lahore were converted into an asylum. Faced with huge costs, some people were of the opinion that the cost should be spread over twenty-five years to get a true estimate, and the running of asylums should 'reflect the highest credit upon the Government for work of such great importance'. As Lahore became the 'Paris of the east' the suburb

[29] Annual Report of the asylum at Delhi, 1883, V/24/1720, India Office Library, London.
[30] Annual Report of the asylum at Delhi,1876–77, V/24/1720, India Office Library, London.
[31] W. Lodge Patch, *A Critical Review of the Punjab Mental Hospitals 1840–1930*, Punjab Record Office, Monograph 13, V/27/858/9, India Office Library, London.

of Anarkali became fashionable and the asylum had to be moved away. By the end of the nineteenth century, financial prudence was strictly enforced. In 1896, it was ordered that only two meals per day were to be served, salaries were reduced and it was regretted that a 'secretary had compared the cost of mental health care across India and decided that it was too expensive'. As early as 1867, an opinion was expressed that large central asylums were a mistake and the separation from the family was not good. It was also suggested that admission to an asylum was necessary for a brief period to establish diagnosis, after which patients could be sent back to 'village colonies'. Though sceptical of home and family care, there was some merit seen by now in emulating the Italian reforms of 1905 that recorded cases of insanity and monitored their care in the community.[32] These and other instances across India perhaps show some of the earliest thoughts about adapting community care for the mentally ill.

Reports from the asylums in the Madras presidency (1877–80) suggest that up to 20 per cent of the inmates were Europeans or Eurasians. Unlike Delhi, it was reported that caste and religious prejudices were as yet too powerful. A number of patients suffered from an acute mania, which had an excellent prognosis, with almost a third recovering entirely. It was again pointed out that 'insanity more often [arose] out of depraved bodily condition, rather than overstrained mind'. In proportion to moral causes, twice as many patients with physical causes were admitted in Madras (289:100), as compared to rates in Europe (129:100). It was suggested that the 'European ... is subject to restless mental activity, keen sensibility and susceptibility to emotion ... to which [the Hindu] ... is a stranger'.[33] It was also suggested that 'asylums should be for cure, and harmless imbeciles and lunatics can be cared for in huts, attached to mofussil [district] dispensaries, and under medical supervision'. Drugs used by now included bromide of potassium, hydrate of chloral, morphia, digitalis tincture, etc, but it was also mentioned 'that wine or a little arrack often proves to be a good hypnotic, and avoids use of opiates'.[34] The sameness of mental symptoms was emphasized, and the point made that different classes and different nations have identical symptoms. Depression was found to be very common, especially with respect to moral causes. Entertainment at the asylum included a fortnightly band and 'open house' where patients mingled with the public; cricket twice a week for Europeans and once a week for Indians, and the occasional circus. It was commented that the native warders were generally indifferent, and on occasion some were suspended for striking a patient. The guiding principle was provided by a quote from Maudsley that 'the true treatment of the insane lies in still further increase in their liberty'.

[32] *Ibid.*
[33] Records of the Madras Asylum, V/24/1704/Madras, India Office Library, London.
[34] *Ibid.*

On the west coast, asylums in the Bombay presidency[35] also had diverse experiences. Inmates of the asylum in Ahmedabad showed significant evidence of caste prejudice, and never entered each other's rooms. The asylum in Poona was 'utterly devoid of the most evident requirements of a medical institution'; the condition in Dharwar 'no credit to Surgeon Major MacKenzie'; while in Haiderbad (Sind) there was an increase in population after 'having seen for themselves how kindly and carefully the patients are treated, to the credit of Dr Holmstead'. Costs in the asylums in Bombay itself reflected the differences in care. Europeans were budgeted at Rs 400 per annum, Parsis and Jews at Rs 263, while the Hindus and Muslims at Rs 213. Annual diet costs were Rs 64 for Hindus and Muslims, but Rs 200 for Europeans. These were the holding asylums, earlier described by Ernst.[36] A superintendent here was to report that 'the Europeans are not inclined to work . . . and it would be difficult and not without danger to employ them in the same shed as natives . . . as insane people are almost always full of prejudices and conceits, and are possessed of irritable and hasty tempers'.[37]

The annual reports of the asylum in Bangalore[38] during the same period show a gradual increase in the number of admissions, and the size of the asylum. It ultimately offered accommodation for 260 patients, at approximately 50 feet per person. The buildings were described as being 'simple, but airy'. The asylum was at an elevation, close to a lake; and adequate water supply and dry earth conservancy were provided. The annual reports repeatedly emphasize the importance of 'moral influence', and the 'dreary misery enlivened by amusements suited to their condition and capacity'. Work was emphasized, and a number of opportunities like gardening, rope weaving and domestic work were offered. The asylum was administered by doctors of the Indian Medical Service, with a number of Indian assistants. After the transfer of power to the kingdom of Mysore, in 1881, it became the only asylum that was supported by a native kingdom.

Work at the Bangalore Asylum was given enough prominence by the administration. Difficulties were frequently encountered. The *pettah* (old city) hospital and the asylum were three miles from the cantonment, and Dr Henderson, the superintendent of the asylum, complained in 1871[39] that it was difficult to complete rounds of all the establishments, as he was also in charge of the general hospital. In addition, he was also expected to see European and Eurasian patients at home. The chief commissioner ordered that 'since duties at the asylum

[35] Records of asylums in the Bombay presidency V/24/1708, India Office Library, London.
[36] W. Ernst, *Mad Tales from the Raj: The European Insane in British India* (London, 1991).
[37] Records of asylums in the Bombay presidency V/24/1708, India Office Library, London.
[38] Annual Report of Special Hospitals in Mysore (1877).
[39] Medical 1870, 1/1870 1–17, Reorganization of Civil Medical Establishment at Bangalore, Karnataka State Archives, Bangalore.

are of a very different nature – moral and disciplinary to a much larger extent than purely medical, Dr Henderson could decide on his own time for rounds'. This was a significant departure from rules, as morning and evening rounds by the doctor, were compulsory. It was suggested that the arrangement 'was practicable without in any way compromising the interests of the lunatics'.[40]

Overcrowding became evident very soon. In 1868, the number of lunatics in the Asylum had reached one hundred, against a projected maximum at that time of 150. Staff shortages was a frequent complaint. Dr Oswald complained in a letter to the government in April 1868 that though the Madras Presidency Asylum had one peon for every three to five lunatics, the Bangalore Asylum had five permanent and two temporary peons for one hundred patients. The seriousness with which this complaint was viewed is reflected in the speed of decision-making. The Viceroy in faraway Calcutta sanctioned more posts in June 1868. It was also observed in 1872 that a large number of paupers were being admitted for humane reasons. Going through the records reveals the famous diversity of India. Patient's religious and national identities were recorded, and Armenians, European Catholics, Italians, Irish, English and people from all parts of India were represented in the patient register. Although it 'may be advisable to provide additional accommodation for caste patients . . . [it] should be done without prejudicing the interests of those who look of European mind', suggested one official communication.[41] While cognizant of local social mores, the administrators were also becoming aware of the changes occurring in Indian society by the advent of western medicine. The diagnoses were very varied, but were consistent with those in use in asylums in the UK at this time. Although the bulk of the patients were classified as having one form of mania or the other, there were a few diagnosed as morally insane (mainly Europeans). Alcohol and cannabis (ganja) are listed as common physical causes.

Academic responses

Although the East India Company doctors were supposed to have some knowledge of Indian languages,[42] the doctor–patient contact would often have to be through interpreters. The official recognition of Indian languages, and by extension, indigenous knowledge, was still evolving.[43] Administrative records could be faulted for not paying enough attention to the voices of the mentally ill, especially those of the 'native'. However, since there are no known first-person accounts from the nineteenth century, these records provide at least some insight

[40] *Ibid.*

[41] Medical 1870, 1–17, letter from secretary to chief commissioner, 29 February 1872; Karnataka State Archives, Bangalore.

[42] D. G. Crawford, *History of the Indian Medical Service 1600–1913* (London, 1914).

[43] B. S. Cohn, *Colonialism and its Forms of Knowledge* (Delhi, 1997).

into the workings of institutional psychiatry in India. Given this handicap, it is quite interesting that the details of the diagnosis and the clinical vignettes contain psychopathological material at all.

Doctors who were to serve in India, had been asked by the company to 'produce a certificate of having diligently attended, for at least three months, the practical instructions given at one of the asylums for the treatment of the insane', as well as to acquire a knowledge of Hindustani before coming over.[44] By 1855, accounts of insanity in India began to be published.[45] This account, written by the superintendent of the Dacca Asylum, was replete with case notes, and suggestions about the etiology. It was also to set the tone for subsequent discussions. One of the aims of psychiatry was 'to determine whether the dark races of man are susceptible to the mental and moral influences necessary to the production of various forms of insanity... and cured by the same plan of treatment as in Europe'. Excessive studying, and a rapid change from a 'less civilized' state, were attributed as causes. It was also observed that, contrary to expectation, significantly smaller numbers of the mentally ill were seen in a 'country where the mental faculties were so less cultivated'. A preliminary attempt at calculating rates was also made, and the total of 157 lunatics under treatment in the Dacca circle (population 9.8 million) was much smaller than the 13,400 under treatment in England and Wales (population 17.9 million). The much higher incidence in Europe was caused by the spread of education and the generally more nervous temperament. Winslow Forbes, the editor of *Psychological Medicine and Mental Pathology*, acerbically pointed out 'many of the facts presented are interesting, but full of error... views that are unsubstantiated'.[46] Forbes disagrees with Wise that 'the Hindoos [*sic*] are perhaps in a lower state of mental development than even the rudest savages'. He found this untenable, as evinced by the literary and architectural achievements in India. Indeed, he was critical of the major role mercantile and military interests were playing in India, and the neglecting of the development of the liberal perspective.[47]

Case reports and reviews in the nineteenth century also began discussing issues related to psychiatry in India. Organic causes were often identified, as in the case of a chronic mania who recovered after developing symptoms of effusion of the brain,[48] and a soldier who developed symptoms of paranoia after developing an ischio-rectal abscess.[49] A soldier who drove a nail into his head

[44] Crawford, *History of the Indian Medical Service*.
[45] T. A. Wise, *On Insanity in Bengal* (Edinburgh, 1852).
[46] Forbes, 'Review of Practical Remarks on Insanity in India'.
[47] W. Forbes, 'Moral Sanitary Economics', *Psychological Medicine and Mental Pathology* 6 (1853).
[48] *Indian Medical Gazette* 13 (1878), 140.
[49] D. M. Moir, 'Mental Depression, Hallucinations and Delusions Associated with Ischiorectal Abscess', *Indian Medical Gazette* 26 (1891), 7–9.

while intoxicated with cannabis, and died after a delirium lasting two weeks, was found to have a clot on post mortem.[50] Case reports, for example, the one by C. K. Swaminath Iyer[51] of an acutely ill twenty-year-old male, who recovered after passing a roundworm, suggests that Indian medical personnel were also beginning to contribute to the scientific literature.

Psychological issues were also described, as in the case of a man who developed a brief psychosis after watching a float that had actors masquerading as being decapitated during a Moharram procession.[52] Chetan Shah, an Indian assistant surgeon, gave an account of hysteria in a fourteen-year-old boy, who could not walk and complained of pain at regular times everyday.[53] Since the boy seemed to be devout, 'an attempt was made during an intermission to produce a deep impression and to invoke the Guru's help'. Dr Shah opined that hysteria in young men was not as rare as mentioned in the textbooks, and felt that faith had a significant role in its cure. Another Indian,[54] Dr Pandurang, reported a case of hysteria that was helped by deva-rishis (native faith healers, but 'sorcerers' in the original report) after his treatment with various drugs and a wine-and-egg mixture had failed. Dr Ram C. Mitter, at the Arrah Charitable Dispensary, treated a case of acute mania in a fourteen-year-old married girl with blistering of the head, purgatives, and cold baths with complete recovery over a week.[55]

The emphasis was on physiological and organic causes of insanity. This was in keeping with the tenor of psychiatry in England in the nineteenth century.[56] There was an ambiguous approach to neurology, but simultaneously an unwillingness to view mental disorders other than manifestations of a brain disease. There was a reluctance to explore psychological models, and thus the absence of much of this in writings from India is not surprising. Emphasis was placed on moral therapy, and that is the predominant theme in the asylums in India.

All these anecdotes, and administrative reports notwithstanding, the initial impetus for providing services was not maintained. Discussing the possibility of employing native staff, an editorial comment in the *Journal of Mental Science* regretfully observed that the 'race prejudice had become the most important fact in the social state of India . . . a conquered country, ruled by a dominant race', not unlike the relation between the races in the citizens of America.[57] The imperial expansion, and wars in the Crimea, Afghanistan and various parts of India needed large amounts of money. There was also widespread famine in

[50] *Indian Medical Gazette* 15 (1880), 71.
[51] *Indian Medical Gazette* 19 (1884), 78.
[52] Dr B. L. Rice, *Indian Medical Gazette* 13 (1878), 112.
[53] Chetan Shah, *Indian Medical Gazette* 23 (1888), 302.
[54] Dr Pandurang, *Indian Medical Gazette* 4 (1869), 55–6.
[55] R. C. Mitter, *Indian Medical Gazette* 2 (1867), 225.
[56] W. F. Bynum, 'Theory and Practice in British Psychiatry from J. C. Prichard (1786–1848) to Henry Maudsley (1835–1918)', in Teizo Ogawa (ed.), *History of Psychiatry* (Tokyo, 1979).
[57] 'Reports on East Indian Asylums', *Journal of Mental Science* 5 (1859), 218–22.

the 1870s. In an order in 1879, it was stated that financial exigencies forced the government to cut back on non-essential expenses.[58]

By the end of the century, things were not in a good shape. An effort to tabulate the services revealed that there were 3,246 insane patients in British India, in twenty-one asylums, and conditions were apparently somewhat better than earlier.[59] The presidential address by T. W. McDowall[60] to the fifty-ninth meeting of the Medico-Psychological association focused on the insane in India and their treatment. Dr McDowall regrets that only 4,311 places for patients exist in the asylums of British India, for a population of 23 million. Even more disturbingly, apart from Mysore, none of the other native states, with a total population of 75 million, had an asylum. Rather than a low rate of insanity, he feels it is neglect of patients and want of services that are revealed in these figures. There was no lunacy board; army medical officers with no particular training in psychiatry administered the asylums, there were frequent changes of staff, the pay was deficient, work irksome and full of petty detail. In general, there was a systemic failure of the administration, annual reports had become worthless and there was no attempt to develop an efficient policy for treatment.

Census reports – early attempts at developing an epidemiology of psychiatry

The Indian census, one of the largest demographic exercises in the world, was initiated in 1872, and conducted every ten years, with the exception of the war years. It was meant to assist the government in planning services and provide insight into social conditions all over India.

Since 1881, the census listed the mentally ill as a separate category, and a statement of number of persons of unsound mind by religion, age and sex (form XIV, census 1881) was to be provided by the returning officers. The governor general of India had requested in 1867, that the number of insane in the provinces be counted.[61] Between 1881 and 1951, the census reports included estimates of the number of mentally ill. Doubts were frequently raised about the quality of information, as, for example, by Dr Deakin of the North-West Province who felt that non-professional enumeration might not identify mild or periodic forms of insanity.[62] However, despite these drawbacks, it still provides a window into the status of services available. Huge variations were seen within India, and between India and the UK. In the 1881 census, more than 80,000 insane were identified in British India.[63] Consanguinity, brain disease and 'disappropriate ambition'

[58] 'The Punjab Reference Book for Civil Officers', 14 (1879), 42–5.
[59] H. C. Burdett, *Asylums – History and Management* vol. I: *Hospitals and Asylums of the World* 4 volumes (London, 1891).
[60] T. W. McDowall, 'The Insane in India and their Treatment', *Journal of Mental Science* 53 (1897).
[61] Patch, *A Critical Review of the Punjab Mental Hospitals 1840–1930.*
[62] 'Census of British India, 1881' (London, 1883). [63] *Ibid.*

and 'intense application to study' were listed as causes. At the same time, the fact that rates were a sixth of those in England and Wales (but almost the same as Italy, a less developed European country) was consequential to the fact that 'mental work (and) intense competition of an active civilization is completely unknown'. In 1881, the census officer of Mysore suggested that some amount of insanity could be attributed to the habit of marrying with relatives, 'which was a compulsory obligation in certain classes and castes'.[64] Addressing this question, the census officer of Assam in 1921 reported that this was not likely, as rates of insanity were the same in exogamous and endogamous tribes.

Geographical, religious and cultural differences were explored in several subsequent census reports, and a ten-fold difference in rates between Coorg in southern India and Burma was observed. By 1921, it was evident that the role of these factors was not substantiated. More importantly, as per estimate, 14 per cent of the insane were already housed in twenty-three asylums of British India. This was important, as it was felt that in the community 'the lunatics' [lives are] not happy . . . [they] receive little sympathy . . . [are] bound hand and foot or [have] a heavy log fastened to the ankle'.[65] Till this point, mental hospitals were to be the mainstay of psychiatric care in India. No data was available for the native states and it was feared that most mentally ill were confined to gaols.

These census reports provide very crude data, but at the same time reflect a concern for establishing the nature of the burden of mental illness, and matching the provision of services to the numbers expected to utilize them. Several epidemiological studies were conducted after Independence, to establish the same issues, with equally disparate results.

Increasing amounts of admissions to the asylums was now causing significant overcrowding. A significant development was the establishing of the hospital for the European insane in 1918 in Ranchi. Though the most modern, its superintendent, Colonel Berkely-Hill noted 'that those responsible for the original design were obsessed with its custodial function so as to sacrifice most, if not all, of its remedial potentialities . . . it has anything but an agreeable appearance'.[66] Occupational therapy, psychoanalysis, amusements, organotherapy and, rarely, hypnotics were used. A follow-up study of discharged patients was attempted, and some effort made to study whether patients recovered sufficiently. One of the best accounts of the state of institutional care in India in the early part of the twentieth century can be found in the reviews of Mapother (1938)[67] and Moore Taylor (1946).[68]

64 B. L. Rice: 'Census of Mysore'.
65 'Census of British India, vol.1, 1921' (Calcutta, 1921).
66 O. Berkely-Hill, 'The Ranchi European Mental Hospital', *Journal of Mental Science* 52 (1924).
67 Report of Professor Edwin Mapother to Sir John Migaw, the president, medical board, India Office, 1938; Archives of the Bethlem Hospital.
68 Summarized in 'Quality Assurance in Mental Health, National Human Rights Commission', 1999.

Mapother report

Professor Edwin Mapother, was requested to visit Ceylon in 1937 and suggest reform of the psychiatric services. For this, he visited India, and submitted a report, which the medical board of the India House decided was not to be published but used exclusively as a background to suggestions for improving services in Ceylon. In the years before this, he had been instrumental in establishing the Institute of Psychiatry at the Maudsley Hospital in London, one of the principal responsibilities of which was to develop services in the British Empire.

'It would be difficult to affirm that with respect to psychiatry, the bearing of the white man's burden has been adequate', notes Professor Mapother at the beginning of his report.[69] In London, there was a psychiatric bed for every 200 individuals, while in India there was one bed for 30,000. Within British India, while in Bombay presidency there was one bed for every 12,000, in the Bengal, Bihar and Orissa region, there was only one bed for every 57,000 individuals. While there were five psychiatric beds for every eight beds for 'physical disease' in London, there was only one bed for psychiatry to every seven in India. There was overcrowding in almost every asylum, and a general shortage of staff. The 'inadequacy was increased by the ignorance and indifference of most medical men' and a 'tactful reticence . . . about defects that cruder persons might publicly call scandalous' he remarked. The asylum buildings, Professor Mapother caustically notes, 'were a permanent monument to brutal stupidity', perhaps 'guided by a PWD[70] concept of a lunatic . . . (one ward) a replica of the accommodation for tigers at the Regent's park Zoo' and some a 'desolate waste, based on the assumption that the insane are indifferent to discomfort and ugliness, and are destructive'. He rated the asylums on a grade of 'badness', with only the Asylums of Ranchi (for Europeans) in British India, and the one in Bangalore in the Kingdom of Mysore having anything to commend them. However, he wondered at the waste of money on an asylum for Europeans, recently established by Berkely-Hill with much triumph, 'based on a concept of race that in practice is unreal, and does not correspond to education, mode of life or any valid claim'. On the other hand, the asylum in Bangalore, he told Sir Sikander Mirza, the dewan of Mysore, was a 'monument to the vision and wisdom of all those responsible for the mental defectives in the East. The Institution is almost unique among mental Hospitals in India . . . it is quite evident that modern methods of diagnosis and treatment are available and freely used'.[71] The impending transfer of power into India hands was of no great concern, indeed many British psychiatrists stated that it was easier to obtain money

[69] Report of Mapother to Migaw.
[70] The Public Works Department (PWD) that was responsible for the design, construction and maintenance of government buildings.
[71] Sir Mirza Ismail, *My Public Life: Recollections and Reminiscences* (London, 1950).

from provincial governments than when the health services were under direct British control. In most places, Indian doctors were managing the asylums, and several had received training in England or the USA.

Professor Mapother[72] was also well aware of the complexity of the Indian social and political situation. While admitting the need for more trained specialists, he was sceptical about the possibility of bringing adequate numbers of Indians to train in the UK or USA in view of the colour prejudices. There was therefore an urgent need to develop a school in India, and the asylum 'at Bangalore was [is] structurally the only center which yet exists that is fit to house a post-graduate school'. In addition to its professional capabilities, it had the benefit of an enlightened native administration, religious harmony and an appeal to nationalism by being established in a native kingdom, Mapother said. Another asylum could be established in Delhi in the future, under British control. The post-graduate school could serve the entire region for training specialists. He also suggested reforms for psychiatric services in India. Easier access, reduction of legal procedures, setting up of visitors' committees and an urgent need to increase the number of beds, irrespective of all pressures, were the major recommendations. He also suggested that psychiatric wards be provided in all general hospitals, and only chronic cases be sent to the asylums. The quality of undergraduate education needed to be improved, and training in psychiatric social work and rehabilitation was to be introduced.

These suggestions, unfortunately, could not be executed at it was felt that 'other needs must have priority and that economic reasons forbade these defects being rectified'.[73] Mapother regretted that any criticism of the system was countered with the need for financial prudence, and the need to maintain the security and prestige of the British Raj. As an example of misplaced priorities, he wonders how an expense of £18 million (of a total budget for India of £60 million) can be justified for building New Delhi for 'ceremonial entertainment'. It was quite evident by now that reform and improvement would not be carried out in British India.

The case of Mysore native asylum in the British Empire

The kingdom of Mysore was administered directly by the British between 1831 and 1881, after which rule reverted to the maharaja of Mysore. At the time of rendition, a policy paper was prepared concerning medical services under the new administration. 'The future medical arrangements must partake of a European character, because there is no native system to fall back upon... the Principal charges are the Civil Hospital, Lunatic Asylum and Medical

[72] Report of Mapother to Migaw.
[73] *Ibid.*; Conversation between Mapother and Megaw.

Stores, and experienced and well-trained men be placed'[74] was the considered advice.

The asylum continued to provide services to the Indian population and the British residents of the army cantonment of Bangalore. Until the early part of the twentieth century, Indian and European women were housed together, but the overcrowding of female European lunatics necessitated the setting up of separate wards for European women in 1913. A gradual increase in the number of patients led to additional wards being constructed, but by 1914 no further expansion was possible. It now accommodated 200 patients, including twenty-seven Europeans and Eurasians. The number of people being admitted every year continued to increase, so that by the second decade of the twentieth century, more than a hundred admissions were made every year.[75] Exclusively Indian staff managed the asylum by now. By 1920, it was evident that 'a new building for the Lunatic Asylum is absolutely necessary... there will have to be specialists in nervous diseases'.[76] Dr Francis Noronha had recently been deputed to train in England, where he worked at the Maudsley Hospital with Dr Mott, from where he returned in 1921. Work was deferred for almost a decade because of lack of funds, but a new building was ready by 1932 in a sprawling campus on the outskirts of the city. Modelled on the plans of the Bethlem Asylum at the Lambeth site, it had four large pavilions, an interior courtyard garden and extensive lawns.

Dr M. V. Govindswamy, a medical graduate from the Mysore Medical College also began working at the mental hospital, and was also sent abroad – to the USA and to the Maudsley Hospital, for further training in psychiatry. In London, he met Professor Willi Mayer Gross, who had been brought over from Germany under the Rockefeller programme. The two shared common interests in philosophy and medicine, and this acquaintance was to guide the development of academic psychiatry in India. Upon his return to India, Dr Govindswamy was an active researcher. He began using cardiazol induced convulsions,[77] insulin coma,[78] and later, psychosurgery,[79] almost as soon as these were available in Europe. A scholar of Sanskrit and English, he also taught himself some German to read the original texts. He was instrumental in maintaining high standards of care, and systematic notes and medical evaluations became a routine at the hospital. Laboratories, rehabilitation services and psychological testing was also

[74] Medical 1880–1, file 1, series 1–2, Karnataka State Archives, Bangalore.

[75] Report of the Mysore State Asylum, Bangalore, 1916; medical 49/17/2 3452–3453 (December 1917), Karnataka State Archives, Bangalore.

[76] Medical 42/22, serial 1–5, 1922, Karnataka State Archives, Bangalore.

[77] M. V. Govindswamy, 'Cardiazol Treatment in Schizophrenia', *Lancet* (1939), 506.

[78] M. V. Govindswamy, 'Insulin Shock and Convulsion Therapy in the Tropics', *Lancet* (1939), 1232.

[79] M. V. Govindswamy, 'Rao BN Bilateral Frontal Leucotomy in Indian Patients', *Lancet* (1944), 466.

introduced. He also felt the need to apply concepts of Indian philosophy to the description of psychopathology, over and above the practice of ayurvedic and other traditional forms of medicine.[80] After Independence, the recommendations of the Sir Joseph Bhore committee in the preceding years to establish a centre for post-graduate education were to be executed.

The only centres thought adequate were the ones at Bangalore and the erstwhile European Asylum in Ranchi. Professor Mayer Gross, who had recently retired in the UK, was invited as a visiting Professor, to Bangalore. Here he helped develop a curriculum for post-graduate training. Dr Govindswamy was convinced that basic neurosciences were crucial to understanding disorders of the brain and mind. He developed a programme that included clinical services in neurology and neurosurgery (in addition to psychiatry, psychology and psychiatric social work), and basic sciences. This hospital was designated as the All India Institute for Mental Health, and began training students for a diploma in psychological medicine, and in clinical psychology in 1956. Unlike the western, especially American experience, psychoanalytical viewpoints were not reflected in the development of psychiatry. Dr Govindswamy himself felt that psychoanalysis was 'a strain on one's credulity',[81] as did Edwin Mapother, who said of a certain analyst that 'he represented the greatest danger to the development of psychiatry in India'.[82] This Institute was redesignated as the National Institute of Mental Health and Neurosciences in 1974. It was indeed ironical, and a tribute to Sir Mapother's perspicacity, that a native-administered asylum, rather than one of the colonial establishments, proved to be the most adept at synthesizing western and Indian approaches, and developing a comprehensive approach to neurosciences and psychiatry.

Case notes at the Bangalore Mental Hospital

Records at this hospital in Bangalore extend to the beginning of the twentieth century. We have tabulated these registers, and have tried to profile the clinical details. In 1903, the asylum had 258 patients (201 males, fifty-seven females) who had been there for an average of seven years. While a few were there from 1865, most had been admitted in the decade after 1895. *Mania acuta* and *Mania longa* were the commonest diagnosis and accounted for almost 60 per cent of the patients. Melancholia was also frequently diagnosed. Organic causes such as epilepsy, and acute and chronic dementias accounted for sixty-two admissions, almost 25 per cent of the total cases. Rarer diagnoses included chronic mania, chronic delusional states and idiocy. Seven individuals were declared not insane,

[80] M. V. Govindswamy, 'Need for Research in Systems of Indian Philosophy and Ayurveda with Special Reference to Psychological Medicine', *Journal of the Indian Medical Association* 18 (1949), 281–6.

[81] *Ibid.* [82] Report of Mapother to Migaw.

but not before they had spent an average of fourteen months in the asylum. Following the records of these patients, it was seen that eighty-seven (34 per cent) died. More than half of those admitted with idiocy, chronic dementia or epileptic dementia died. A significant number of those with mania recovered entirely, although a fourth of these patients also died over the next seven years. Of all the individuals admitted between 1895 and 1903, at the end of 1910 only thirty-five were still in the asylum. Eighty-eight had been discharged as cured or improved, while fifty-five had died.

We also analysed records of the new patients admitted in the years 1903–4. Relatively small numbers were admitted afresh – forty-two in 1903, and thirty-seven in 1904. This had remained relatively static for several years, for instance there had been thirty-eight admissions in 1878. Their average age was in the early thirties and a significant number had sought treatment earlier from the asylum. We could chart the outcome of these new cases through the casebooks of the successive years. *Mania acuta* and *Mania longa* were still the most common diagnoses. The large majority of these recovered or were discharged to the care of the family, and only five patients stayed on till 1910. Half of the new admissions stayed in the asylum between six and seven months, and *mania acuta* had the best recovery rate. Some died soon after admission, but most of these were suffering from epilepsy or idiocy.

Religion, caste and social background were recorded, and were representative of the population of Bangalore. Hindus accounted for 70 per cent of admissions, Muslims 21 per cent and Christians 8 per cent (including Europeans and native Christians). While most new cases who were discharged were from the city of Bangalore, a larger proportion of those who stayed in the asylum for longer periods were from more distant places in the kingdom.

In 1878, there were only eight diagnostic categories, but by 1904, nineteen diagnostic categories were in use. The case notes were reviewed, and quite often the diagnosis would be changed a few months after admission. New categories in 1904 included hypochondriac melancholia, and several categories of dementia. This probably reflected a better understanding of the causes of dementia by this time in medicine.

Changes in diagnostic practice are quite evident in cannabis-related psychosis. In 1879, ganja was identified as a cause in 75 per cent of the admissions.[83] Of the patients resident in 1903, ganja use was a factor in ten cases of mania, and a few of dementia. However, after 1900, ganja-induced psychosis as a diagnosis decreases substantially in the records. The closing years of the nineteenth century had seen a huge interest in cannabis. From the initial curiosity regarding its possible use in treatment,[84] there had been growing concern about its

[83] Annual Report on special hospitals in the province of Mysore for 1879.
[84] W. B. O'Shaughnessy, 'On the Preparations of Indian Hemp or "*Gunjah*"' (Calcutta, 1839).

role in causing madness.[85] The final report of the Indian Hemp commission, after interviewing a number of Indian and European experts, stated that there was insufficient reason to identify ganja as a cause of psychosis. By 1900, this opinion was widely shared, thus accounting for the rapid decline in rates of diagnosis.

Case notes from the 1930s included detailed psychopathological observations, family history, social functioning and a thorough medical review. Patients were seen everyday for the first few days after admission, and less frequently later. Laboratory tests such as the Wasserman reaction, blood counts and x-ray were available. Drugs in use included opium, chloral, paraldehyde, bromides, antipyrin and Jamaican dogwood. The residency surgeon, from the British Army, justified the expense in a letter to Dr Govindswamy, stating that 'a large number of cases are due to organic causes... the more patients are cured, the less will be the recurring expenses. In other words, it is better to spend money on drugs that cure, rather than on maintenance, that does not'.[86]

The development of the asylum in Bangalore encapsulates the various trends in institutional care in India. It started as a ward in a general hospital for civilians, as part of the services by the British Army in the first half of the nineteenth century. It became an institution, acquired a building, full-time staff and, after 1881, was administered by an Indian kingdom. Western medicine by now had gained social and intellectual acceptance, and Indian doctors managed the asylum well. Advances in medicine were incorporated quickly, and there was only minor evidence of any deliberate attempt to maintain social distinctions. Other asylums were not so fortunate.

The Royal Indian Medical Psychological Association

India by this time was well on its way to independence, and the Second World War was looming. A growing number of specialists in psychiatry were now practising in the hospitals and asylums. In 1936, a move to establish an Indian division of the Royal Medical Psychological Association (RMPA)[87] had been initiated, allegedly the first in the Empire, outside the UK. Dr Banarsi Das, superintendent of the Agra Mental Hospital, wrote to Dr R. Worth, the president of the RMPA, with a plan for the association and an estimate of the costs involved. It was also suggested that all those who had worked in asylums for a

[85] J. H. Tull-Walsh, 'On Insanity Produced by the Abuse of Ganja and other Products of Indian Hemp', *Indian Medical Gazette* 29 (1894), 333–7, 369–73.
[86] Mysore Residency Files 621/1, 1937, correspondence regarding a grant to the Mental Hospital, Bangalore for purchase of European medicines for the treatment of the mentally ill patients of the Civil and Military station, Bangalore. Karnataka State Archives, Bangalore.
[87] Records of the Indian Division of the Royal College, Royal College of Psychiatry, London.

long time but had not acquired specialist degrees (at that time, this was possible only from the UK) be allowed to become members. This was not permitted by the RMPA. Eventually, the Indian division came into existence and held two meetings in Agra (1938) and Lahore (1941). At the first meeting, Dr Thomas, the superintendent of the Hants County Mental Hospital in England, represented the RMPA, thus signifying some degree of co-operation between the psychiatric professions in the two countries.

The issues discussed were overcrowding of the hospitals, training of hospital attendants, improved undergraduate education and opportunities for postgraduate study, and the design of single cells best suited for use in India. The need for reform and expansion was thus acutely felt, both by the practitioners in India and visitors from abroad. The members of this association were the superintendents (by now largely, but not exclusively, Indian) and the growing number of psychiatrists in general hospitals and medical colleges.

After the death of Dr Banarsi Das in 1943, Lt. Col. Moore Taylor, superintendent of the European Mental Hospital at Ranchi, took over as president. By this time the war and the Indian political unrest was well on its way. In 1946, moves had been made to establish a separate Indian society. In April 1947, Taylor resigned as he felt that the Indian division was being allowed to die. The Indian Psychiatric Society with Col. Davis as its secretary had already been established, and the Indian division of the RMPA had 'ceased to function as such', as Dr Davis told the RMPA during a visit.[88] By November 1947, a few months after Independence, the Indian division was dissolved. Despite its short life, this association affirmed the close links between the Indian and the British medical professions, and their similar preoccupations.

Bhore Committee and the Moore Taylor Report

The Health Survey and Development Committee 1946 (Sir Joseph Bhore Committee) included reform of psychiatric services in its ambit and Colonel Taylor was asked to survey the mental hospitals. His report was based on his observations of nineteen mental hospitals with 10,181 beds. His findings were quite similar to those of Mapother a decade earlier. Asylums were designed for custodial care and not for cure. 'The worst of them were the Punjab Mental Hospital, the Thana Mental Hospital, the Agra Mental Hospital and the Nagpur Mental hospital . . . conditions of many hospitals in India today are disgraceful and have the makings of a major public scandal.' Increasing bed capacity, without concomitant increase in personnel, lack of attention to training and education at all levels, inadequate provision for rehabilitation, and poor liaison with medical services were pressing problems, and it was time he felt 'for Government to

[88] *Ibid.*, undated note.

take account of stock, overhaul resources, and rechart the course for the next 30 years'.[89]

The Bhore Committee chronicled the dismal state of health services in India.[90] There were only 73,000 medical beds in the whole of British India (0.24/1,000), the doctor – population ratio 1/6,000, and the nurse – population ratio 1/43,000. Life expectancy was only twenty-six years, compared to above sixty years in other parts of the empire, like Australia and New Zealand; and infant mortality rates were five times higher. However, the committee made sweeping suggestions for the development 'in forty years', of 'an integrated, preventive and curative National Health Service embracing within its scope institutional and domiciliary provision for health protection of a reasonably high order'. Loosely planned on similar reform in the UK, these suggestions had been hinted at by Dr Dalrymple-Champneys[91] (an adviser to the Bhore Committee) and Professor A. V. Hill[92] in the early 1940s. The Committee envisaged the setting up of a health administrative unit for every three million population, with primary health centres for every 20,000 and a specialist general hospital with 2,500 beds that would include care of the psychiatrically ill. The estimated cost would be Rs 2 per annum. However, as a unit, the costs were several times lower than those budgeted for similar services in England, prompting some to question the feasibility of it all.[93]

Suggestions for increasing the number of asylums, and beds for psychiatric services were made. However, progress was slow. By 1980, the number of mental hospitals had been increased to thirty-seven, but there were only 18,918 beds. The post-Independence expansion of services in India coincided with the introduction of pharmacological treatments. These became available widely in India very quickly, and were the mainstay of treatment by the end of the 1950s. Indeed, the first workshop of medical superintendents on improving mental hospitals called for a restraint in the use of tranquilizers! The growing awareness of the drawbacks of aslylum-based long-term care was also evident. As a result of all these diverse influences, between 1951 and 1961, only five more asylums were added, with approximately 2,500 beds.[94] However, the number of admissions increased several fold, as did the number of discharges.

[89] As quoted in 'Quality Assurance in Mental Health: National Human Rights Commission' (New Delhi, 1999).

[90] Lt. Gen. Sir Bennett Hancie, 'The Development and Goal of Western Medicine in the Indian Sub-Continent (Sir George Birdwood Memorial Lecture)', *Journal of the Royal Society of Arts* 25 (1949).

[91] Sir Weldon Dalrymple-Champneys, Health Review of India, GC 139/H2, Wellcome Library, London.

[92] A. V. Hill, 'Health, Food and Population in India,' *International Review* 21 (1945), 40–50.

[93] *Ibid.*

[94] S. Sharma and R. K. Chadda, *Mental Hospitals in India: Current Status and Role in Mental Health Care* (Delhi, 1996).

In the first two decades after Independence, the emphasis on asylum-based care for the mentally ill continued. Hospitals were added in Amritsar, Hyderabad, Srinagar, Jamnagar and one of the last in Delhi in 1966. Surprisingly, the one in Delhi was finally built at a site identified almost a century earlier to replace the asylum destroyed during the Indian Mutiny. Institutional care in India now consists of these forty-odd hospitals, with a total of 20,000 beds. Dr Vidyasagar in the Amritsar Mental Hospital introduced one of the most remarkable innovations in mental hospital care in the early 1950s.[95] He erected tents in the grounds, and encouraged families to live with the patients, until they recovered. He shared almost all his working hours with the patients and their families. Principles of mental health, derived from religious and medical sources, were shared. This significantly reduced the stigma of mental illness, and demonstrated the feasibility of community care.

The rapid availability of pharmacological treatments for psychiatric disorders allowed the government to envisage that care for mental disorders could be successfully amalgamated into the general health services, as perhaps suggested by the Bhore Committee.

A series of public interest litigations in the 1980s has led to sporadic attempts at reform. A review by the National Human Rights Commission[96] in 1999 pointed out the deficiencies in the system. This report again highlighted the exceedingly disparate standards of care, commented upon a century-and-a-half earlier.[97] The cost of maintaining a patient varied from Rs 19 ($ 0.3) to Rs 275 ($ 7) per day, an average of Rs 106 ($ 2.5). More than a third were still housed in converted gaols, with all the custodial trappings of a century ago. Twenty per cent lacked any investigation facilities at all, and the wide range of services for psychosocial intervention and rehabilitation were woefully inadequate. 'Despite the increase in budget... utilization is so variable... no appreciable improvement in many hospitals' observed the review.

Conclusions

Although hospitals are an 'article of faith' by several historians, there is little to suggest that they were widely available before the advent of European, and specifically, British influences.[98] Medical care was provided by trained doctors at patients' homes, and social divisions perhaps precluded any creation of a common public space for care. However, the choice of physician was often

[95] M. K. Isaac, 'Trends in the Development of Psychiatric Services in India', *Psychiatric Bulletin* 10 (1995), 1–3.
[96] 'Quality Assurance in Mental Health: National Human Rights Commission' (New Delhi, 1999).
[97] Ernst, *Native Asylums in Colonial India.*
[98] A. L. Basham, 'The Practice of Medicine in Ancient and Medieval India', in C. Leslie (ed.) *Asian Medical Systems* (Berkeley, 1976).

very eclectic – and Ayurvedic, Unani and European doctors would be consulted with equal felicity. Traditional medicine also suffered from a lack of acceptance of insanity. It has been suggested that the insane lost all caste distinctions, and were considered defiling, and pious householders and Brahmins were advised not to look at insane persons.[99] Islamic societies (and medieval India was administratively an Islamic society) did not make a specific provision for public institutions and services for the poor. Although the notion of charity allowed the setting up of poorhouses, these were often run on private donations and not systematically supported.[100] Troublesome lunatics were often locked into gaol, while harmless ones wandered the streets and joined the poor and vagabonds near the mosques and temples.[101]

Medicine was often outside the traditional social systems, as doctors, by the nature of their profession, had to handle unclean substances. The practice of medicine – both by the professions and the people – did not conform to the rigid demands of religious dogmas. The origins of European medicine, and its use by a wide section of the population in India, were thus no surprise. In essence, in public approaches to illness, whatever was empirically effective, was used. Charles Smith, at the Hospital for Peons, Paupers and Soldiers referred to earlier, was able to document 23,406 consultations between 1836 and 1849, and in 1849 alone had 4,336 admissions through the year, from a population of only 100,000 in Bangalore. And this despite the fact that rich Indians and Brahmins seldom used the hospital. Despite other allegations of colonial imposition, hospitals and asylums thus proved quite popular and acceptable to the population of India.

Medical colleges were established in 1835 in India, and created a large body of Indian professionals trained in western medicine. Leaving service conditions and administrative rules aside, this implied that western notions of hospital care became a part of social and intellectual life. Rich businessmen offered to fund special facilities, such as the special wards for Parsees in the Pune Asylum which was a 'charming villa for 40 patients',[102] or donations to the asylums. Medicine was seldom seen as a tool of Empire, unlike the railways.[103] There have been suggestions to the contrary, but there is little evidence that colonizing the mind was as useful (or successful) an enterprise as colonizing the body.[104] The growing Indian medical elite identified themselves closely with the Raj, as

[99] M. Weiss, 'History of Psychiatry in India: Towards a Culturally and Historiographically Informed Study of Indigenous Traditions' *Samiksa* 40 (1986).
[100] M. W. Dols, *Majnun: The Madman in Medieval Islamic Society* (Oxford, 1992).
[101] B. Pfeiderer, 'Mira Datar Dargah: the Psychiatry of a Muslim shrine', in Imtiaz Ahmed (ed.), *Ritual and Religion among Muslims of the Sub-Continent* (Lahore, 1995).
[102] Report of Mapother to Migaw.
[103] D. R. Headrick, *The Tools of Empire* (New York, 1981).
[104] D. Arnold, *Colonising the Body: State Medicine and Epidemic Disease in 19th-Century India* (Delhi, 1993).

seen in the attempts to create an Indian association aligned to the Royal Society, just years before Independence.

The East India Company passed laws regarding the detention of the insane in its territories several years before similar Poor House Acts were enforced in England. The nineteenth century was marked by a frenzy of asylum building. Although it has been suggested that these were symbols of imperial domination, their actual utilization by the Indian people was quick. The prevailing ideas about the causes of insanity were extrapolated to the region. Though racial issues were recognized, it was equally evident that a considerable degree of effort to understand and improve the services was made. There is little evidence that a systematic denial of the psychological space of 'natives' was attempted. This was a reflection of the trends in psychiatric care in the UK in the nineteenth century.

Other issues in medical science and technology are also important. Until the early part of the nineteenth century, there was a significant give and take between the healing traditions of India and the British. However, scientific advances increased the distance between the two approaches. Unlike Canada and Australia, a comprehensive techno-scientific education was not provided, but one more akin to achieving technical skills and a 'PWD type' of education.[105] In the absence of this broad scientific background, progress in medicine was slow. The lack of adequate sharing of scientific knowledge was to prompt a severe rebuke by A. V. Hill.[106] This was quite apparent in medical services, and perhaps equally true of psychiatric care.

By the early twentieth century, there was an increasing dependence on Indian professionals, and provincial governments in any case were responsible for health care. This perhaps prevented the kind of formal analysis of the issue of race as a factor in mental illness that was to bedevil African psychiatry. The first asylums in Africa were established only towards the end of the nineteenth century and the early years of the twentieth, and social contacts between the two cultures were not as complex as had been established in the Indian subcontinent over the past 300 years. There is seldom any use of metaphors of race in describing the Indian insane, nor is there a difference in their symptoms. The sameness is repeatedly emphasized, although differences on account of geography, climate and organic disease are often suggested.

In the nineteenth century, moral treatment was sought to be extended to all the citizens of British India. Although initiated as an exercise to reduce public nuisance, it was soon regarded as a 'noble work'. However, by the end of the nineteenth century, increasing reparations to the UK, and the costs involved,

[105] D. Kumar, *Science and the Raj* (Delhi, 1997).

[106] A. V. Hill, 'India-Scientific Development or Disaster', in *The Ethical Dilemma of Science and other Writings* (New York, 1960).

proved prohibitive. Endless debates about separate asylums for Europeans culminated in two buildings: one in Berhampore (which was quickly discarded as it turned out to be too much like a gaol), and at Ranchi.

For most of this period, asylum populations remained almost static at below 15,000 beds for a population of several hundred million. Financial and administrative lacunae (parsimony and neglect) were blamed for this appalling state. But the great incarceration simply never happened.

This was to have several consequences for services in India. Unlike the West, where social psychiatry and community care evolved as extensions of the asylum, there were no comparable services. The ancillary professional staff – psychologists, psychiatric social workers, mental-health nursing, etc., were woefully inadequate. Prompted by developments in pharmacology and innovations in community care, asylums began playing a diminishing role in the provision of care, reserved only for the destitute and abandoned. General hospital psychiatry units, established in only half of the medical colleges, attended to acute cases, and chronic cases fell into the background. Sporadic attempts at reform have been partially successful, and a few of the asylums have been made autonomous, and provided increased funds to improve the quality of care. It is quite likely that no new facilities will be established, though the need for long-term care is quickly being filled up by private asylums and halfway homes that were permitted under the revised Indian Mental Health Act of 1987. Whether these will go the way of the private madhouses of the eighteenth and nineteenth centuries remains to be seen. Economic reforms have increased the role of the private sector in health provision, and have been accompanied by reduced funding for public health. This raises questions about the retreat of the state from the responsibility of care for the chronically ill, and these are likely to intensify in the future as families become smaller, society more 'industrial' and the demands for care more complex.

Colonial institutions in India include the railways and the parliament, as well as the asylums. Though setting up of each of these was prompted by the needs of the colonial administration, they have been incorporated into all aspects of contemporary Indian life. There is constant debate about the relevant adaptations of each of these to the needs of the Indian society. As perceptions about the nature of psychiatric disease and care changed over the past two centuries, so did attitudes towards institutional care. The sheer paucity has sometimes been viewed as an advantage, as the ills of 'chronic institutionalization' were avoided. The needs of the chronic mentally ill are still woefully neglected, and a more responsive institutional care service will perhaps be necessary. Asylums in India will necessarily have to reinvent themselves to continue to be relevant.

12 Confinement and colonialism in Nigeria

Jonathan Sadowsky

In a recent article Shula Marks has asked, what is colonial about colonial medicine?[1] The answer, of course, depends in part on what one considers 'colonial' to mean. One of the benefits – perhaps unexpected – of the growth of studies of colonial medical institutions in recent years has been a growing appreciation of the diversity of colonial contexts, the recognition that colonialism was not the same in all places. This chapter seeks to contribute to that understanding by posing the question, what was distinctively colonial about the confinement of the insane in Nigeria, with an emphasis on institutions in the southwest of the country?

The history of Nigeria's asylums re-enacted developments common in the comparative history of psychiatric institutions, but also illustrates themes peculiar to the politics and priorities of colonialism. In the beginning, the institutions were, like many colonial imports, already obsolete by metropolitan standards, replicating many of the faults British psychiatry had come to pride itself on overcoming. For most of the early twentieth century, colonial officials in Nigeria lamented the state of the asylums and planned fitfully to reform them. But when reform was achieved in the late 1950s and early 1960s, it was contemporary with Nigeria's gradual shift to independence, and the reform was largely accomplished through the initiatives of Nigerians.

Victorian Britain enacted a series of dramatic changes in lunacy policy, including increased institutionalization, the rise of 'moral treatment' and other optimistic therapies. There was, however, a growing disillusionment with institutional options by the early twentieth century. With each of the changes, the psychiatric establishment radiated an image of progress. Michel Foucault and others have, of course, argued that the reforms of modern psychiatry have spawned cultural hegemonies and forms of social control all the more insidious

This chapter is based on my monograph, *Imperial Bedlam: Institutions of Madness in Colonial Southwest Nigeria* (Berkeley, 1999), especially chapter three. The research was funded by the US Social Science Research Council. All the acknowledgments made in the monograph apply here.

[1] S. Marks, 'What is Colonial About Colonial Medicine? And What has Happened to Imperialism and Health?', *Social History of Medicine* 10 (1997), 205–19.

for seeming benign. But for most of the colonial period in Nigeria this view would have little applicability. The colonial lunatic asylums of Nigeria were simply not benign enough to be insidious. Nor would a view of colonial asylums as 'panoptic' really be apt.

Roughly, colonial asylum policy in Nigeria can be periodized as follows: asylums were established in the first decade of the twentieth century, shortly after the establishment of a colonial state incorporating most of the country now known as Nigeria – a country whose borders were drawn according to agreements made in Europe without regard to the natural or human geography of Africa.[2] For the first two decades, the asylums were used as purely custodial institutions, with colonial officials having no higher aspiration for them. By the late 1920s, there began to be calls for a reformed, curative hospital, calls that were received with scorn from most in the government at first, but with more sympathy starting in the mid-1930s. Once the government determined that a hospital would be desirable, though, inertia carried the day, until near the end of the Second World War. Development of therapeutic facilities quickened in the mid-1950s, as Nigerian psychiatrists began to staff the institutions, which were then re-named hospitals. By examining the material conditions in the asylum, the social processes of admission and discharge, the ideological conflicts in debate over asylum reform and, finally, the process of reform itself, we can see how colonial policies for the confinement of the insane reflect the contradictions of the colonial policy of indirect rule.

The beginnings of Nigeria's asylums

Neither insanity nor psychiatric care were new to Nigeria when the colonial era began. Yoruba is the dominant language in the region that is the focus of this chapter, and the language contains not only a word for madness, *were*, but a number of more specific diagnostic categories used by mental healers.[3] Traditional psychiatric healing among the Yoruba resembled western psychiatric care of the late twentieth century more than that of the time of the European occupation in a number of key respects. For example, Yoruba healers used an effective psychotropic drug derived from the plant *rauwolfia* that is chemically similar to the phenoziathines used in western countries to treat psychosis since the 1950s. Yoruba healers also combined therapeutic ritual with the use of drugs, similar to the combination of psychotherapy and medication now highly esteemed in western psychiatry, and recognized heredity as a major risk factor for mental illness. For these reasons, traditional Yoruba psychiatric care can be said to

[2] See A. I. Asiwaju, *Western Yorubaland Under European Rule, 1889–1945* (London, 1976), on some of the effects of European boundary-making in Africa.

[3] See Sadowsky, *Imperial Bedlam*, chapter three, and R. Prince, 'Indigenous Yoruba Psychiatry', in A. Kiev (ed.), *Magic, Faith, and Healing* (New York, 1964).

have been at least as 'developed' as that practised in Europe a century ago – certainly more so than the strictly custodial model that was imported under the colonial regime. Yoruba healers have employed chains and other forms of restraint for patients, but the region had no tradition of large-scale confinement prior to colonialism.

Other European imports had been familiar near the Bight of Benin for centuries before formal colonialism, and European medicine made its first significant intrusions with the Christian missions in the middle of the nineteenth century. The asylum came only in the first decade of the twentieth century, after the establishment of formal British rule. The growth of the colonial state was accompanied by a growth of the urban centres of the country, and by the late nineteenth century these centres had growing numbers of vagrant insane. Both Africans and the European administrators began to argue that the state needed to take some responsibility for them.

Large institutions for confining the insane have existed in Europe since the Middle Ages.[4] The British diffused these institutions with the growth of their global empire in the modern era. There were, for example, asylums in operation in India by the mid-eighteenth century.[5] These housed the European insane, for fear they would become vagrant or otherwise compromise British prestige. The asylums in British Africa, by contrast, mostly confined Africans. Asylums were used in Southern Africa by the early nineteenth century; these included ones at Robben Island, a predecessor of the notorious prison in which Nelson Mandela was confined.[6] West African asylums in Kissy (Sierra Leone) and Accra (the Gold Coast) were in use by the late nineteenth century.

The southern provinces of Nigeria launched an asylum policy in the first decade of the twentieth century, with the enactment of laws for confining lunatics. The Lunacy Ordinance of 1906 empowered the government to establish a lunatic asylum when necessary, and defined a 'lunatic' as any person of 'unsound mind', including 'idiots'.[7] Any medical officer could detain an individual as a suspected lunatic for up to a month. Local magistrates could also

[4] Midlefort suggests that asylums were not a European innovation, but were developed by Arab societies and imported to Christian Europe by way of Spain. E. C. Midlefort, 'Madness and Civilization in Early Modern Europe: A Reappraisal of Michel Foucault', in B. Malament (ed.), *After the Reformation: Essays in Honor of J. H. Hexter* (Philadelphia, 1980).

[5] See W. Ernst, 'The European Insane in British India 1800–1858', in D. Arnold (ed.), *Imperial Medicine and Indigenous Societies* (Manchester, 1988), and W. Ernst, *Mad Tales from the Raj: The European Insane in British India* (London, 1991).

[6] J. Iliffe, *The African Poor: A History* (Cambridge, 1987), 100; H. J. Deacon, 'Madness, Race and Moral Treatment: Robben Island Lunatic Asylum, Cape Colony, 1846–1890', *History of Psychiatry* 7 (1996), 287–97. Institutions would also later be used in Nyasaland and Kenya. On Nyasaland, see M. Vaughan, 'Idioms of Madness: Zomba Lunatic Asylum, Nyasaland, in the Colonial Period', *Journal of Southern African Studies* 9 (1983), 218–38.

[7] Government of Southern Nigeria, *Government Gazette*, 7 November 1906. With the amalgamation of Nigeria in 1916, a colony-wide ordinance was enacted with much the same language.

hold hearings regarding a person's sanity. The district commissioners (local colonial officials) would have ultimate authority for designating lunatics, and were to hold hearings, call witnesses, and appoint medical supervisors to that end. The medical supervisor was required to certify the individual as a lunatic and 'proper subject for confinement', and to specify what warranted the certificate. The first buildings designated exclusively as asylums were the Yaba Asylum in Lagos and the Calabar Asylum in southeastern Nigeria.

Nigeria's colonial prisons and asylums were functionally equivalent: they confined deviant or troublesome individuals, and refrained from cure, rehabilitation, or otherwise normalizing their inmates for return to the outside.[8] Yet administrators distinguished carefully between the two. Asylums lay in the middle of a spectrum between prisons and hospitals, reflecting an ambiguity in colonial rhetoric as to whether lunatics were essentially a health or social control problem.[9] Literal overlaps between the asylums and the prisons existed as well. By 1915, for example, with the Yaba asylum already overcrowded, several cells in Lagos prison were converted into a lunatic asylum.[10] In 1906, the Yaba Asylum was developed in the former headquarters of the Nigerian Railways.[11] Dr Crispin Curtis Adeniyi-Jones, a Nigerian physician who served in the Lagos Medical Service became the first director of the asylum. Adeniyi-Jones wrote two reports about the Yaba Asylum in November and December of 1907, which provide impressions of what it was like at that early date.[12] As of the 31 October, fourteen lunatics: eight women and six men, had been admitted, and an additional female inmate was admitted in November.[13] Adeniyi-Jones noted that 'The classification as to forms of insanity has not been decided in every case.' The asylum already had physical defects, including broken locks, gutters and fences. Adeniyi-Jones wrote a number of letters to superior medical officers during his tenure as head of the asylum; they show him to be diligent but already embattled by the lack of resources the government was willing to devote. A European, Thomas Beale Browne succeeded him in 1909, and Europeans would direct the asylum until Nigeria's transition to independence in the late 1950s.[14]

[8] The number of people confined in colonial asylums in Nigeria was always much smaller than the number in prisons.

[9] In the Gold Coast, in fact, lunatics were generally placed in prisons until the 1940s. R. C. Brown, *Report III On the Care and Treatment of the Mentally Ill in British West African Colonies* (London, 1938).

[10] Nigerian National Archives, Ibadan (hereafter NAI) AI CSO N1210/1916. These were cells that had previously housed European convicts.

[11] T. Asuni, 'Aro Hospital in Perspective', *American Journal of Psychiatry* 124 (December 1967), 763–70.

[12] For more on Adeniyi-Jones, see A. Patton, Jr, *Physicians, Colonial Racism, and Diaspora in West Africa* (Gainesville, Fla., 1996), 192–3.

[13] Yaba Lunatic Asylum, Monthly Report for November, 1907, from personal collection of Dr Alexander Boroffka.

[14] Adeniyi-Jones's letters are available at the Contemporary Medical Archives Centre, Wellcome Institute for the History of Medicine, London, GC/146/1.

Material conditions

From the inception of the asylums, critics drew attention to their revolting conditions, referring mainly to dirt and overcrowding. In 1928, in the first major report to the government on lunacy in Nigeria, visiting alienist Bruce Home described dark, congested cells, poor bathing facilities, lack of basic supplies, and the use of chains.[15] He added that the asylums in Calabar and Yaba were little better than the prisons. In Calabar, he said 'the unfortunate patients are exposed to view, and are objects of amusement to the public'. Later reports by other visiting alienists such as Robert Cunyngham Brown (in 1938) and by J. C. Carothers (in 1956) echoed Home's with regard to material conditions. The disgust repeatedly led to calls for reform, but little action.

Although asylums in other regions of the country did not suffer quite as much overcrowding as in Yaba and Calabar, most of them were not much more attractive. Some officials thought lunatics were better off left in the streets![16] The Port Harcourt prison extension was especially dreadful; an official witnessed 'several violent lunatics shivering naked in damp, dark cells, chained like animals to a ring in the floor; others also naked, wandered aimlessly around a barbed wire enclosure'.[17] Brown observed that the asylum in Kano resembled 'a fortress rather than an asylum'. The one exception was in Zaria, where 'The reporter found the asylum clean, tidy and, according to native requirements, proper in every way'.[18]

In order to deal with overcrowding, the Yaba and Calabar asylums tried to exclude so-called 'harmless lunatics.' This did not mean that all the inmates were criminal. On the contrary, the distinction between 'criminal' and 'civil' lunatics was a carefully kept marker of the inmates' identity. Criminal lunatics were those who had been arrested, usually for violent crimes, and found unfit to stand trial because of their mental state. Civil lunatics were usually people who were considered public nuisances, but who had not committed any crime, or at least no serious crime. Whenever possible, the government preferred to leave these people in the care of relatives.[19] The material conditions were made more alarming by the expectation that many more lunatics would need to be confined. In 1928, Home 'calculated' that provision would have to be made for 4,000 cases of insanity, one quarter of which would require urgent care.[20] The word 'calculated' requires quotation marks because there was no basis for his numbers. He was probably trying to stress that a lot of beds were needed, to draw attention to the problem.

[15] B. Home, *Insanity in Nigeria* (Lagos, 1928), 5.
[16] NAI New File MH (Fed) 1/1 3313, 69–70.
[17] NAI New File MH (Fed) 1/1 3313, 'Lunatics, Care of', 31, inspection notes by J. G. C. Allen, Senior Resident, 18 July 1955. See also MH (Fed) 1/2 MH 59 vol 2, 54.
[18] Brown, *Report III*, 40.
[19] See, for example, NAI CSO 26 26793, 'Abeokuta Mental Hospital', vol. 1, 7.
[20] Home, *Insanity*, 6.

Therapy was obviously not a priority in these conditions. It is possible that the physical health of some inmates may have improved, as the Annual Medical Report for 1927 claimed.[21] This was plausible, given that many of the inmates were impoverished, undernourished vagrants, who received regular meals upon admission. Also, medical examinations were given every three months.[22] Still, a number of factors may have worsened the health of inmates. Exposure to rain and unsanitary conditions, for example, may have undermined some of the benefits.

Regardless of attempts to maintain the physical health of the inmates, there were barely any attempts to cure any psychiatric problems they faced for most of the colonial period. Very modest occupational therapy was all Yaba offered until independence.[23] In the 1950s the asylum began selling the inmates' handiwork, with dual goals of raising money and reducing the stigma attached to the mentally ill.[24]

The contradictions of Indirect Rule

Official colonial policy in British Africa called for Indirect Rule, which entailed the recognition of local political leaders (though in some cases these were installed by the colonial government itself) and an official ethos that there should be as little interference in African ways of life as possible. But the policy was contradictory, since colonialism was also intended to benefit the metropolitan country financially, and bring the alleged benefits of civilization for the colonized people, two goals that made interference inevitable. Asylums – even dreadful as they were – were associated with the benefits of civilization. The expense of a truly modern asylum, though, was incompatible with the economic goals of colonialism. At the same time, financial restraint was justified by the goal of preserving the African way of life.

The directive to preserve tradition was followed more rigorously in the north of Nigeria; as a result the northern provinces relied on small native administration asylums throughout the colonial period. In 1928, the acting secretary of the northern provinces attached a scathing review to Home's call for expanded therapeutic services. He invoked the authority of a colonial medical official named Cameron Blair to support his disdain for Home's plan, quoting at length

[21] Public Records Office (hereafter PRO) CO 657 20, 419. In 1936, Brown also reported that the physical health of inmates was 'quite satisfactory', even finding that the asylum patients in the south were generally more robust than their counterparts in the small native administration asylums in the north; see Brown, *Report III*, 28 and 35.

[22] PRO CO 657 24, 466–67.

[23] NAI CSO 01507/ S.2, 'Yaba Asylum', 11–13.

[24] PRO CO 657 24, 467; PRO CO 657 69, 27. Traditional healers in the region also sometimes employed patients in farm work, but this was more typically as payment *after* treatment, not for any possible curative effect.

from a world-weary 1926 letter written by Blair. Blair, an alienist, wrote with cynical acumen; his comments merit extended quotation, to provide a sense of the tone with which asylum reform was derided:

> I hear a dreadful rumour to the effect that the question of adopting a systematic lunacy policy has cropped up again and is being seriously entertained but trust this not be true; for if it be, the results are likely to be disastrous. Here, I know what I am writing about; for I was an alienist myself for over nine years. If such a policy be adopted, this is the sort of thing that will happen – To begin with, the Director of Medical and Sanitary Services will have to devote, at least, a paragraph of his annual report to Lunacy. In due course his report will as usual, be passed on by the Secretary of State to the Medical and Sanitary Advisory Committee. When that body arrives at the Lunacy section of the report, the Chairman will call a halt and deliver himself after this fashion: 'Gentlemen, I trust some of you know something about Lunacy; for I am sure I don't.' The long odds are that the other members will declare themselves in the same boat. Then the Chairman will say: 'Well, gentlemen, I propose we have this section extracted and pass to our colleagues, the Commissioners of Lunacy . . . for the favour of their comments.' This motion will be carried unanimously: and then the fact [*sic*] will be on the fire with a vengeance. The commissioners will sputter over it, looking at the question from home standards; their comments will establish a state of panic; the Secretary of State will share in it; and the panic will find relief in imposing extravagant expenditure on unfortunate Nigeria. In a very few years we shall have some twenty-five thousand certified lunatics under public control, at a minimum charge of ten shillings a week per head . . . You can imagine what this means with Nigeria's revenue; and the only people who will profit will be the native lawyers and the food contractors who will charge extravagant prices for the supply of food-stuffs, in not a few cases for the feeding of their own insane relations whom they can well afford to feed at their own cost.[25]

As the appendix continues, it becomes clearer that this debate was not merely over the need of resources for accommodations, but cut to the heart of what colonialism was about. The secretary added that the 'loyal cooperations of the local authorities'

> might be construed as a polite euphemism for dragooning natives of the Northern Provinces into alien ways and ignoring what is serviceable in their own, – or in other words, pursuing the gospel of Direct Rule . . .
>
> The assumption that there is no middle course between native administration gaols and the expenditure of some £500,000 a year on providing European comforts and architecture is wholly illogical and wrong.[26]

These responses to Home's recommendations embodied the contradictions of indirect rule. They justified parsimony in the name of cultural preservation, and assailed reform as an imposition on the native way of life. Underlying this stance was the crucial self-deception that one could, somehow, have colonialism

[25] Home, *Insanity*, appendix, 2. [26] *Ibid*., appendix, 3.

without making any impositions. The government – more by deferring decision than through articulated policy – followed a middle course, avoiding both a significant investment in treatment, and a strict 'hands-off' policy. Nigeria therefore ended up with institutions that had none of the potential advantages Home called for, and all the disadvantages his opponents predicted.

Brown's 1936 report, though similar to Home's, had a different reception. There is no record of his recommendations receiving the derision that Home's report did. Instead, the tenor of his report was echoed in virtually all subsequent memoranda and reports.[27] In just ten years, the recognition of the need for expanded accommodation, and the desirability of a curative institution, became conventional wisdom. This change reflected shifts in both colonial and psychiatric thinking. The Colonial Development Act of 1929 signalled the beginning of a move to greater planning and centralization from the colonial state. At the same time institutional psychiatry in Britain itself was becoming more pro-active, increasingly admitting voluntary patients from the 1930s.[28] But these changes in thinking were not reflected in significant institutional reform in Nigeria's asylums. As Falola notes, the 1929 Act achieved little in its eleven-year history because the depression and the Second World War inhibited capital investment, and so the intense emphasis on promoting exports continued.[29] But the arguments against expanded and improved psychiatric facilities no longer ridiculed such reforms as incompatible with the very nature of colonialism itself, but rather more defensively, they cited budgetary restraints.[30]

There were several financial obstacles to a mental hospital. The salaries needed for a trained staff and the large amount of water supplies required were seen as prohibitive expenses. Besides the inherent expense, there were competing considerations when the first calls for a curative programme were made in the late 1920s – on the eve of the worldwide depression. In the 1940s, of course, the British government was preoccupied with the Second World War and its aftermath. A mental hospital was also more expensive than other public-health projects. In 1930, the Nigerian government requested a public-health budget from the Colonial Office in which the cost of a mental hospital (£33,700) was almost a third of the total (£107,000).[31] The other items for which funds were requested were training centres for midwives and sanitary inspectors, ambulances, grants to native dispensaries, the leper colony in Benin, general sanitary improvements and research. None of these items alone approached the cost of the mental hospital.

[27] See, for example, PRO CO 657 47, 231.
[28] K. Jones, 'The Culture of the Mental Hospital', in G. E. Berrios and H. Freeman (eds.), *150 Years of British Psychiatry, 1842–1991* (London, 1996), 23–24.
[29] T. Falola, *Development Planning and Decolonization in Nigeria* (Gainesville, Fla., 1996), 17–23.
[30] NAI CSO 26 1507/S.1/T.1 [31] NAI CSO 01507, vol. 4, 412c.

Financial expediency was the main reason a curative institution was delayed, but it dovetailed with ideological inhibitions. One was the belief that a mental hospital was inappropriate for Africans. As Anne Phillips has remarked, 'British colonial practice seemed to pride itself on retarding rather than hastening change.'[32] This pride was misplaced, since there was no way the economic goals of colonialism could be met without hastening change. The theory of indirect rule acknowledged that economic, political, kinship and religious systems were interrelated parts of African life. One goal, therefore was not only to support those they acknowledged as traditional rulers, but also to preserve systems of land tenure.[33] But the encouragement of cash crop cultivation to serve the economic goals of colonialism, along with missionary activity and western education, were causing significant disruptions.[34] The ideology of indirect rule – what Freund has called the 'cult of tradition'[35] – nevertheless had significant effects; one result of the contradiction between ideology and practice was half-measures like asylums – measures which dimly recognized the social changes colonialism incurred, but also denied responsibility for them.

Getting in, getting out

The social processes of determining who was a lunatic can be studied by looking at the processes of confinement and discharge. Cases where Africans and Europeans disagreed outright on the madness of a case seem to have been rare. Most of the confined seem to have been people who caused considerable distress and confusion to those around them. For the most part, patients were not brought in voluntarily by themselves or their families, but involuntarily by police or other relative strangers. Asylum administrators were usually willing to release patients if someone would care for them, though there were exceptions to this.

Some officials were aware that the labelling of insanity was problematic, especially in a colonial situation. In 1950, an official named J. H. Pottinger provided guidelines for detention. A careful diagnosis must be made, he said, because lunatics should not be lightly deprived of liberty. Pottinger also argued that asylum life would worsen the chance of recovery for a patient:

[32] A. Phillips, *The Enigma of Colonialism: British Policy in West Africa* (London, 1989), 3. It may not yet go without saying that this 'African way of life' was frequently imagined and constructed according to colonial convenience.

[33] The political organizations of indirect rule did not so much 'preserve' existing forms as transform them and even substantially create them in many places. This is a very well documented aspect of colonial history, but for one treatment see M. Chanock, 'Making Customary Law: Men, Women, and Courts in Colonial Northern Rhodesia', in M. J. Hay and M. Wright (eds.), *African Women and the Law: Historical Perspectives* (Boston, Mass., 1982).

[34] My discussion here draws on J. Coleman, *Nigeria: Background to Nationalism* (Berkeley, 1968), 53.

[35] B. Freund, *The Making of Contemporary Africa* (London, 1984), 79.

Life under detention is regular, food and drink appear at stated times without effort on the part of the individual and shelter and clothing are available. The patient therefore tends to lose initiative... He gets out of touch with the life he has previously known, and after a prolonged detention becomes positively unfit for any other kind of life.[36]

Family members wrote letters to administrators requesting the release or confinement of patients.[37] There were very few requests to confine family members; despite occasional colonial complaints about families 'dumping' unwanted lunatics into the asylums, this was not typical. There were apparently occasional instances of families or communities who, frustrated with attempts to manage chronic cases, sought the state's help, but this was exceptional. Most of the letters request the release of a patient, and of these, the majority concede the insanity of the confined but urge home care instead. The file also contains letters from administrators containing requests for release, and in most cases the requests were granted.

Another striking aspect of these letters is the rarity of dispute over the madness of the inmate in question. Rather than deny that the patient was insane, most letters proposed that a local healer should care for the patient. This may have been a rhetorical gambit; letter writers may have thought their chances of getting someone released were better if they conceded the madness, and this may have even become known as a formula for procuring release. But I think it more likely that the concession to the inmates' madness was sincere. Asylum inmates were likely to be the most unambiguous cases – that is, people whose dramatically anomalous behaviour caused widespread apprehension.[38]

While Nigerians did not always question the ability of colonial institutions to recognize madness in Africans, they did frequently question the institutions' ability to care adequately for mad persons, and traditional treatment was sought as an alternative. In many cases, authorities agreed that the patients would be better served by traditional healers and released them.[39] Official support for local healers was partly a matter of convenience; the more patients who were handled by healers, the fewer who burdened the asylums. There was, though,

[36] NAI Oyo Prof. 1015, 'Lunatics, Oyo Province', vol. 2, 147. Pottinger thus claimed a 'secondary gain' which inhibited recovery. As for the regularity of food, there was at least one recorded case of a patient starving to death while under detention; Elizabeth A. was confined in Osogbo prison, due to lack of asylum space, and died in May 1956, as a result of 'chronic starvation, self-neglect, and cardiac failure'. Her starvation may have been a form of protest, but this can only be speculated. NAI Oshun Div. 1/1 86/13, 15.

[37] Starting in the 1970s a number of studies have stressed the importance of the complex social processes by which patients and their families determine their therapeutic options in pluralistic medical settings. Landmark treatments include J. Janzen, *The Quest for Therapy in Lower Zaire* (Berkeley, 1978); S. Feierman, 'Struggles for Control: The Social Roots of Health and Healing in Modern Africa', *African Studies Review* 28 (1985), 2–3 and 73–147; and L. Mullings, *Therapy, Ideology, and Social Change: Mental Healing in Urban Ghana* (Berkeley, 1984).

[38] See NAI Comcol 1 735/S. 1, vol. 1, 'Lunatics, General Matters Affecting', 406.

[39] See NAI Comcol 1 735/S. 1, vol. 1, 'Lunatics: General Matters Affecting', 18–20.

some concession to the abilities of healers. Conversely, when petitions for release were turned down, medical officers sometimes claimed that native cure had been tried and failed, even if this was not the only reason for the denial.[40] Officials based the release of patients more on their harmlessness than on their health. If someone assumed the care of the patient, and the patient was not considered a public danger, release was granted even when the symptomatic behaviour was still present.

Criminals were probably only a small minority of asylum inmates. Many of the so-called 'civil lunatics', were also detained by police or other authorities for being nuisances, if not exactly criminals. Often these were people who walked naked in public, urinated or defecated in the open, or threatened people for no apparent reason. While the behaviour could be simply odd, such as giving away money at random, it was usually bothersome to other people. Sometimes, entire communities petitioned to confine a threatening person.[41]

Decolonizing psychiatry

While Nigeria's colonial asylums were scandals even to the very people responsible for them, their successors – the mental hospitals of the independence period – developed a number of innovations that have made them famous in world psychiatry. These therapeutic innovations and their timing perhaps highlight all the more the negligence that characterized the colonial period.

By the 1950s, there was official recognition that the custodial model of care that had characterized colonial policy needed to be abandoned. In the early 1960s Lagos, Yaba Lunatic Asylum became Yaba Mental Hospital. A German psychiatrist, Alexander Boroffka, came to oversee the transformation and Yaba began employing a battery of therapies, including drugs and electro-convulsive therapy, derived from western psychiatry. But it was the Aro Mental Hospital which had paved the way in the 1950s, and which deserves close examination because of its innovative treatment plan.

Funds were first committed, and planning begun, for the development of Aro in the 1930s.[42] Brown noted then that 'the Alake [king] of Abeokuta is actively interested in the welfare of the insane of Abeokuta...';[43] the Alake had, in fact, offered a lease of land at nominal rent, for the hospital as early as 1929.[44] But some officials continued to believe that a mental hospital for Africans was an extravagance, some arguing for public-health initiatives directed more

40 NAI Comcol 735/ S. 1, vol. 2, 'Lunatics: General Matters Affecting', 615–27.

41 A 1950 letter requesting someone's incarceration came from a compound in Ibadan with fifty-seven signatures and several thumb-prints. The man, the petitioners complained, wielded cutlasses against pedestrians. NAI Oyo Prof. 1 1015, vol. 2, 'Lunatics – Oyo Province', 129. For another example, see NAI Oshun Div. 1/1 86/10, 1.

42 NAI Comcol 1 735 vol. 1, 'Lunatic Asylum', 130.

43 Brown, *Report III*, 60.

44 NAI CSO 26 01507, vol. 4, 496.

towards infectious diseases.[45] The Second World War provided the essential catalyst for the Aro project, because of the repatriation of soldiers who served with the Allied forces.[46] West African soldiers received free medical care for disabilities resulting from service, as well as employment assistance, upon their repatriation.[47] Rehabilitation centres were set up in Lagos, Freetown and Accra, which dispensed artificial limbs manufactured by Italian prisoners of war. Upon repatriation, five mentally ill Nigerian soldiers were transferred to Abeokuta from Yaba, where they could not be accommodated due to overcrowding.[48]

The responsibility the government assumed for insane soldiers highlights the element of bad faith in the prior claim that Europeans could not treat mental illness in Africans. The returning soldiers were too many to stay in Yaba. And it would have been an impressive feat of ideological rationalization to deny them treatment altogether. After their experiences fighting for the Allies in Asia, preservation of their 'African way of life' could hardly be seen as a pressing aim.

The soldiers were lodged in a building formerly used as the Abeokuta convict prison – at that time also an overcrowded institution – the name of which was changed to Lantoro Lunatic Asylum.[49] Lantoro, which became the nucleus for Aro, was a curative institution, and excluded all cases 'associated with crime or violence'.[50] Nigerian psychiatrist Tolani Asuni has referred to the 'unfortunate circumstances of Lantoro's origin', which, he said resulted in a stigma being attached to the patients there.[51] Lantoro was, however, kept as a relatively 'closed' extension of Aro, for what Asuni described as the most 'disturbed and uncooperative' patients.[52]

Civilian patients were admitted to the Lantoro site beginning in 1946.[53] Construction of Aro began in the early 1950s about seven miles from the Lantoro site. The initial staff of Aro consisted of thirteen attendants transferred from

[45] NAI CSO 52898, 'Mental Hospital (Aro Site) Abeokuta', 5–19.

[46] Several thousand Nigerian soldiers, most of them from the south of the colony, served with Allied forces in the Second World War. The majority served in Burma, some in the Middle East, and a small number in the European theatre. See also PRO 657/56, 934.

[47] NAI Oyo Prof. 2/3 C.311, 'Resettlement of Ex-Soldiers After the War'.

[48] Geoffrey Tooth, a psychiatrist working in the Gold Coast, noted that in Accra, an 'unduly large' proportion of lunatic asylum admissions in the post-war period were ex-soldiers. NAI MH (Fed) 1/1 2nd Accession 5280B, 'Research in Mental Illness and Juvenile Delinquency', A1. See also NAI Oyo Prof. 2/3 C.311, 'Resettlement of Ex-soldiers After the War', vol. 1, 90.

[49] NAI Oyo Prof. 1. 2188, 'Unification and Staffing of the Colonial Prisons Service, Ibadan Native Authority System', 9. 'Lunatic soldiers' were transferred to Lantoro from Yaba; see NAI MH (Fed) 1/1 3420, 23.

[50] NAI CSO 52898/S. 1, 'Lantoro Lunatic Asylum (Abeokuta)', 1.

[51] Asuni, 'Aro Hospital in Perspective', 769.

[52] T. Asuni, 'Development in Mental Health Care in Africa with Special Reference to Western Nigeria', *Proceedings of the IV World Congress of Psychiatry* (Madrid, September 1966), 1067–8.

[53] Aro Mental Hospital, *40th Anniversary of the Establishment of Aro Neuro-Psychiatric Hospital Complex* (Ibadan, 1984), 9.

Yaba.[54] Although there were still no psychiatrists in attendance, medical officers from the Abeokuta General Hospital supervised. The professional staffs of both Aro and Yaba at this time were still dominated by Europeans. In 1951, Abraham Ordia, the first Nigerian trained as a psychiatric nurse, began working at Yaba. Ordia, who had spent thirty-two years in mental-health nursing in Holland, England, Sweden and Switzerland, helped oversee the beginning of the transition of Yaba from asylum to hospital.[55]

Work on Aro Mental Hospital proceeded slowly, and it was not completed until 1958. In the early 1950s, reflecting the continued sense of alarm about the number of psychiatric cases, T. A. Lambo decided to take up psychiatry as a specialty. He studied psychiatry at Maudsley Hospital in Great Britain and assumed authority over Aro in 1954. He remained at Aro until becoming chair of the department of psychiatry at the University of Ibadan in 1963. Aro's physical plant was not yet completed in 1954, but Lambo, whose enthusiasm and commitment were becoming legendary, was not to be stopped. He devised the Aro village scheme, an outpatient care system which permitted the beginning of treatment.

Lambo's roots in Abeokuta were crucial. He persuaded a number of families near the Aro site to allow mental patients to live with them in exchange for work, mostly help in farming, and a lodging fee. The patients could go to the hospital for treatment in the morning, and work in the farms in the afternoon. In addition to the rents the villagers were able to charge, they received several public-health benefits: piped, purified water, pit latrines, and a mosquito control squad.[56] Nurses were available to the patients on a 24-hour basis.[57] Landlords were able to borrow money from the hospital in order to build extensions to their houses so they could accommodate patients.[58]

Lambo was acutely aware of the ways in which the mental hospital was alien to African cultures. Understanding how cultural familiarity was critical to mental-health interventions, and also strongly convinced of the therapeutic efficacy of local medicines, he travelled across Nigeria and handpicked fifteen traditional healers from different cultural backgrounds to serve as mediators between the hospital staff and the patients. The hospital's treatment programme, though, strongly emphasized interventions which were then state of the art in

[54] Asuni, 'Aro Hospital in Perspective'.

[55] According to Alexander Boroffka, 'Two psychiatrists were working with him for one year each (Cameron 1955 and Campbell Young 1958) but did not contribute much to this task, lacking Mr Ordia's optimism and zest.' Boroffka, 'History of Psychiatry in Nigeria', *Psychiatry* 8 (1985), 709–14.

[56] See T.A. Lambo, 'Patterns of Psychiatric Care in Developing African Countries: The Nigerian Village Program', in H. David (ed.), *International Trends in Mental Health* (New York, 1966), 149.

[57] Aro Mental Hospital, *40th Anniversary*, 11.

[58] Ben Park, 'The Healers of Aro' (16 mm film, New York: United Nations, 1960).

western psychiatry: electro-convulsive therapy (ECT), insulin-coma therapy, psychotropic drugs such as largactil, and interactive psychotherapy, as well as an expanded occupational therapy programme.[59] The village system became a model for mental-health care in a number of African countries, including Ghana. At a time when the world psychiatric community was searching for alternatives to institutionalization, the Aro plan offered an imaginative compromise – out-patient care which could help the patients re-integrate into society, in close proximity to the medical technology of the hospital.

The Nigerianization of psychiatric institutions provides an example of the creative energy for which independence provided greater scope. At the most basic level, the transformation of the asylums from near-prisons into hospitals, while indeed a change in the form of 'social control' of the mentally ill, was a departure from the most crudely coercive and dehumanizing form of control. And this change needs to be understood in the context of Nigeria's independence, not only because it provided more opportunity for Nigerian physicians to use their expertise publicly, but because the welfare initiatives which characterized colonialism in the 1950s represented, as many in the government themselves understood, a collapse of the basic logic of colonialism itself. The Aro village plan merits particular recognition as a significant achievement. The village system was not completely novel; it resembled, for example, Belgium's Geel Colony, where lunatics travelled in search of cure and lived in the community, and which was a model for some fitful experiments in Britain in the nineteenth century.[60] Out-patient care was a growing trend in European psychiatry when Lambo was studying psychiatry, as anti-psychotic medications came into use; in Britain, this trend was represented by the development of Maudsley Hospital, which was a reaction to the large, long-term custody mental hospital.[61] There was also an African precedent in the out-patient care developed by Dr Tigani El-Mahi at the mental hospital in Khartoum, which Lambo had observed first-hand. The use of traditional healers in active conjunction with biomedical care was also not completely innovative; Alexander Leighton, Lambo's collaborator on *Psychiatric Disorder Among the Yoruba*, promoted a similar plan in the 1940s among the Navaho.[62] But while there were precedents for the Aro

[59] Asuni, 'Aro Hospital in Perspective', 765.

[60] W. Ll. Parry-Jones, 'The Model of the Geel Lunatic Colony and Its Influence on the Nineteenth-Century Asylum System in Britain', in A. Scull (ed.), *Madhouses, Mad-Doctors, and Madmen: The Social History of Psychiatry in the Victorian Era* (Philadelphia, 1981).

[61] See P. Allridge, 'The Foundation of Maudsley Hospital', in Berrios and Freeman (eds.), *150 Years of British Psychiatry*. See also D. Bennett, 'The Drive Towards the Community', in the same volume.

[62] *New York Herald Tribune*, 27 December 1942. See the Alan Mason Chesney Medical Archives, Johns Hopkins University, biographical file on Alexander Leighton.

plan, Lambo executed it with a practical conscientiousness that made it justly famous.

A measure of this care can be gained by comparing the Aro plan with the de-institutionalization movement in the United States. De-institutionalization was no doubt well intentioned in some aspects, and in some respects successful. It too frequently, though, failed to think through exactly what 'community care' should mean – what was the community? What was in it for the community? Indeed, the combination of a desire for fiscal restraint and an often blithe invocation of the merits of 'community care' in the de-institutionalization movement inadvertently echoed the rhetoric of Nigeria's colonial policies. If this seems a harsh charge against de-institutionalization, consider how specifically criticism of Home's report anticipates the language of the de-institutionalization movement. Advocates of de-institutionalization also saw hospitals as impositions, as a form of cruelty worse than the disorders they were supposed to correct. In the United States, the critique of institutions dovetailed conveniently with the desire to cut government budgets. Many advocates of de-institutionalization also rejected disease models of mental illness to begin with, just as Home's opponents sought to decouple 'mental abnormality' from sickness. And, just as the critic hoped to solve the problem more simply with medication, de-institutionalization became viable in part because of developments in psychotropic drugs during the 1960s which offered the hope of less involved, less expensive care away from large hospitals.[63] Many critics of de-institutionalization have noted that while the rhetoric of the policy called for 'community care', in only a few instances was a 'community' carefully identified, consulted and articulated with the planning in a meaningful way.[64] This was precisely the merit of the Aro-village plan, and a principal reason why the village system there has been maintained and even expanded to the present.[65]

If colonial asylums were not subtle enough in their social control function to merit a Foucaultian approach, it is in the reforms of the late colonial period that one might think to adopt such an approach. But especially in light of the negligence of the asylums under colonialism, the reforms of the late colonial and early independence period are more appropriately described straightforwardly as achievements. Warwick Anderson has recently suggested that in a sense, all modern medicine is colonial.[66] He presumably is referring to the intrusiveness of modern medicine, the ways it comes to dominate more and more spheres

[63] See A. B. Johnson, *Out of Bedlam: The Truth about De-Institutionalization* (Boston, 1990).

[64] For a scathing indictment of de-institutionalization, published before such critiques became common, see A. Scull, *Decarceration: Community Treatment and the Deviant – A Radical View* (Englewood Cliffs, NY, 1977).

[65] From the 1970s onward, other areas near the hospital have taken part in the housing of patients.

[66] W. Anderson, 'Leprosy and Citizenship', *Positions* 6 (1998), 707–30.

of life, 'colonizing' our consciousness, as we 'medicalize' practices and adopt medical idioms to understand our sufferings and predicaments. Discussions of such processes often ascribe to psychiatry a central place. Yet what is striking about colonial asylums in the Nigerian context is how little they were colonial in this sense. Foreign rule resulted in a reluctance to provide medical care even from administrators who overtly recognized the need for it.

13 'Ireland's crowded madhouses': the institutional confinement of the insane in nineteenth- and twentieth-century Ireland

Elizabeth Malcolm

In 1921 the popular Irish novelist known as George A. Birmingham[1] published a short story entitled 'A Lunatic at Large'. This concerns a young Englishman working as a dispensary doctor in the west of Ireland.[2] Although only recently appointed to the post, Dr Lovaway has already been startled at the number of lunatics he is being called upon to certify, especially by the police. Indeed, as the story begins he is in the process of attempting to compose an article entitled, 'The Passing of the Gael. Ireland's Crowded Madhouses' for the *British Medical Journal*. He ponders various explanations for this startling phenomenon.

> He balanced theories. He blamed tea, inter-marriage, potatoes, bad whisky, religious enthusiasm, and did not find any of them nor all of them together satisfactory as explanations of the awful facts. He fell back finally on a theory of race decadence. Already fine phrases were forming themselves in his mind: 'The inexpressible beauty of autumnal decay.' 'The exquisiteness of the decadent efflorescence of a passing race.'[3]

The doctor's musings are interrupted by the arrival of a member of the Royal Irish Constabulary, announcing that, according to information received from his aunt, a young labourer on a distant and isolated farm is in immediate need of committal.

Accompanied by the police constable and also the local sergeant, Dr Lovaway makes an arduous journey in pouring rain to the farm. There, however, he finds the labourer exhibiting no signs of mental illness at all, or of violence either,

I would like to thank the Wellcome Trust for funding much of the research upon which this chapter is based.

[1] G. A. Birmingham was the pen-name of Canon J.O. Hannay (1865–1950), who was born in Belfast and served as Anglican rector of Westport, Co. Mayo, from 1892 to 1913. He worked as an army chaplain in France during the First World War and then spent much of the rest of his career in various English parishes. Between 1905 and 1950 he wrote nearly sixty novels, most of them about Ireland and most of them comic.

[2] For the Irish dispensary system, by which government-paid doctors provided free medical care to the poor, see R. D. Cassell, *Medical Charities, Medical Politics: The Irish Dispensary System and the Poor Law, 1836–72* (Woodbridge, Suffolk, 1997).

[3] G. A. Birmingham, 'A Lunatic at Large', in P. Haining (ed.), *Great Irish Detective Stories* (London and Sydney, 1994), 110.

although the doctor concludes that the boy is 'evidently of weak mind'.[4] Yet the labourer and his aunt, strongly supported by the sergeant, are extremely anxious for his committal. The sergeant assures the doctor that the boy will be out of the asylum in two weeks, and his uncle chimes in to announce that he has to be, as he is needed at home for the spring planting. The doctor is puzzled by their demand that he certify when it is obvious that the boy is not insane. He therefore refuses to sign the committal papers.

In the *denouement* to the story it is revealed that the police, supported by the community generally and even with the connivance of Dr Lovaway's Irish predecessor, have been 'scouring the country . . . searching high and low and in and out for anyone, man or woman, that was the least bit queer in the head'. The doctor is paid a guinea for every lunatic certified. The community want to encourage the doctor to stay and his predecessor, who is serving in the army medical corps, aids them in this endeavour. Thus there is concern and bewilderment when the doctor refuses in a number of instances to sign the medical certificate and so collect his fee. The 'English is a queer people', concludes one of the locals.[5]

Birmingham's story humorously proposes that it is the Irish who are more canny and less 'queer' than the English, contradicting Dr Lovaway's eloquent 'scientific' theories of race decay. Over and above this, the story touches upon a number of key issues concerning the confinement of the insane in nineteenth- and early twentieth-century Ireland. It suggests that there were high rates of committal to mental institutions in the west of Ireland by the First World War – when the story is set. This phenomenon, moreover, was causing alarm and there was speculation in the British medical press as to the reasons for it, with eugenicist theories of racial degeneration being employed. Further, according to the story, committals were generally instigated by families and carried out by the police,[6] with doctors playing an essential but basically subsidiary role. Yet, at the same time, the story subverts this picture by portraying doctors and institutions as being manipulated by the community for its own purposes. Thus the message conveyed is that Irish committal rates are not necessarily a reliable index of the levels of mental illness in the society. Through an examination of the confinement of the insane in Ireland over the past two centuries, this chapter will test the validity of each of these assertions.

[4] *Ibid.*, 111.

[5] *Ibid.*, 119.

[6] The role of the police in the committal process is not examined in any detail in this chapter, although in Ireland it was an important one. Other chapters in this volume look more closely at police involvement, in particular Coleborne on Australia. Given that Australian colonial police forces were modelled on the Irish police, it is not surprising to find them playing leading roles in committal. For Irish influence on Australian policing, see E. Malcolm, '"What would people say if I became a policeman": the Irish Policeman Abroad', in O. Walsh (ed.), *The Irish Abroad: Politics and Professions in the Nineteenth Century* (Dublin, 2003), 95–107.

I

Although set in a remote part of the province of Connacht, Birmingham's story shows that there was a local asylum. In addition, the community was well aware of its function and seemingly not at all averse to having family members committed, at least for short periods. This is a fair reflection of the fact that Ireland supported a well-developed asylum system from the early nineteenth century and that large numbers of people were confined in it, frequently at the instigation of their families. And, in fact, this system – with the same high committal rates, often to the very same hospital buildings – continued to function into the latter half of the twentieth century.[7]

There were some private asylums in eighteenth-century Ireland which received government grants as well as a certain amount of haphazard local-authority provision for the insane.[8] However, growing public and political alarm at an apparent rapid increase in mental illness after 1800 led the British government[9] to begin building a network of state asylums, well in advance of the comparable English system. The first of these hospitals was the Richmond Asylum opened in Dublin in 1815 with 250 beds.[10] A select committee, initiated by the young and innovatory Irish chief secretary and later British Tory prime minister, Robert Peel, investigated relief of the lunatic poor in 1817 and recommended the establishment of a system of district asylums throughout Ireland. The system, moreover, was to receive central-government loans and was also largely to be regulated centrally.[11] Thus it was not permissive or decentralized as was the English model of the time.[12]

[7] For the architecture of Irish asylums and the history of the buildings, see M. Reuber, *Staats-und Privatanstalten in Irland: Irre, Arzte und Idioten, 1600–1900* (Cologne, 1994); M. Reuber, 'The Architecture of Psychological Management: The Irish Asylums, 1801–1922', *Psychological Medicine* 26 (1996), 1179–89; F. O'Dwyer, *Irish Hospital Architecture: A Pictorial History* (Dublin, 1997), 10–13, 33–4.

[8] E. Malcolm, *Swift's Hospital: a History of St Patrick's Hospital, Dublin, 1746–1989* (Dublin, 1989), 16–103.

[9] Ireland was ruled by Britain as part of the United Kingdom from 1801 until 1922, when the south of the country achieved independence. A local administration based in Dublin Castle, and generally referred to as 'Dublin Castle', carried out British policies. It was composed of a civil service and was headed by a lord lieutenant and a chief secretary, representing respectively the crown and the political party in power in Britain.

[10] For histories of the Richmond Asylum, subsequently re-named St Brendan's Hospital, Grangegorman, see J. Reynolds, *Grangegorman: Psychiatric Care in Dublin since 1815* (Dublin, 1992); B. O'Shea and J. Falvey, 'A History of the Richmond Asylum (St Brendan's Hospital), Dublin', in H. Freeman and G. E. Berrios (eds.), *150 Years of British Psychiatry*, volume II: *The Aftermath* (London, 1996), 407–33.

[11] *Report from the Select Committee on the Lunatic Poor in Ireland*, H. C. 1817, viii, 430.

[12] For the public asylum system in England, before the establishment of asylums was made mandatory in 1845, see L. D. Smith, *'Cure, Comfort and Safe Custody': Public Lunatic Asylums in Early Nineteenth-Century England* (London and Washington, DC, 1999).

As a result of the select committee's report nine asylums with a little over 1,000 beds were built between 1824 and 1835, while a further twelve with nearly 3,500 beds were erected between 1852 and 1869. Together with the Richmond in Dublin, the Irish public asylums had some 7,600 beds by the early 1870s and, after major extension, by the mid-1890s they had 13,600 beds, which rose to 16,600 beds shortly after the turn of the century. In 1904 the asylums in Belfast, Cork and Ballinasloe had well over 1,000 beds each, while the Richmond dwarfed them all with over 3,000 beds, or some 19 per cent of all Irish public asylum accommodation.[13]

Yet these massive increases in asylum provision have to be set against a background of limited industrialization and urbanization, and as well, from the late 1840s, a steadily declining population.[14] The Irish population, according to census figures, fell from 8.2 million in 1841 to 6.5 million in 1851, largely due to famine, but continued to decline over the next half century reaching 4.4 million in 1901. This of course meant that the number of persons per asylum bed fell dramatically: from 713:1 in 1871 to 270:1 by 1904.[15]

Therefore, although Dr Lovaway may have been misled into believing that mental illness in Connacht was worse than it actually was, he was nevertheless right to be startled at the level of committals to Irish asylums. And his speculations as to the causes of this phenomenon accurately reflect much of the contemporary debate.

II

Before considering this debate, it is necessary to look a little more closely at the workings of the Irish asylum system. To begin with, beds were by no means evenly distributed throughout the country. The province of Leinster with 26 per cent of the population in 1901 had nearly 36 per cent of asylum beds; while in the province of Ulster these percentages were almost reversed: Ulster had

[13] M. Finnane, *Insanity and the Insane in Post-Famine Ireland* (London, 1981), 227. The present chapter deals with the patient populations of Irish public asylums, but not all 'lunatics and idiots' were held in these institutions. According to the 1891 census only 58 per cent of them were confined in public and private asylums, while 23 per cent were 'at large' and 19 per cent were in workhouses. In 1851, 3 per cent of this group had been in Irish prisons. The chapter also focuses on 'lunatics', whereas in 1891, 8 per cent of the total asylum population were classed as 'idiots'. *Census of Ireland for the Year 1891. General Report. Part II*, 45 [C.6780] H.C. 1892, xc, 61.

[14] That Ireland experienced a massive growth in asylum accommodation without substantial industrialization or urbanization contradicts some of Andrew Scull's theories as regards the development of the asylum system in England. For a discussion of this issue, see E. Malcolm, '"The House of Strident Shadows": The Asylum, the Family and Emigration in Post-Famine Ireland', in G. Jones and E. Malcolm (eds.), *Medicine, Disease and the State in Ireland, 1650–1940* (Cork, 1999), 177–8.

[15] Finnane, *Insanity and the Insane*, 227; W. E. Vaughan and A. J. Fitzpatrick (eds.), *Irish Historical Statistics: Population, 1821–1971* (Dublin, 1978), 3.

36 per cent of Ireland's population but only 27 per cent of asylum beds. The other provinces, Munster and Connacht, had asylum facilities largely in line with their proportions of the general population. The differentials in Leinster and Ulster reflected in particular the situations in the cities of Dublin and Belfast. The Richmond Asylum in Dublin, as we have seen, was by far the largest in the country and had in 1904 a bed to district population ratio of 1:179. Yet the comparable ratio for the Belfast Asylum was 1:482.[16] Indeed, the Belfast Asylum was notorious for its overcrowding throughout the latter part of the nineteenth century.[17]

The less generous provision in Belfast highlights, among other issues,[18] local concerns about the funding of Ireland's extensive asylum system. Asylums were costly to build and expensive to operate. It has been estimated that the twenty-one asylums built between the 1820s and the 1860s cost some £1.14 million. To put this figure in perspective: a comparable sum of money was spent in the late 1830s and early 1840s to build 130 workhouses in Ireland.[19] Central government advanced the capital to erect asylums, but the thirty-two Irish counties were obliged to share the cost of repaying these advances within fourteen years. Such outlays were clearly a substantial financial burden upon Irish ratepayers. While some towns had initially appreciated the employment and trade that a local asylum generated,[20] as the century advanced complaints about the cost

[16] *Ibid.*

[17] In 1889, for instance, Belfast Asylum had 103 patients more than the 550 it had been built to house. Yet, despite this serious overcrowding, Belfast's complement of beds remained static between 1886 and 1896, while nationally the number of beds increased by nearly 47 per cent. The workhouse was used more extensively to house the insane in Belfast than in any other Irish asylum district. Close to 50 per cent of 'lunatics and idiots' were in the workhouse, where they were cheaper to accommodate, compared to a level of around 20 per cent for the country as a whole. H. C. Burdett, *Hospitals and Asylums of the World*, vol. I (London, 1891), 245; Finnane, *Insanity and the Insane*, 227; Malcolm, 'House of Strident Shadows', 189n.21, 179.

[18] Sectarianism was also clearly a feature of the operation of the Belfast Asylum. The governors, who were overwhelmingly protestant, preferred, as already indicated, to keep many lunatics, who were disproportionately catholic, in the workhouse. At the same time they opposed the appointment of a catholic chaplain for many years. O. Walsh, '"The Designs of Providence": Race, Religion and Irish Insanity', in J. Melling and B. Forsythe (eds.), *Insanity, Institutions and Society, 1800–1914: A Social History of Madness in Comparative Perspective* (London and New York, 1999), 228–33; Malcolm, 'House of Strident Shadows', 179; D. V. Griffiths and P. M. Prior, 'The Chaplaincy Question: The Lord Lieutenant of Ireland Versus the Belfast Lunatic Asylum'. I would like to thank Dr Prior for providing me with a copy of this unpublished paper.

[19] Finnane, *Insanity and the Insane*, 33.

[20] Even well into the twentieth century, mental hospitals could make significant economic contributions to Irish provincial towns. The Connacht District Asylum was opened in Ballinasloe in 1833; and this is the asylum to which Dr Lovaway would have been sending patients. It was re-named St Brigid's Hospital after independence in the 1920s. According to the 1951 census, however, the town of Ballinasloe had a population of 5,596, of whom 2,078 were mental-hospital patients and 439 full-time employees of the hospital – in other words, as late as the 1950s, an extraordinary 45 per cent of those resident in the town were confined in or employed by the mental hospital. D. Walsh, 'Mental Health Care in Ireland, 1945–97 and the Future', in J. Robins

and lack of effectiveness of asylums became increasingly common. Costs spiralled after 1850. Spending on patient maintenance, for instance, leapt from £6 per head of the general population in 1852 to £123 by 1914. During the same period per capita spending on the dispensary medical service had increased from £9 to £48 and on the poor-law system from £139 to £238.[21] In 1874, in response to repeated complaints, the British government agreed that the cost of the maintenance of asylum patients should be met in part by the treasury, which would contribute four shillings per head per week for all paupers housed in public asylums.[22] Although the contribution fluctuated somewhat over the years, during the 1880s and 1890s it generally amounted to somewhat over 40 per cent of total maintenance expenses. Nevertheless, by 1898 Irish ratepayers were still contributing nearly £217,000 per annum to support the asylum system.[23] Yet Irish asylums were far from generously funded. A knowledgeable contemporary observer commented in 1890 that boards of governors, being 'non-representative', were 'timid as to expenditure', so that 'a spirit of economy' existed 'which would not elsewhere be considered consistent with due provision for the insane'.[24]

The issue of cost was closely related to the issue of control. The cost was paid out of the county rates, known as the 'cess', which were levied on all occupiers of land. Yet, at the local level, asylums were directed by boards of governors appointed by the lord lieutenant, who headed the Irish administration based in Dublin Castle. Such boards were dominated by local protestant landowners and gentry. As a means of trying to curb expenditure, ratepayers increasingly demanded that they be represented on asylum boards. In 1888 the government conceded that half the membership of boards should consist of the nominees of local authorities. When Irish local government was radically reformed in 1898, the system of asylum administration also changed substantially. The unrepresentative grand juries, which had controlled counties for centuries, were replaced by popularly elected county councils.[25] They were charged with providing and maintaining accommodation for pauper lunatics, with appointing and removing asylum officers and with regulating expenditure. In future, asylum boards of governors were to be selected by local councils. However, despite

(ed.), *Reflections on Health: Commemorating Fifty Years of the Department of Health, 1947–97* (Dublin, 1997), 127.

[21] Finnane, *Insanity and the Insane*, 228.

[22] V. Crossman, *Local Government in Nineteenth-Century Ireland* (Belfast, 1994), 39.

[23] Local funding continued well into the twentieth century. In the south of Ireland it was not fully matched by government until 1947 and local rates were not finally abolished until 1978. Finnane, *Insanity and the Insane*, 230; O'Shea and Falvey, 'A History of the Richmond Asylum', 411; Burdett, *Hospitals and Asylums of the World*, 242, 244.

[24] Burdett, *Hospitals and Asylums of the World*, 244.

[25] For the role of county grand juries in Irish local government, see Crossman, *Local Government in Nineteenth-Century Ireland*, 25–41.

this decentralization of power over asylums, the government's subsidy to the system was still to be determined at the centre by Dublin Castle.

This battle over funding and control illustrates the fact that the asylum system in nineteenth-century Ireland was administered and regulated in a complex, confusing and, it could be argued, counter-productive manner. The asylum expert H. C. Burdett summed up the administrative code regulating Irish asylums thus:

> As far as any glimmer of a general principle can be seen through it, its design seems to be to exemplify the cunning motto of the old Roman, *Divide et impera*, for, unless it has been framed for the express purpose of giving perpetual opportunity for the intervention of the central powers, it is impossible to see why it should contain, as it does, numerous enactments, manifestly intended to produce a perpetual state of unstable equilibrium . . . The rules, besides, have the faults ordinarily found in regulations drawn up by persons entirely ignorant of the matters to be regulated. They often contradict each other.[26]

Government enquiries recommended major administrative restructuring in 1878 and again in 1891, but until 1922, when the south achieved independence, no such reforms were introduced.[27] Aside from the British government, Dublin Castle, local authorities and boards of governors, there were powerful inspectors of lunacy regulating asylums and also asylum medical superintendents represented by their own professional body, the British Medico-Psychological Association. All wielded significant influence and all competed to shape the development of Irish asylum-system.

The lunacy inspectorate grew out of the prison inspectorate established in 1787 with a remit to inspect madhouses as well as prisons. But the prison inspectors were generally military men with little interest in, and even less knowledge of, the insane.[28] By an Act of 1845 a separate two-man lunacy inspectorate was created, based in Dublin Castle, and staffed by medical practitioners. Two inspectors were appointed as it was decided that one should be catholic and one protestant. Only five men filled these positions during the nineteenth century, reflecting the fact that several of them served for thirty or forty years.[29] Such lengthy service did not encourage radical thought or innovation; and especially in the 1880s, due to age and ill health, neither inspector was fulfilling his job properly. Their response to the increasingly overcrowded state of Irish asylums, which was becoming evident as early as the 1840s, was to build more asylums. The inspectors consistently advised government from the 1860s through into

[26] Burdett, *Hospitals and Asylums of the World*, 253.

[27] R. B. McDowell, *The Irish Administration, 1801–1914* (London, 1964), 174.

[28] O. MacDonagh, *The Inspector General: Sir Jeremiah Fitzpatrick and Social Reform, 1783–1802* (London, 1981), 320.

[29] Dr George Hatchell served as an inspector from 1847 until his death in 1889, while Dr John Nugent served from 1857 until his retirement with a knighthood in 1890. Hatchell was ill for a number of years prior to his death, while Nugent was eighty-four at the time of his retirement.

the 1910s that institutionalization was the only answer to Ireland's apparently high levels of mental illness.

The appointment of medical practitioners as inspectors in the 1840s hastened the medical takeover of patient management in asylums. From the 1830s medically trained asylum managers, who were in the minority, began to campaign for the abolition of lay managers. By the 1860s all Irish public asylums were headed by medical superintendents appointed by the lord lieutenant. Yet, while boasting a medical degree, few of these men had any specialist training in the treatment of the mentally ill. Their appointments were also often the result of patronage rather than merit. Pointing out that the Irish medical profession over the previous forty years had made not one significant contribution to the literature on insanity, Burdett claimed in 1890 that the office of asylum medical superintendent was 'given away almost wholly from political considerations' and that, in fact, 'there is no office under the Crown for which there is more canvassing, direct and indirect, personal and political'. Like the lunacy inspectors, many asylum medical superintendents served for lengthy periods. For instance, at the Richmond Asylum in Dublin, later re-named St Brendan's Hospital, only five men filled the position of medical superintendent from the 1850s up to the 1990s.[30]

Visiting physicians to asylums continued to be appointed in Ireland into the 1890s, forty years after the office had disappeared in England. Their appointments were in the gift of boards of governors, who were jealous of their powers of patronage. In 1890 about half of Irish public asylums had in addition assistant medical superintendents, also appointed by boards of governors. As regards their appointment, according to Burdett: 'Canvassing, instead of being forbidden, as it ought to be, is expected, if not demanded, and influence decides the matter.'[31] Clearly patronage played a significant role in the appointment of asylum medical personnel. Thus it is not surprising to discover that Dr George Hatchell, the longest-serving lunacy inspector of the nineteenth century, whose own appointment was the result of patronage, had two sons both employed as asylum medical superintendents.[32] Severe financial constraints, confusing regulations, divided authority and corrupt appointments inevitably produced a

[30] The doctors, with their periods of service, were Joseph Lalor (1857–86), Conolly Norman (1886–1908), John O'Conor Donelan (1908–37), John Dunne (1937–66) and Ivor Browne (1966).

[31] Burdett, *Hospitals and Asylums of the World*, 254, 256, 258. The asylums, however, were not unusual in this regard. Patronage and nepotism were the rule rather than the exception in Irish hospital appointments throughout the nineteenth century and beyond. In the earlier years especially, such jobs were openly bought and sold. For the career of a leading catholic doctor, who made his way in a protestant-dominated profession partly through buying and selling appointments, see E. O'Brien, *Conscience and Conflict: A Biography of Sir Dominic Corrigan, 1802–80* (Dublin, 1983), 67–8, 79, 82–3, 95–6.

[32] Finnane, *Insanity and the Insane*, 52n.108, 64.

far from satisfactory asylum system. This state of affairs became most evident visually: in the deterioration of the buildings themselves.

By the 1890s some asylum buildings were seventy years old, while even the newest were between twenty and thirty years old. There were suggestions that some of the oldest were so inadequate that only demolition and the erection of new buildings on totally different designs could provide satisfactory accommodation. Despite the large sums spent on erecting them, asylums were often shoddily built and, due to funding restrictions, their maintenance left a lot to be desired. Burdett condemned them roundly.

> That they are plain to the verge of wanton ugliness might perhaps be defensible on the ground of frugality, if they were at the same time suitable in other respects; but this, unfortunately, cannot be said at least of most of them. Everywhere the windows incline to be too high from the ground: in many places they are too small . . . Flagged corridors are not unknown, nor were flagged single rooms till very recently. All internal stone and wood work is of the rudest and poorest description. Means of ventilation are quite primitive. In most asylums there are no means of heating single rooms and dormitories . . . The sanitary appliances in some of the institutions have recently been found to be in a truly shocking state. Whitewash is freely used . . . and communicates to everything around its own inevitable chill and pauperising tone. Kitchens, laundries, and other offices are usually on a par with the wards in faulty construction and defective fittings.[33]

Burdett was also struck by the meagre furnishing of wards and day rooms. Plain deal tables and deal forms without backs comprised the main furniture of day rooms; floor coverings were rare; and windows seldom had blinds or curtains. He concluded: 'It is probable that no workhouse in England presents nowadays so gaunt and cheerless an appearance as may be found in many Irish asylums.' Indeed, 'perpetual reminders of the prison' were everywhere.[34]

Younger and more active inspectors were appointed in 1890 and, as already mentioned, elected local councils took over the running of the asylums after 1898. But the inspectors were still committed to asylum building and in the late 1890s new asylums were erected in Antrim town and Portrane, Co. Dublin, to take pressure off the large Belfast and Dublin asylums. Moreover, local councils, dominated by farming and shopkeeping interests, were even more determined than the old gentry-dominated grand juries to keep down the cost of asylums. In the first decade of the twentieth century a number of county councils and asylum boards petitioned the British government requesting that the state take full responsibility for the funding of asylums. But in each case the government refused. Thus, despite the creation of a more representative form of asylum administration after 1898, buildings continued to deteriorate, facilities were not improved and overcrowding increased during the first two decades of the twentieth century. As Mark Finnane has written:

[33] Burdett, *Hospitals and Asylums of the World*, 243. [34] *Ibid.*, 243–4.

The balance of power had tipped from Dublin Castle to popularly-elected local government authorities. With this change the emphasis of asylum politics shifted, from the mid-century obsession of the inspectorate with confining *all* the insane, to the provincial preoccupation of the early twentieth century with reducing numbers of inmates and certainly costs. The interests of the confined themselves were largely ignored in the course of this shift of power.[35]

Yet, despite the best efforts of penny-pinching local councils, the numbers of patients in Irish public mental hospitals continued to rise during the first half of the twentieth century, reaching a peak in 1958. In that year there were 21,046 mental hospital patients, comprising 0.5 per cent of the population of the Irish Republic. In 1961 about one in seventy residents of the country aged twenty-four and over was a patient in a mental hospital.[36]

III

How then were lunatics committed to Irish asylums? Again, Birmingham's short story offers a good deal of accurate information. Complaints, often about violent behaviour, were made to the constabulary, usually by relatives, sometimes by friends or neighbours. The person complained of would then be taken into police custody, summoned before magistrates and, up until the 1870s at least, probably committed to prison.

Before the mid-1840s asylum inspectors were also prison inspectors. Burdett in 1890, as already mentioned, drew a parallel between the Irish asylum and the prison. This all points to the close connection that existed in Ireland during the nineteenth century between institutions to house the insane and institutions to house criminals. Or, as one historian has commented, rather than insanity being associated with poverty through the poor-law system, as in England, in Ireland there was 'an intimate link between insanity and criminality'.[37] This link was apparent in the legislation that governed Irish committal procedures for over a century.

Committal to an asylum was regulated by two major Acts: one passed in 1838 (1 & 2 Vic. c.27) and one in 1867 (30 & 31 Vic. c.118). The latter superseded the former, but, significantly, both carried the same title: the Dangerous Lunatics Act. Their provisions continued to govern Irish committal procedures, at least in the south of the country, until 1945. The 1838 act empowered two magistrates

[35] Finnane, *Insanity and the Insane*, 82.

[36] Walsh, 'Mental Health Care in Ireland, 1945–97', 126; The Psychiatric Services, *Planning for the Future: Report of a Study Group on the Development of Psychiatric Services* (Pl.3001, Dublin, 1984), 2.

[37] Walsh, 'Designs of Providence', 225. For the English situation, see P. Bartlett, *The Poor Law of Lunacy: the Administration of Pauper Lunatics in Mid-Nineteenth-Century England* (London and Washington, DC, 1999).

to commit anyone they deemed a dangerous lunatic to prison. There the lunatic would remain until discharged by the magistrates or transferred to an asylum under an order of the lord lieutenant. The magistrates in making their decisions could call for a medical opinion, but they were not obliged to do so. In addition, a lunatic could be committed directly to an asylum by a relative or friend. In this instance a medical certificate was required, as well as a declaration from a magistrate or clergyman that the lunatic was a pauper. Also, under this procedure, the person committing had to agree to take the lunatic back when the asylum decided that he or she was ready for release. The same obligation did not apply if the lunatic was committed by magistrates.

As a result of this legislation, over the following thirty years, a large proportion of those committed to Irish asylums were committed as dangerous lunatics. In addition, at any one time, there was a significant proportion of lunatics housed in Irish prisons. The lunacy inspectors were critical of this act, arguing that families used it so as not to be obliged to take unwanted members back, that magistrates made committal orders in a cavalier fashion, and that a period in prison could only further harm a person's mental health. The 1867 act was intended to rectify these shortcomings. It did not, however, achieve this goal. It allowed magistrates to commit dangerous 'lunatics and idiots' directly to an asylum, but a medical certificate from a dispensary doctor confirming a diagnosis of mental deficiency was now an essential requirement. This is the Act that would have governed the actions of dispensary doctors like Dr Lovaway.

Yet, despite the inspectors' intentions, under the 1867 Act, the proportion of the Irish insane committed as criminals actually increased substantially: from around 42 per cent in 1860–2 to fully 66 per cent forty years later in 1900–2. Thus under the new Act, while few of the mentally ill were now actually sent to prison, larger numbers of them were criminalized. They were generally taken into custody by the police after a complaint from relatives; they were certified by one doctor normally untrained in mental-health matters; they appeared in court before local landowners and businessmen acting as magistrates, where an indefinite sentence of committal was passed; and they were transported by the police to an asylum. There they had no right to have their diagnosis or committal reviewed. If and when they were released depended upon the decision of a lay board of governors, advised by a medical officer who might have no training in the treatment of mental illness. All in all the Irish committal process gave enormous powers to the ignorant and the potentially malicious. It criminalized the mentally ill and allowed them few if any avenues of redress. It was a formula for abuse that operated from the 1830s into the 1940s.

Lovaway and doctors like him sought explanations for what they perceived as extraordinarily high rates of mental illness among the Irish in theories concerning race, religion and society. Perhaps they would have been better advised to have investigated the legislation governing Irish asylums and the manner in which the system was administered. The inadequacies and conflicts inherent in these go some way at least towards explaining why so many people were committed so easily and why so few of them were ultimately released.

IV

By the late nineteenth century the overcrowding of Ireland's large asylum system, as well as the apparently high rates of committal to asylums among the Irish abroad,[38] had given rise to a heated debate about Irish mental health. This debate was not restricted to the medical profession and it continued well into the twentieth century. In 1909 the nationalist MP and historian, R. Barry O'Brien, identified the 'three scourges, which afflict Ireland' as 'Emigration, Tuberculosis and Lunacy'.[39] But were the Irish especially prone to mental illness, and, if so, why? Dr Lovaway rehearsed some of the popular aspects of this debate in his proposed article, quoted at the beginning of this chapter. Was Irish mental illness the product of 'race decadence' or was it due to social factors, such as peculiar marital practices, excessive alcohol consumption or oppressive religiosity?

In 1903 a special conference was convened at the Richmond Asylum, composed of representatives from most Irish asylums, to discuss the causes of the apparently high rates of Irish mental illness. Heredity, intemperance and emigration were highlighted. Hereditary mental defects among isolated rural families, accentuated by intermarriage and alcoholism, were identified as especially significant. In addition, emigration was seen as draining away from such communities the brightest and most energetic members, leaving behind those more mentally inadequate or vulnerable. As a solution to the problem, the conference – perhaps not surprisingly, given that most of those attending worked in asylums – came out in favour of further hospital building. It considered it essential that lunatics should be taken out of their communities, where they could have a bad influence on others; and, above all, that they should not be allowed to marry and reproduce freely.[40]

In 1911 one of the lunacy inspectors, Dr W. R. Dawson, delivered a wide-ranging address on Irish insanity and social factors before the annual meeting of the British Medico-Psychological Association, held that year in Dublin. Dawson began by noting that at the present time 23,174 persons were confined in Irish public asylums and workhouses as lunatics. This amounted to

[38] This phenomenon is discussed in E. Malcolm, '"A Most Miserable Looking Object": The Irish in English Asylums, 1851–1901: Migration, Poverty and Prejudice', in J. Belchem and K. Tenfelde (eds.), *Irish and Polish Migration in Comparative Perspective* (Essen, 2003), 115–26.

[39] R. B. O'Brien, *Dublin Castle and the Irish People* (Dublin and London, 1909), 315.

[40] D. Healy, 'Irish Psychiatry in the Twentieth Century', in Freeman and Berrios (eds.), *150 Years of British Psychiatry*, 269–72.

5.3 persons per 1,000 of the total population. In Waterford this figure was an 'enormous' 9.2, or nearly one per cent of the county's population of 84,000.[41] Dawson then proceeded to search for correlations between these figures and rates of population density, poverty, emigration, age, death, tuberculosis, crime and alcoholism in each county. He concluded that: 'in Ireland insanity tends to prevail in the agricultural counties, and has a close relation . . . to pauperism, which also prevails in rural districts'. It also bore some 'little relation' to emigration, criminality and alcoholism.[42] But beyond registering crude statistical patterns, Dawson made no effort to explain causal links. How rural poverty related to mental illness, emigration, crime and alcoholism was not explored. Also, like many writers of the time and since, he regarded institutionalization as an accurate measure of mental illness. That a significant number of those committed might not have been mad and that many 'lunatics' continued to live in the community does not seem to have occurred to him.

By 1900, even supposed experts were convinced that the Irish exhibited a proneness to mental illness. An Irish stereotype that highlighted drunkenness, aggression, fecklessness and poverty had been evident in English colonial discourse for centuries. It was reinforced in the mid nineteenth century by new racial and genetic theories.[43] By the latter part of the century insanity had been added to the traditional list of Irish negative characteristics. The Irish were perceived as being naturally emotional, volatile and irrational. Sometimes these characteristics were seen in a favourable light: they were what made the Irish great poets and soldiers.[44] But, more often, they were interpreted negatively: the Irish had a propensity to mental instability that made it more difficult for them to cope with poverty and hardship. And the high admission rates to Irish asylums, which became evident in the mid nineteenth century and continued for over a hundred years, seemed to offer clear evidence of the truth of this proposition.

[41] W. R. Dawson, 'The Presidential Address on the Relation between the Geographical Distribution of Insanity and that of Certain Social and other Conditions', *The Journal of Mental Science* 57 (1911), 577.

[42] *Ibid.*, 587.

[43] The first clear presentation of this stereotype occurred as early as the twelfth century; see Gerald of Wales, *The History and Topography of Ireland*, trans. J. J. O'Meara (rev. edn, Harmondsworth, 1982). For anti-Irish attitudes during the nineteenth century, see L. P. Curtis Jr, *Apes and Angels: The Irishman in Victorian Caricature* (rev. edn, Washington and London, 1997).

[44] Matthew Arnold in a famous series of lectures 'On the Study of Celtic Literature', delivered in Oxford in 1867, identified the Celts generally, and the Irish particularly, as being different in temperament and culture from the Anglo-Saxon English. They were 'undisciplined', 'anarchical', 'turbulent of nature', 'sensuous' and vehemently in 'reaction against the despotism of fact'. Their temperament made them great poets, but bad politicians. See M. Arnold, *On the Study of Celtic Literature and Other Essays* (London, n.d.). For a discussion of the supposed martial characteristics of the Irish, who were disproportionately represented in the British Army throughout the nineteenth century, see J. Bourke, *An Intimate History of Killing: Face-to-Face Killing in Twentieth-Century Warfare* (London, 1999), 118–20, 125, 137.

V

Although Ireland underwent a political transformation in the 1920s, with the division of the country into two states – the Irish Free State (from 1948 the Irish Republic) and Northern Ireland – little changed as regards the public asylums, or mental hospitals as they came increasingly to be called. Before the country was partitioned Irish political leaders, whether nationalist or unionist, were preoccupied with constitutional questions, and issues connected with health and medicine were not accorded a very high priority. In the first decades after partition, the new government of the south was attempting to consolidate its existence amid extremely difficult economic circumstances.[45]

In the south,[46] therefore, it was not until 1945 that a major new piece of mental-health legislation was passed. This was the Mental Treatment Act (no.19 of 1945) which superseded the 1867 Dangerous Lunatics Act. It allowed for voluntary admission to mental hospitals, the establishment of out-patients' clinics, the 'boarding out' of the mentally ill and the release of involuntary patients 'on trial'.[47] It certainly liberalized committal procedures, but discharge of both voluntary and involuntary patients was still very much at the discretion of the medical superintendent of the hospital concerned.

The act, however, had little success in reducing rates of committal, which, if anything, rose during the 1950s. A commission of inquiry into mental illness, appointed in 1961, later reported that the Irish Republic in that year probably had the highest rate of psychiatric bed usage in the world. Whereas the number of beds per 1,000 of population was 4.6 in the United Kingdom and 4.3 in the United States, in Ireland the comparable figure was 7.3; while in parts of the west of Ireland the figure was a remarkable 11.0 beds per 1,000 of population.[48] In other words, committal rates in Connacht which Dr Lovaway had considered alarming during the First World War were even higher half a century later.

Thus the public debate about Irish mental health, begun in the late nineteenth century, continued well into the late twentieth century. Heredity and social deprivation were still being put forward as explanations, although new theories that emphasized politics were also introduced, as anthropologists, sociologists and historians joined the discussion.

Some psychiatrists identified what they termed an 'epidemic' of schizophrenia in the west of Ireland. This interpretation was presented most controversially

[45] R. Barrington, *Health, Medicine and Politics in Ireland, 1900–70* (Dublin, 1987), 22, 86–8, 110–12; J. Robins, *Fools and Mad: A History of the Insane in Ireland* (Dublin, 1986), 186–90.

[46] Developments in Northern Ireland after 1922 will not be discussed here as they are treated at some length in P. M. Prior, '"Where Lunatics Abound": A History of Mental Health Services in Northern Ireland', in Freeman and Berrios (eds.), *150 Years of British Psychiatry*, 292–308.

[47] Walsh, 'Mental Health Care in Ireland, 1945–97', 127–8; *Planning for the Future*, 2.

[48] Healy, 'Irish Psychiatry in the Twentieth Century', 268.

in 1979 by an American anthropologist, Nancy Scheper-Hughes, in a book entitled *Saints, Scholars and Schizophrenics: Mental Illness in Rural Ireland.* Scheper-Hughes summed up her argument thus:

> I share with other recent ethnographers ... the belief that rural Ireland is dying and its people are consequently infused with a spirit of anomie and despair. This anomie is expressed most markedly in the decline of the traditional agriculture ... and in the virtual dependence of the small communities of the west upon welfare schemes ... The flight of young people – especially women – from the desolate parishes of the western coast, drinking patterns among the stay-at-home class of bachelor farmers, and the general disinterest of the local populace in sexuality, marriage, and procreation are further signs of cultural stagnation. Finally, the relative ease with which a growing proportion of the young, single, male farmers are able to accept voluntary incarceration in the mental hospital as a panacea for their troubles is a final indication that western Ireland, one of the oldest and most continually settled human communities in Europe, is in a virtual state of psycho-cultural decline.[49]

At the time of its publication this argument generated a storm of protest in Ireland, and yet it was far from a new argument. Dr Lovaway had speculated that high rates of mental illness in the west of Ireland were the sign of a 'passing race'; and Irish asylum doctors were analysing social deprivation in rural Ireland as a factor in precipitating insanity at the turn of the century. Professor Scheper-Hughes – even if she did not realize it – was offering an interpretation with a very long pedigree, although her argument was decked out with detailed recent field research and the latest elaborate statistics.

An interpretation of Irish rates of mental illness as dramatic and colourful as Scheper-Hughes's, but with more of a political bent, was put forward in the 1990s by Liam Greenslade, an English psychologist of Irish descent. He looked at the health of the Irish community in Britain during the last quarter of the twentieth century. Much of what he said, however, related to Ireland itself and to a much longer time span. Arguing that illness is a 'socially constructed notion', Greenslade declared that the 'vast majority of mentally ill people are not "ill" in the sense proposed by the medical model of illness ... What such people are, in fact, are the physical manifestation, the "symptoms", of the oppression, contradictions and pathologies in the society in which they live.' In the case of Ireland, Greenslade laid the blame for the country's social – and by implication individual – ills squarely at the door of English imperialism. As a colony, indeed England's first colony, Ireland 'over its history suffered conditions of colonial violence, oppression and expropriation, underdevelopment and clientism' comparable to the worst excesses of imperialism experienced in Africa, Asia and Australasia. Drawing on the work of the French colonial

[49] N. Scheper-Hughes, *Saints, Scholars and Schizophrenics: Mental Illness in Rural Ireland* (Berkeley and London, 1979), 4–5.

psychiatrist, Franz Fanon, who postulated that colonialism created a sense of worthlessness and dependence in the colonized, Greenslade concluded that Irish mental health would not noticeably improve until Irish people 'rid themselves of their historical inferiority'.[50]

Rather more prosaic explanations of the apparently poor state of Irish mental health have been advanced that also stress the nature of British rule. In Ireland, British government was highly centralized and highly interventionist in social as well as political terms, testing out policies some of which were later introduced into England itself.[51] Thus a large national, public asylum system was established in Ireland considerably in advance of a comparable system in England. Some writers have suggested that there were high asylum committal rates in Ireland simply because the British provided so many asylum beds, and continued to provide beds even when the Irish population began to decline substantially from the mid-1840s.[52] However, this is to overlook two key facts: Irish asylums were chronically overcrowded and thus there was an almost constant shortage of beds, not an over-supply; and, secondly, committals were in the main instigated by families, not by government officials or even by doctors.

Another version of this argument suggests that the Irish themselves developed a 'love affair' with the hospital bed, for not only did Ireland have extremely high rates of committal to mental hospitals but also extremely high rates of hospital-bed usage generally.[53] Dr Lovaway found that there seemed to be no stigma attached to short-term committal to an asylum. Some writers have argued that the Irish came to see the mental hospital as a benevolent institution, which could help the family in times of crisis, as opposed to the workhouse which was generally dreaded and shunned.[54]

Burdett commented on the poor quality of Irish asylum superintendents and on the apparent lack of interest of the Irish medical profession in insanity. The claim that Irish training in psychiatric medicine was inadequate and thus, for ill-informed, over-worked and under-paid doctors, committal was the easiest option, has been applied to the twentieth century as well.[55] Irish universities were slow to establish chairs in psychiatry, while Irish psychiatrists remained members of the British Medico-Psychological Association even after political

[50] L. Greenslade, 'White Skin, White Masks: Psychological Distress among the Irish in Britain', in P. O'Sullivan (ed.), *The Irish World Wide: History, Heritage and Identity*, volume II: *The Irish in the New Communities* (Leicester and London, 1992), 202–5; L. Greenslade, 'The Blackbird Calls in Grief: Colonialism, Health and Identity among Irish Immigrants in Britain', in J. MacLaughlin (ed.), *Location and Dislocation in Contemporary Irish Society: Emigration and Irish Identities* (Cork, 1997), 36–60.

[51] O. MacDonagh, *Early Victorian Government, 1830–70* (London, 1977), 178–96.

[52] E. Kane, 'Stereotypes and Irish Identity: Mental Illness as a Cultural Frame', *Studies* 75 (1986), 539–51.

[53] Healy, 'Irish Psychiatry in the Twentieth Century', 283.

[54] Finnane, *Insanity and the Insane*, 129.

[55] Healy, 'Irish Psychiatry in the Twentieth Century', 281–2.

independence, as they showed little enthusiasm for setting up their own independent professional body. In addition, the Catholic Church looked askance at many of the new developments in psychiatric medicine, especially the growing influence of psychoanalysis and behaviourism. Child psychology was considered deeply suspect and there was an outright ban on the use of hypnosis until the 1950s. Thus able and ambitious Irish medical graduates were unlikely to opt for a career in psychiatry. Jobs in mental hospitals carried with them little if any professional prestige.[56]

It was not until the 1960s that a more sympathetic attitude to psychiatry within the Irish church began to become obvious. The editors of an important book on *The Priest and Mental Health*, published in 1962, were at pains to point out that, if the priest could benefit from knowing something of modern psychiatry, equally the psychiatrist could benefit from knowing 'more about the religious dimensions of the patients' problems'. Dismissing the 'facile solutions' of the Jungian school, the editors welcomed the fact that in recent times psychiatrists seemed to have ceased to regard religion itself as a form of mental illness. They certainly endorsed a recent Vatican ruling forbidding priests to practise psychoanalysis. Yet, one of the editors, a professor of logic and psychology at University College, Dublin, who was also a priest, went on to re-interpret Freud, arguing that far from 'explain[ing] away the existence of God', he had in fact thrown light on the complexities of the human attitude to God.[57] Although clearly very sceptical of aspects of modern psychiatry, the book was nevertheless an attempt to demonstrate that some knowledge of psychiatry could be of considerable value to priests in their pastoral work – and even psychoanalysis was not wholly without merits.

VI

Over the last twenty years, after more than a century of almost continuous growth, Irish specialist institutions for the mentally ill have begun to disappear – and to disappear rapidly. The government commission established in 1961 finally reported in 1966, recommending a move away from large, isolated, specialist mental hospitals towards smaller psychiatric units based in general hospitals and community facilities such as day hospitals and clinics, and hostels. The report suggested that if these recommendations were put into effect the

[56] Personal communication to the author from a former medical director of St Patrick's Hospital, Dublin, who entered psychiatric medicine in the 1930s after a bout of TB prevented him from pursuing a career in more prestigious specialities. In this context it is also interesting to note that, in the early decades of the twentieth century, a number of Ireland's first women medical graduates, excluded from leading teaching hospitals, were nevertheless able to secure junior positions in various provincial asylums.

[57] E. F. O'Doherty and S. D. McGrath (eds.), *The Priest and Mental Health* (Dublin, 1962), v–vi, 69–70.

number of psychiatric beds in the Irish Republic could be cut by more than 50 per cent over the following fifteen years – that is to about 8,000 by 1981. In fact this goal was not achieved. In 1984 there were still nearly 13,000 patients in psychiatric institutions, and indeed, admissions had increased from over 18,700 in 1970 to nearly 23,700 in 1982. In the latter year only 4 per cent of psychiatric beds were in general hospitals.[58]

In response to the perceived failure of the 1966 report, a study group was appointed by the Irish government in 1981, which issued a report in 1984 entitled *Planning for the Future*. Like its 1966 forerunner, this report stressed the need to develop community-based residential and day-care facilities, to close large, old mental hospitals and to rehabilitate their long-stay patients. Unlike its forerunner, however, this report was acted upon. By 1995 the resident population of public and private Irish psychiatric hospitals and units was down to a little over 5,800, and plans were well in train to close all specialist psychiatric hospitals by the year 2005 at the latest.[59]

VII

Birmingham's short story is quite accurate in its depiction of the remarkably high asylum-committal rates that prevailed in large parts of Ireland during the early years of the twentieth century. These rates had been increasing for well over half a century and, indeed, would continue to increase for another half century. More significantly, however, the story is astute in its recognition that committal was a complex process, involving a variety of parties and sometimes having little to do with actual mental illness. Asylums had social as well as medical uses.

Serious social and economic deprivation certainly existed in rural Ireland during this period. A declining population, poverty, physically taxing labour, isolation, the threat of famine and destitution, an authoritarian family structure, high levels of emigration especially among young women, low marriage rates, a puritanical church and recreation largely devoted to heavy drinking, all helped create a lifestyle that was often bleak and unfulfilling, if not a positive threat to both physical and mental health.[60] In addition, as has been demonstrated, the legislation that regulated asylums and later mental hospitals from the 1830s into the 1940s made committal all too easy. Local authorities certainly struggled to limit the numbers committed, essentially in order to curb costs, but their efforts were largely in vain. Families, police, magistrates, clergy and doctors

[58] Walsh, 'Mental Health Care in Ireland, 1945–97', 131–2; *Planning for the Future*, 143–53.

[59] The progress achieved since 1984 has been reviewed in *Green Paper on Mental Health* (Pl.8918, Dublin, 1992) and *White Paper. A New Mental Health Act* (Pn.1824, Dublin, 1995).

[60] See T. W. Guinnane, *The Vanishing Irish: Households, Migration and the Rural Economy in Ireland, 1850–1914* (Princeton, 1997).

co-operated to take advantage of lax procedures so as to rid their communities of those deemed troubled or troublesome.

Thus the picture that emerges from Ireland is in line with recent research into the committal process in a number of countries.[61] It portrays a system of negotiation between communities and authorities that resulted in large numbers of individuals being institutionalized over a long period of time. What is perhaps most distinctive about Ireland is that the numbers and time period concerned were probably greater than almost anywhere else. While grand theories speculating about racial decline and colonial trauma are unconvincing, the question of why Ireland used asylums/mental hospitals more intensively and for longer than most other countries still awaits a wholly satisfactory explanation.

[61] See, for example, several chapters in the present volume, especially Murphy on England and Prestwich on France.

14 The administration of insanity in England 1800 to 1870

Elaine Murphy

Introduction

The social history of insanity has proved a seductive paradigm for students of the management of the dependent poor in nineteenth-century England. The insane have been perceived as 'casualties' of class and gender power relations during the transformation from a paternalistic rural economy into an industrialized capitalist state. While the Elizabethan Poor Law legislation of 1601 was the administrative foundation on which the system of care was constructed, until recently two other themes dominated the historiography of mental disorder: first, that of the rise of psychiatry and psychiatrists; and second, the expansion of the Victorian asylum as society's preferred response.[1] A reappraisal of the 'revisionist' interpretation of events is now underway, however, and a more complex picture is emerging. Mad paupers are no longer so readily annexed to political dogma.

Scull's 'deeply researched and provocative account of the growth of public asylums'[2] in nineteenth-century England, published as *Museums of Madness* in 1979,[3] attributed the expansion of 'asylumdom' to the emerging commercial market economy and the consequent extrusion of inconvenient non-working people from the mainstream of family and community life. Scull interpreted the growing interest in madness by specialist mad-doctors as an unattractive bid for power and status by a group of financially insecure members of a profession still on the threshold of respectability.

Looking back twenty years later, Scull[4] acknowledged that his work was stimulated in part by Foucault's brilliant but flawed essays on power relations,

[1] J. Walton, 'Poverty and Lunacy: Some Thoughts on Directions for Future Research', *Bulletin of the Society for the Social History of Medicine* 34 (1984), 64–7.

[2] J. Melling, 'Accommodating Madness', in J. Melling and B. Forsythe (eds.), *Insanity, Institutions and Society, 1800–1914* (London, 1999), 3.

[3] A. Scull, *Museums of Madness: the Social Organization of Insanity in Nineteenth Century England* (London, 1979), 254–66.

[4] A. Scull, 'Rethinking the History of Asylumdom', in Melling and Forsythe (eds.), *Insanity, Institutions and Society*, 295–395.

Madness and Civilisation.[5] Scull's Marxist historical sociology proved to be a red rag to the mainstream British psychiatric bull and to fans of the asylum like Kathleen Jones, who viewed Scull's interpretation as a challenge to the legitimacy of the psychiatric profession, which it was, and a late flowering of the 1960s anti-psychiatry movement, which perhaps it was not. In the early 1990s Jones complained that the social history of mental-health services had become an 'academic minefield', smarting perhaps from Scull's heavyweight criticism of her analysis of events as a story of progress and social enlightenment thwarted by 'backsliding, misunderstandings and incompetence'.[6]

The effect of Scull's challenge was to provoke further analyses of the meaning of the asylum as a solution to managing derangement, facilitated by the wealth of archival material from the institutions and the county magistracy that administered them.[7] The place of the insane in social welfare provision was located by Jones and Scull in their early works within the reforming zeal of the county magistrates, the mid-Victorian Lunatics Acts and the central inspectorate responsible for policing the Acts, the Commissioners in Lunacy.[8] The literature underplayed the legal and administrative context of the poor law within which lunacy was managed and paid only glancing attention to the influence of the changing role of the state and the growth of nineteenth-century government administration.

Porter meanwhile was excavating an earlier seam of eighteenth-century evidence, which challenged the notion that the nineteenth-century response to madness was discontinuous and different from previous centuries. He rescued the humanity of earlier attempts to care for and cure the mad from the overwhelmingly dismissive accounts of eighteenth-century 'care' as barbaric. He further challenged the notion that the curative ideal of 'moral treatment' was a solely nineteenth-century phenomenon.[9] Just as importantly though, he reasserted the

[5] M. Foucault, *Madness and Civilisation: A History of Insanity in the Age of Reason* (New York, 1965).

[6] K. Jones, *Asylums and After* (London, 1993), 4.

[7] The literature is too large to quote more than a handful of studies here but see A. Digby, *Madness, Morality and Medicine: A Study of the York Retreat 1796–1914* (Cambridge, 1985); W. F. Bynum, R. Porter and M. Shepherd (eds.), *The Anatomy of Madness* (London, 1985); M. Finnane, *Insanity and the Insane in Post-Famine Ireland* (London, 1981); C. MacKenzie, *Psychiatry for the Rich: A History of Ticehurst Private Asylum* (London, 1992), chapter four; D. J. Mellett, *The Prerogative of Asylumdom: Social, Cultural and Administrative Aspects of the Institutional Treatment of the Insane in Nineteenth-Century Britain* (New York, 1982); E. Showalter, *The Female Malady: Women, Madness and English Culture 1830–1980* (New York, 1985); N. Tomes, 'The Anglo-American Asylum in Historical Perspective', in J. Giggs and C. Smith (eds.), *Location and Stigma* (1988), 3–20; J. Walton, 'Lunacy in the Industrial Revolution: A Study of Asylum Admissions in Lancashire 1848–50', *Journal of Social History* 13 (1979), 1–22.

[8] K. Jones, *Lunacy, Law and Conscience 1744–1845* (London, 1955); Scull, *Museums of Madness*.

[9] R. Porter, 'Medicine and the Enlightenment in Eighteenth-Century England', *Bulletin of the Society for the Social History of Medicine* 25 (1979), 27–41; R. Porter, 'Was There a Moral Therapy in the Eighteenth Century?', *Lychnos* (1981–2), 12–26; R. Porter, 'The Rage of Party: A

value of a pragmatic analysis of events in the context of a broader cultural approach. The madman, his family, the parish and the poor law moved centre stage.

Over the past fifteen years, largely through the work of Bartlett, Wright, Smith, and Forsythe and Melling,[10] the asylum and 'mad-doctors' have been repositioned on the periphery of a target that places the administration of the poor law at its centre. The author's own studies in east London supplement this recent literature using administrative sources from a metropolitan geographical patch a world away from Melling's leafy Devon and the Middle England where Bartlett's, Smith's and Wright's studies are set.[11] This new generation of historians, released from the imperative of chasing Foucault's shadow, have continued the search for an understanding of institutions in the management of the poor and disadvantaged during the process of social and political development of the modern state. What emerges is that even at their peak of expansion, asylums were only part contributors to a broad spectrum of institutional and domestic 'supervisors of care', orchestrated by the multi-layered poor law administrative system.

Detailed local studies have unpicked a complex and geographically diverse weave of relationships and motives. Old certainties about the social class of inmates, the economic and commercial drivers of institutionalization, the central role of a status-crazed embryonic psychiatric profession and the tidy chronological progression from local parish bumbling to central state control and regulation must now be abandoned, or at least substantially modified.

The important role of kinship and family ties in the admission, discharge and negotiation of patients' care are revealed by Wright's work on the Buckinghamshire asylum, harking back to Walton's earlier work on the importance of family preferences in the committal process. Placement is now regarded as far more the outcome of contractual and bargaining negotiations between family, poor-law officials and doctors, between the 'community' and 'authority', than an imposed medical solution. In this respect the care of the insane does not look so very different from the care of the dependent aged in Richard Smith's exposition of the balance of domestic and institutional care driven by rational

Glorious Revolution in Psychiatry?', *Medical History* 29 (1983), 35–50; R. Porter, *Mind Forg'd Manacles: A History of Madness in England from the Restoration to the Regency* (London, 1987).

[10] P. Bartlett, 'The Asylum and the Poor Law: The Productive Alliance', in Melling and Forsythe (eds.), *Insanity, Institutions and Society*, 48–67; D. Wright, 'Getting Out of the Asylum: Understanding the Confinement of the Insane in the Nineteenth Century', *Social History of Medicine* 10 (1997), 137–55; L. Smith, 'The County Asylum in the Mixed Economy of Care', in Melling and Forsythe (eds.), *Insanity, Institutions and Society*, 33–47; B. Forsythe, J. Melling and R. Adair, 'The New Poor Law and the County Pauper Lunatic Asylum: The Devon Experience 1834–1884', *Social History of Medicine* 9 (1996), 335–55.

[11] E. Murphy, 'The Administration of Insanity in East London 1800–1870', PhD thesis, University of London (2000).

economic necessity.[12] Furthermore for many patients, incarceration was not permanent and could be a fairly short-lived affair. Wright found that approximately 50 per cent of all Buckinghamshire asylum patients were eventually discharged, and that of those who were discharged, three-quarters left the asylum having stayed fewer than twelve months.[13]

The birth of county asylums

Len Smith's detailed studies of the first county asylums established following Wynn's Act of 1808 (which enabled county justices to construct county asylums) but before the comprehensive Lunatics Acts of 1845 (which obliged county justices to build public asylums), suggest that a competitive mixed economy of provision existed long before 1808 and continued to be important long afterwards.[14] Porter's eighteenth-century melting pot of commercial competition, soaring scientific ambitions and multi-faceted world of personal motives and flexible public rules survived far longer into the nineteenth century than has been acknowledged. The language of entitlement, rights and contractual commitments of the old poor law gave way only slowly to the Victorian new poor-law rhetoric of virtue, vice and exclusion. Smith's work documents in fascinating detail the daily life in early asylums of keepers, patients, managers and the internal/external relationships of administrators and parish officials that is also evident in the constant negotiation over pauper patients in east London.[15]

The first county asylums were no larger and usually considerably smaller than the existing pauper madhouses used by the parishes. Smith suggests that the early public asylums were a natural development of embryonic quasi-public hospitals like St Luke's Hospital and Bethlem rather than a new phenomenon.[16] Porter had also concluded there was greater continuity of customs in managing the mad between the seventeenth, eighteenth and nineteenth centuries than was generally acknowledged.[17] Smith points to the difficulties the new asylums had in finding a place in a market economy of parish purchasers, private providers and voluntary hospitals.[18]

[12] R. Smith, 'Charity, Self Interest and Welfare: Reflections from Demographic and Family History', in M. Daunton (ed.), *Charity, Self Interest and Welfare in the English Past* (London, 1996), 23–8; R. Smith and P. Horden, 'Introduction', *The Locus of Care* (London, 1998).

[13] D. Wright, 'The Discharge of Pauper Lunatics from the County Asylums in Mid-Victorian England: The Case of Buckinghamshire', in Melling and Forsythe (eds.), *Insanity, Institutions and Society*, 93–108.

[14] L.D. Smith, *Cure, Comfort and Safe Custody: Public Lunatic Asylums in Early Nineteenth Century England* (London, 1999), Introduction, 1–11.

[15] Murphy, 'Administration of Insanity,' 84–118.

[16] Smith, *Cure, Comfort and Safe Custody*, chapter one.

[17] Porter, *Mind Forg'd Manacles*, 278–80.

[18] Smith, *Cure, Comfort and Safe Custody*, chapter one; L. D. Smith, 'The County Asylum in the Mixed Economy of Care 1808–1845', in Melling and Forsythe (eds.), *Insanity and Institutions*, 33–47.

When the new Middlesex county asylum at Hanwell opened in 1831, even though Bethlem was fast removing itself from the pauper market, the asylum was hardly well placed to compete in the east London parishes with the long-established local licensed houses and local voluntary St Luke's Hospital. The flexible charging system whereby parishes 'topped-up' family resources to fund places in St Luke's or a licensed house was an added bonus of the old system. Hanwell only became an acceptable alternative when the cost dropped significantly below the licensed houses. Local scandals seem to have made remarkably little difference to the overall use of the private sector; cost was the prevalent determinant until the 1845 Act made the use of county asylums obligatory. For the overseers of the poor in east London the new asylum at Hanwell was by no means the obvious preferred choice in the mixed economy of care between 1808 and 1845. The London unions were initially just as reluctant to fill up the second county asylum at Colney Hatch when it opened in 1851, to the desperation of the magistrates trying to run an economical institution.[19] Reducing the charge was the only way to attract patients since parish supervisors were happy to leave the insane in the workhouse unless they posed a serious risk. These London findings concur with those of Melling and Forsythe in Devon and Smith in the Midlands that county asylums had to compete for trade largely on price but also by astute marketing of the advantages of an asylum over the private trade.[20]

One of the main aims of the 1834 Poor Law Amendment Act was to impose national consistency of practice in poor relief. The Act introduced the 'New Poor Law', to be administered by boards of guardians of unions of several parishes. Stricter rules made relief available only on condition of admission to a workhouse; outdoor relief was meant to be abolished, although the more punitive aspects of the law were short-lived and largely unworkable. The New Poor Law was to be monitored by a new central government agency, the Poor-Law Commission. Central guidance should have produced uniformity in dealing with the mad. Prior to the Amendment Act there were striking differences in east London, for example, between the neighbouring Holborn parishes of parsimonious St Andrew's and generous St Sepulchre. There were however equally striking differences in the decade after the Act between the policies of the poor-law guardians of sage Stepney, punitive Poplar and generous St George-in-the-East.[21] The guardians of impoverished St George-in-the-East for example spent double what most of their neighbouring parishes and unions spent per 1,000 population on specialist lunatic placements in the early 1840s.

[19] Circular letter from Benjamin Rotch to Metropolitan Union Guardians, Whitechapel Union Board of Guardians Minutes BG/Wh/13, 174.

[20] Smith, 'The County Asylum', in Melling and Forsythe (eds.), *Insanity, Institutions and Society*, 33–47.

[21] Murphy, 'Administration of Insanity', 84–118, 119–85.

Similar diversity between neighbouring boards of guardians is described in rural and county town Devon by Forsythe, Melling and Adair and by Bartlett in Leicestershire.[22] Some rural areas relied heavily on boarding out single patients with nurses or keepers, a practice disapproved of by the Commissioners in Lunacy on the grounds that such patients were outwith the oversight of the central inspectorate, although they had no direct evidence that the practice was unsatisfactory.[23]

The high cost of placing individuals in special asylums had the positive effect of obliging the first New Poor-Law guardians to consider individual paupers and their families just as their predecessors had. However, once the responsibility for making judgements on where to send people was removed from the guardians by the growth of county asylum places and the obligation to use them, there was less reason to consider cases in such depth. The magistrates' capital solution of building county asylums tied the guardians into an inflexible system that, while it had the merit of being cheap per individual case at the outset, proved expensive in the long term. Bartlett points out that the justices of the peace who built the county asylums were merely the top tier of poor-law administration.[24] Bartlett's work on the complex relations in Leicestershire and Rutland between the justices, the guardians, the asylum administrators and patients' families resonates with Forsythe, Melling and Adair's reading of the shifting nuances of administrative power at the Devon asylum.[25] Melling's group concluded that the axis of power was balanced between the magistrates and poor-law officials, the Lunacy Commission playing only a small part.[26] Walton's study of the admission process in Lancashire and Wright's on discharges from the Buckinghamshire asylum also stressed the role of poor-law officials in the lunatics' life career.[27] A similarly complex picture emerges from the early days of the New Poor-Law period in east London where the guardians, officials and parish doctors clearly regarded the union pauper lunatics as 'theirs'. Pauper

[22] Forsythe, Melling and Adair, 'The New Poor Law', 335–55; P. Bartlett, *The Poor Law of Lunacy* (London, 1999), 151–96.

[23] Fifteenth Annual Report of the Commissioners in Lunacy (1860) Single Patients, 53–66, 69–70.

[24] P. Bartlett, 'The Asylum, the Workhouse and the Voice of the Insane Poor in 19th-Century England', *International Journal of Law and Psychiatry* 21 (1998), 421–32.

[25] P. Bartlett, 'The Poor Law of Lunacy: The Admission of Pauper Lunatics in Mid-Nineteenth-Century England with Special Reference to Leicestershire and Rutland', PhD thesis, University of London (1993), 278.

[26] Forsythe, Melling and Adair, 'New Poor Law', 3; B. Forsythe, J. Melling and R. Adair, 'Politics of Lunacy. Central State Regulation and the Devon Pauper Lunatic Asylum', in Melling and Forsythe (eds.), *Insanity, Institutions and Society*, 68–92.

[27] Walton, 'Lunacy in the Industrial Revolution, 6–7; J. Walton, 'Casting Out and Bringing Back in Victorian England: Pauper Lunatics 1840–1870', in W. Bynum, R. Porter and M. Shepherd (eds.), *The Anatomy of Madness: Essays in the History of Psychiatry*, vol. II: *Institutions and Society* (London, 1985); L. Ray, 'Models of Madness in Victorian Asylum Practice', *European Journal of Sociology* 22 (1981), 229–64; D. Wright, 'The Discharge of Lunatics from County Asylums in Mid-Victorian England', 93–112.

lunatics might be 'on loan' to the county asylum or to the private Bethnal Green Asylum. The guardians might take advice from the asylum doctors on discharge, although they might not and the first generation of Lunacy Commissioners was nothing like as much trouble to them as they might have feared.

The tussle of wills between the guardians, officials and parish doctors about the disposal of pauper lunatics was similarly played out by county asylum officers and the county magistrates (justices) who served on the asylum governing committees. Parishes and unions resented the justices' greater powers and generous budget creamed off from their own resources without their sanction but also wanted to shift the management burden of difficult-to-manage paupers. 'Dangerousness' was the language of negotiation used by all interested parties in east London to convince others of the need to act. Adair, Forsythe and Melling have also remarked on the importance of the concept of 'dangerousness' as an admission bargaining criterion between poor-law officials, doctors and asylum staff in the Devon county asylum.[28] Overseers and guardians attempted to match expense to the pauper's perceived level of dangerousness and behavioural nuisance.

Debates about whether the post-1845 Lunacy Commission was effective or influential as an inspectorate depend on whether the question pertains to their local visitorial or central policy role. Hervey set the Lunacy Commission's work within the context of changing conceptions of the role of government, the development of a central administrative bureaucracy and the rise of supervisory central agencies designed to oversee and 'police' the implementation of central government policy through local government.[29] Hervey judged the commissioners were effective locally in Kent in their early years, within the narrow confines of their remit.[30] Mellett thought their remit so constrained it prevented them doing very much at all and Bartlett found their role to be largely conciliatory and weak in the East Midlands.[31] Forsythe, Melling and Adair in contrast found the Lunacy Commission 'authoritative and successful' in Devon.[32]

Local Commissioners had only as much influence as individual members could exert through force of personality, negotiating skill and tenacity. Commissioners were generally far more constrained in their relationships with poor-law officials and guardians about conditions in workhouses than with public asylums and magistrates. They failed miserably to get major improvements

[28] R. Adair, B. Forsythe and J. Melling, 'A Danger to the Public: Disposing of Pauper Lunatics in late Victorian and Edwardian England: Plympton St Mary Union and Devon County Asylum', *Medical History* 42 (1998), 1–25.

[29] N. Hervey, 'The Lunacy Commission 1845–60 with Special Reference to the Implementation of Policy in Kent and Surrey', PhD thesis, University of London (1987), 50–5.

[30] Hervey, 'The Lunacy Commission', 455–64.

[31] D. Mellett, 'Bureaucracy and Mental Illness: The Commissioners in Lunacy 1845–1890', *Medical History* 25 (1981), 243; Bartlett, *The Poor Law of Lunacy*, 266–7, 294.

[32] Forsythe, Melling and Adair, 'Politics of Lunacy', 68–92.

in insane wards of the eastern metropolitan workhouses, compared with the self-appointed trio of doctors who comprised the 'Lancet Commission', whose dynamite journalistic prose battered the complacent London guardians through a series of well-publicized fortnightly articles in the *Lancet* through 1865–6.[33] The Lunacy Commission chairman Shaftesbury, however, had sufficient standing and parliamentary clout to ensure that for twenty-five years the collective commission influenced central government aspirations and the central ideology of care. Ultimately though, the Lunacy Commission's aspirations to annexe the universe of imbeciles to their lunatic empire ran aground, although their annual reports set the moral tone and care standards for a generation of asylums.

While it is clear that first the county magistrates, then central government in the form of the Poor-Law Board gradually appropriated the care of the insane in England, the old mixed economy system survived far longer than has been recognized. Private licensed houses played an important and respected part in the grand scheme until the last years of the nineteenth century and very often received better reports from the commissioners than the county asylums. The transition from private to public provision occurred in step-wise fashion, the private sector quick to adopt new strategies as old markets were denied them. Ultimately public providers triumphed because the legislature was on their side. Scull retains today his earlier interpretation of the rise of public asylums as a 'side effect' of the growing capitalist state in which the suspect motives of nineteenth-century alienists aid and abet the state's convenience.[34]

Lunacy in east London

Public asylums were not only used for paupers. The Devon asylum, Forsythe and his colleagues found, sheltered the middling class of insane from the stigmatizing workhouse, with the acceptance of the guardians. While officially the Middlesex county asylum at Hanwell, serving London, took only paupers, and a declaration to that effect was made by the parish officers, in fact the parish or union was charged by the asylum and then the parish clerks pursued close relatives for the cost of the placement. Considerable energy was expended in devising ways of shifting or reducing the costs of lunatic placements through extracting costs from relatives. In Bethnal Green parish for example, the guardians' minutes for July 1839 record 'Mr William Young of 76 Church Street came before the Board to advocate of Mr Mann late of Pollard Row, chairmaker who was now in a state of insanity and about to be removed to Messrs Warburton'(Bethnal Green Asylum).[35] Mann had a weekly pension of 10*s*. per week

[33] Lancet Sanitary Commission, *Lancet* II (1865), 14–22; Metropolitan Workhouses, *Lancet* II (1865), 131–3; *Lancet* I (1866), 104–6; *Lancet* I (1867), 215–16.

[34] Scull, 'Rethinking the History of Asylumdom', 295–315.

[35] Bethnal Green Minutes of the Board of Guardians (Be BG) 5: 1 July 1839, 309.

from the East India Company to which the guardians had a legal claim to pay Mann's asylum costs.

Mr Young wished a certain portion to be applied for the benefit of the wife but in as much as she did not apply for relief and as her husband was not at present removed to Hanwell asylum the board suggested the matter had better remain as it did. Mann finally got his place in Hanwell some two months later, reducing the cost burden on the parish. Mr Young, accompanied by Mann's son, renewed his application on behalf of Mann's wife. The guardians agreed that the pension would be used to pay Warburton's oustanding bill of £1 17*s*. 6*d*. and that the sum of 5*s*. 6*d*. a week would be paid from the pension towards the cost of Hanwell. Mann's wife could keep the 4*s*. 6*d*. remainder 'to which son and Mr Young assented'. In another instance, when Sally Bartlett died in the Bethnal Green asylum in 1839, the guardians placed a charge over her £30 residual estate held in the local savings bank in order to cover Warburton's bill of £14 15*s*. 0*d*. The remainder was handed over to her son-in-law.[36]

The notion that institutionalization was a manifestation of social control finds little echo in recent work. The insane with financial resources were also admitted to workhouses if parish officers decided it was in the individual's best interests and the relatives were incompetent. Confused old James Lock was admitted urgently to Stepney Union's Mile End workhouse in the winter of 1838 because Mr Story, the parish medical officer and Mr Warren, the relieving officer

> found James Lock sitting by the fire. It seems he had gone to the necessary [*sic*] and having stopped long his daughter went to look for him and found him lying on the stairs in the yard quite exhausted from cold... The only clothing he had on him at the time was a coat and one shoe and it would seem he was in the habit of going about almost in a state of nudity. It was a respectable sort of house but there was no vestige of furniture in his room except a little flock in one corner although his daughter is understood to be in receipt of £50 per annum. The Relieving Officer adds that he considered it safer to remove him at once to the workhouse than to trust him to the care of the daughter who by the accounts given her by her neighbours, appears to be addicted to drinking.[37]

While broadly adhering to the revisionist notion of the asylum exercising a 'social control' function, Bartlett sets the care of the insane within the competing administrative models of the punitive New Poor Law and the high moral paternalism of the Lunatics Acts, illustrating with his studies in Leicestershire and Rutland that there was no hierarchical power structure but a complex web of influence between the guardians, Poor-Law Commissioners, Lunacy Commissioners and asylum 'professionals'. Workhouses remained an important part of the mixed economy of care since crucially, between a quarter and a third of

[36] Be BG 5: 30 September 1839; 14 October 1839.

[37] Stepney Union Minutes of the Board of Guardians, Response to Complaint to the Poor Law Commission. St BG/L/3, 145.

those designated lunatics and idiots was cared for in local workhouses. Insanity was therefore not only of peripheral interest to the guardians.[38]

The parishes of east London had opted for an institutional solution for the care of the insane long before the beginning of the nineteenth century. Parish workhouses and vast urban 'pauper farms' provided care for 'harmless' idiots and the chronically mad and the huge private madhouses in Hoxton and Bethnal Green took the most difficult to manage. The individual parish was a small and feeble unit of administration and was unable to respond to any unusually heavy financial burden, but it had humanity and flexibility. The handful of insane people that each overseer and later, the paid assistant, had to deal with annually meant that each case was handled on its merits. Inconsistency of practice also allowed adaptability; if the quality of care or costs of a specialist institution changed in an unattractive way, then it was a relatively simple matter to move one or two paupers elsewhere. When for example Mr Boak, the parish beadle of the City of London parish St Andrew Undershaft visited Jonathan Tipple's pauper farm house on Christmas Day 1807 to see the parish handful of dull-witted, incompetent dependent paupers he 'found the accommodations very bad and thought it expedient they be removed'.[39] The beadle consulted with 'some gentleman of the parishes of St Peter and St Michael Cornhill' and found that they maintained such folk at a house in Bethnal Green. St Andrew's paupers were moved when negotiations on terms were completed a month or two later.

The metropolis was unusual in having a workhouse classification system in place long before the new poor law introduced it as ideology. Pauper farms have largely been excluded from earlier debates, yet were a crucial part of the provider system for managing some species of mentally incompetent poor in the metropolis up to 1834.[40] The London pauper farms were part of a continuum of types of refuge, which also included the licensed houses, refractory and idiot wards of parish workhouses, houseless refuges for casuals and local prisons. The shift away from pauper farms to private asylums for lunatic placements in the first two decades of the nineteenth century probably reflects what Porter identifies as the emergence of a 'cadre of specialist entrepreneurs of madness' and increasing willingness of the overseers to regard madness as requiring special expertise that could only be had in asylums.[41]

Funding systems and revenue cost comparisons with workhouses closed the city pauper farms that took the foolish and simple unproductive pauper and transferred their clients to the asylum. When Edward Byas, who kept a four-hundred-place pauper farm at Grove Hall in Bow, east London, could not get

[38] Bartlett, *The Poor Law of Lunacy*, 197–237.

[39] Minutes of the Trustees of the Poor, St Andrew Undershaft. Ms. 4120 vol. 6, 6 January 1808.

[40] E. Murphy, 'Mad Farming in the Metropolis: Part I. A Significant Service Industry in East London', *History of Psychiatry* 12 (2001), 245–82.

[41] Porter, *Mind Forg'd Manacles*, 228.

sufficient paupers of this type, he turned in the early 1840s to the conventional lunatic trade to fill the gap. Then, when he could no longer compete financially with the county asylum system, Byas lost most of his union trade and in 1856 turned to the military for their mad trade to fill the void places.[42] Grove Hall was cheap, far cheaper than any other asylum at about 5–6*s*. a week in the early 1840s, although the charges went up as it turned more into an asylum and less a standard pauper farm. By 1842, Whitechapel union was using Byas routinely for the care of idiots in preference to the workhouse, although John Liddle the parish doctor responsible for visiting them was dissatisfied with their care: 'The house is not clean or orderly.'[43]

Whitechapel Guardians continued to place a handful of insane paupers at Grove Hall at a cost in 1845 of 11*s*. a week, the same cost as Hoxton House and Peckham private asylums and pricier than Hanwell's charge of 8*s*. 9*d*. Bow was after all very convenient geographically and the parish staff knew the institution officers well. In Stepney, when the Poor-Law Commissioners made it clear that there was a requirement under the new Acts of 1845 to remove treatable lunatics from workhouses to asylums, the woman workhouse 'Master' Mrs Megson and her close ally, medical officer Daniel Ross, pressed the guardians hard for the thirteen lunatics in Wapping workhouse to be sent to a proper asylum. They were shipped off to Grove Hall because there was no room either at Hanwell or at any of the other local institutions.[44]

Control over capital

The story in east London fits Scull's interpretation only in part. It was not capitalism but the *control* of capital that was at the heart of shifting institutional policies. Early restriction of outdoor relief in many poorer London parishes established an 'asylum' solution to managing diverse kinds of human incompetence long before the machinations of meddlesome magistrates. The burgeoning poor-relief bill in London consequent upon economic recession after the Napoleonic wars had a more direct effect on the parishes' and unions' increasing resort to institutional solutions than the rise of the industrial economy. Green's parallel arguments for the transformation of 'artisans to paupers' in the later recession of the 1860s also stresses that an institutional response to the burden of the poor was an attempt to set limits on relief expenditure.[45]

Under the old poor-law system the control of capital expenditure and the size of the capital resource were in the hands of independent madhouse entrepreneurs

[42] Murphy, 'Administration of Insanity', 73–84.
[43] Whitechapel Union Minutes of the Board of Guardians StBG/Wh/3: April 1839, 143.
[44] Stepney Union Minutes of the Board of Guardians StBG/L/10: 11 September 1845.
[45] D. R. Green, *From Artisans to Paupers: Economic Change and Poverty in London 1790–1870* (London, 1995), 181–209.

who largely determined the service configuration. By the mid nineteenth century England was awash with capital available for public works, the fruit of the burgeoning capitalist economy. The availability of cheap capital allowed the magistrates and later the central government machine to invest in an attractively 'global' solution. As the agency where capital expenditure was controlled became more remote from the lives of the mad and their families, the larger and cheaper the institutional solution became. Emotional and geographical distance between the decision-makers and the recipients of the service facilitated an administrative solution that ignored local families', parish officers' and union doctors' individually devised solutions and substituted instead 'benefits' from the point of view of the central administration.

Having constructed the asylums it was imperative to keep them full to hold unit costs down. Financial incentives were rapidly introduced to do just that. The main increase in institutionalization rates for insanity in Middlesex beyond that predicted by population growth occurred after the Union Chargeability Act and the Metropolitan Common Poor Fund in the late 1860s produced irresistible financial incentives for guardians to use asylums. It is important not to underplay other reasons for the growth of numbers of patients. There was for example an inevitable accumulation of chronic cases throughout the nineteenth century.[46] It has already been noted that there was a significant turnover of acute cases. There was also a surprisingly high institutional mortality rate and in east London at any rate, a continuing merry-go-round after 1862, when the Lunatics Amendment Act made possible the transfer of chronic insane cases from asylums to workhouses, of omnibuses full of 'incurables' being swapped for disruptive workhouse 'recents'. Nevertheless these made only a marginal difference to the long-term resident population.[47] It was the remorseless train of patients, and by 1860 they often were on a train, making their final 'passenger station stop' that seemed never-ending that was a direct result of financial policies. There was no concerted campaign to query the wisdom of further enlargements and additional asylums. Medical superintendents were ambivalent; they wanted to respond to the suffering masses currently without benefit of their asylum and there must have been a modest satisfaction in being indispensable. Professional standing rises when a growing trade is knocking at the door.

The official reasons for the increase in numbers of those requiring asylums was that there had been a miscalculation of the numbers in official statistics and

[46] There is an extended discussion of the chronicity problem in A. Scull, *The Most Solitary of Afflictions: Madness and Society in Britain 1700–1900* (New Haven, Conn., 1993), 267–93.

[47] For coverage of the dynamic shifts in and out of the Victorian asylum see L. Ray, 'Models of Madness in Victorian Asylum Practice', *European Journal of Sociology* 22 (1981), 229–64; Walton, 'Casting Out'; J. Crammer, *Asylum History: Buckinghamshire County Pauper Lunatic Asylum – St John's* (London, 1990), chapter five; MacKenzie, *Psychiatry for the Rich*, chapter four.

that hidden cases were emerging from the community as detection and services increased.[48] A likely significant cause of the apparent increase was the widening of the definitions of insanity and in particular, what passed as a suitable case for institutionalization produced by funding incentives.[49] The 'no-man's-land' of disputed cases was peopled with epileptics, idiots and imbeciles, people with traumatic brain damage, cases of degenerative brain disease and senile dementia, all groups who in metropolitan London at any rate, if they were not in an asylum would be in other forms of institutional care.

The tight reciprocal bond between the parish rate-payers contribution and union expenditure was weakened for lunatics before any other group of paupers by a clause in the 1853 Lunatics Amendment Act which made unions rather than parishes the accountable units of administration responsible for paying asylum fees.[50] It was another twelve years before the Union Chargeability Act of 1865 applied the same ruling to costs of relief for all other classes of pauper. A second clause in the 1853 Act obliged parish medical officers to visit and report quarterly to the overseers and guardians every pauper who in their judgement might be properly confined in an asylum. This repeated reporting of cases gave the parish doctors considerable influence over their guardians as local 'moles' of the Lunacy Commission. The doctors had no special interest in keeping the costs of asylum placements down but a very strong interest in reducing the burden on workhouse staff and their own time.

The Union Chargeability Act of 1865 provided an even greater advantage, from the point of view of the poorer parishes, of a more equitable rating system across rich and poor unions. This enhanced the spending power of unions without drawing further on their beleaguered ratepayers. Two years later the 1867 Metropolitan Poor Act severed the link in London between asylum funding and rate-payers' pockets by the creation of the common poor fund, a central pot on which unions could draw to place any number of designated cases. The 1867 Act produced 'immense and disproportionate growth in poor law lunatic asylums and other forms of poor relief'.[51] Furthermore, the extra 4*s.* per week subsidy for every lunatic placement made available after 1875 was sufficient to reduce the real cost to the guardians in London of a placement to almost nothing. It is not surprising to find that almost any pauper with a hint, a suspicion of eccentricity, indecorous habits or behavioural inconvenience was a candidate for the asylum.

[48] Scull, *Most Solitary of Afflictions*, 340, quotes the fifteenth Lunacy Commissioners' report of 1861, 79; the Select Committee Reports of 1859 and 1877 and J. Bucknill and D. H. Tuke, *A Manual of Psychological Medicine* (London, 1858), 48, 58.

[49] Scull, *Most Solitary of Afflictions*, 344–52.

[50] Lunatics Amendment Act (1853), 16/17 Vict. c 96 & 97.

[51] D. Cochrane, '"Humane, Economical and Medically Wise": the LCC as Administrators of Victorian Lunacy Policy' in Bynum, Porter and Shepherd (eds.), *The Anatomy of Madness*, vol. III (London, 1988), 251.

Funding systems designed to facilitate one social policy frequently have an unwanted effect of stimulating unforeseen changes elsewhere. The drive in the 1860s to make funding systems fairer and more equitably burdensome across rich and poor unions had the entirely unexpected effect of inducing further institutionalization in the poorer unions. Contemporaneously in London the weekly horrors of the Lancet Sanitary Commission reports on appalling conditions in workhouse infirmaries, the regular but ineffectual pressure from poor-law inspectors and the marginally more persuasive Commissioners in Lunacy urged the extrusion of insane people from the workhouse. The objective of Gathorne-Hardy, the President of the Poor-Law Board and main champion of the Metropolitan Poor Act, was to provide sufficient financial incentives for the guardians to cede power to the central poor-law bureaucracy. The common poor fund was the wooden horse that lured the guardians into Troy. Lunatics became proxy parcels of cash through which centralism was achieved.

The main changes in east London from 1800 to 1871 were transinstitutional shifts from pauper farms to asylums in the old poor-law period and from workhouses and private asylums to county and imbecile asylums in the New Poor-Law period. There is no suggestion in the guardians' minutes or parish and union doctors' letters that they thought the nature or rate of insanity was changing. Alienists and asylum inspectors puzzled their heads over the rising rate of lunacy through the nineteenth-century but the guardians did not.[52] Some interesting questions remain however about nineteenth-century changes in the epidemiology of syphilitic general paralysis,[53] the role of alcohol abuse consequent on cost fluctuations in alcohol[54] and the changing age demography. A little-explored area is the possible increase in dementia that would probably have accompanied the increasing life expectancy through the century and the rising proportion of indoor paupers who were elderly.[55] Modest shifts in the prevalence of these disorders could have had a major impact on institutionalization rates. There is insufficient evidence of a change in the epidemiology

[52] Report of the Metropolitan Commissioners in Lunacy to the Lord Chancellor 1844 PP 1844(001) XXVI.1, 30; W. A. Browne, *What Asylums Were, Are and Ought to Be* (Edinburgh, 1837), 51–5; Bucknill and Tuke, *A Manual of Psychological Medicine*, 48–58; discussion in Scull, *Most Solitary of Afflictions*, chapter seven.

[53] E. H. Hare (posthumous, edited by J. L. Crammer), *The Origin and Spread of Dementia Paralytica* (London, 1998), 36–71.

[54] The rising availability and fluctuations in the cost of alcohol had a marked effect on rates of alcohol abuse in the twentieth century and may also have had a similar effect in the nineteenth. M. Sutton and C. Godfrey, *The Health of the Nation Targets for Alcohol: A Study of the Economic and Social Determinants of High Alcohol Consumption in Different Population Groups* (York, 1994).

[55] While the size of the population of sixty-five years and over as a percentage of the total rose only from 3.0 per cent in 1851 to 5.0 per cent in 1901, as a percentage of indoor paupers the sixty-five years and over rose from 19.8 per cent in 1851 to 36.5 per cent in 1901. K. Williams, *From Pauperism to Poverty* (London, 1981). Statistical appendix table 4.24, 'Aged Indoor Paupers 1851–81', 205.

of insanity in England during the nineteenth-century to draw firm conclusions. The most likely explanation for the increasing use of the public asylums was the guardians' desire to use available finance as cost effectively as they could.

The tradition of sending mentally ill people away from their home for 'treatment' in the hope that they would be restored, was established by the sixteenth century and became increasingly popular as beliefs about mental disorder were reframed from the religious/mystical to the realm of personal suffering and 'illness'. Porter cites from numerous examples of parishes as well as private families seeking expert help in the centuries before the nineteenth.[56] By the seventeenth century there were numerous madhouses offering care for all classes and all purses. Parry Jones pointed out that the cheaper institutions were huge enterprises catering for hundreds rather than tens of people and that patients of all classes were sent many miles away from home as a matter of course.[57] There was nothing strange or unusual about the magistrates' notion of sending the insane to asylums in pleasant far away places for care.

The early asylums were not built primarily in rapidly industrializing areas where it might be conjectured that rapid social change might throw out more human incompetence to be mopped up by the welfare system. They were built largely because of the determination of a handful of powerful magistrates; in Devon a small group of Tory land-owners, in Bedford the Brewer-philanthropist Whitbread, in Middlesex a trio of ambitious humanitarians of diverse political and religious persuasion. These were men who knew how to raise the necessary capital finance and knew that the greater the capital investment the cheaper the running costs, so long of course as there were sufficient patients to fill the buildings. The first asylums were built where individual reformers could get hold of sufficient capital, usually in fact in rural areas, not because of any special characteristics of the local poor.

The planned size of the early county institutions reflected the social entrepreneurs' perception of the population at which the single institution was targeted. The Middlesex magistracy made the same global 'one capital project' decision about its convicted felons and remand prisoners until it became convinced of the wisdom of separating the young remand prisoners from the contagion of evil recidivists. Values that shifted the capital spend from the solution which would be cheapest on revenue to a more expensive one had to be convincing and widely held, deeply embedded within the culture of the corporate public body. The magistrates were not exposed to a satisfactory alternative to the asylum and the only vision they had to provide them with their mission was

[56] Porter, *Mind Forg'd Manacles*, 120.

[57] W. L. Parry Jones, *The Trade in Lunacy: A Study of Private Mad-Houses in England in the Eighteenth and Nineteenth Centuries* (London, 1971), 281–92.

that provided by the growing handful of asylum specialists and the evangelical campaigners' detest of the profit motive.[58]

Concluding remarks

Over the course of the nineteenth century, the subtle complexities of relations between the family, the poor-law parish, the magistracy and the central agencies gradually gave way to a dominant centre so removed from the patients that it ceased to rely on a mission of service provision other than to warehouse huge numbers of people. The county magistrates were capable of creating sweeping cheap global solutions in their county asylums. Nevertheless, the shameless creation of megalithic human warehouses did not really become apparent until the creation of the imbecile asylums built by the Metropolitan Asylums Board, the central agency under Poor-Law Board Control established by the Metropolitan Poor Act. The remoteness of the board from day-to-day concerns of the parishes and unions facilitated the establishment in 1871 of the apotheosis of cheap barn-like eighty-bed identical dormitories and 'living rooms' designed for 150 people in the vast and anonymous 2,000-bed institutions at Caterham and Leavesden.

The public asylum was always only a partial solution to managing the insane, even at the peak of asylum expansion. The policy of confinement of the insane in England began long before the nineteenth century, having its origins in the old poor-law parish poorhouses, the widespread practice of farming out the poor for private care from urban parishes too small to sustain their own poorhouse and also in the private madhouses, which had grown up to serve the wealthier classes in the seventeenth and eighteenth centuries. The role and influence of families and the fundamental importance of the local poor-law authorities and increasing power over public policy by central government redefines the importance of the embryonic psychiatric profession and the role of the Lunacy Commission. Neither seem quite so influential in the grand scheme of management as their own reports and trade journals imply. For patients of course, the care and treatment they received from their workhouse and asylum doctors and keepers and the influence the visiting commissioners could exert on their behalf were vital to determining the quality of the regime in which they lived.

[58] Shaftesbury's view as expressed to the Select Committee of 1859: 'When I look into the whole matter I see that the principle of profit vitiates the whole thing; it is at the bottom of all these movements that we are obliged to counteract by complicated legislation, and if we could but remove that principle of making a profit we should confer an inestimable blessing upon the middle classes, getting rid of half the legislation and securing an admirable, sound and efficient system of treatment of lunacy.' Quoted in D. H. Tuke, *Chapters in the History of the Insane in the British Isles* (London, 1882), 193.

Index

Printed in the United Kingdom
by Lightning Source UK Ltd.
134530UK00001B/303/A

9 780521 802062